AGRICULTURE AND FOOD
SECURITY IN CHINA

AGRICULTURE AND FOOD
SECURITY IN CHINA

WHAT EFFECT WTO ACCESSION AND REGIONAL TRADE ARRANGEMENTS?

CHUNLAI CHEN AND RON DUNCAN (EDS)

ANU

THE AUSTRALIAN NATIONAL UNIVERSITY

E PRESS

Asia Pacific Press
The Australian National University

ANU E PRESS

Published by ANU E Press and Asia Pacific Press
The Australian National University
Canberra ACT 0200 Australia
Email: anuepress@anu.edu.au
This title is available online at http://epress.anu.edu.au/afsc_citation.html

National Library of Australia Cataloguing-in-Publication entry

Title:	Agriculture and food security in China : what effect WTO accession and regional trade agreements? / editor, Chunlai Chen.
Publisher:	Canberra : Asia Pacific Press, 2008.
ISBN:	9780731538171 (pbk.)
	9781921313646 (online)
Notes:	Includes index.
Subjects:	World Trade Organization. Food supply—China. Agriculture and state—China. International trade—China. Foreign trade regulation—China
Other Authors/Contributors:	Chen, Chunlai.
	Duncan , Ron.
Dewey Number: 338.1851	

Cover design: Teresa Prowse, ANU E Press
Cover photo: Rob Broek. iStockphoto, File Number: 4570704

CONTENTS

TABLES

FIGURES

SYMBOLS

..   not available
n.a.   not applicable
-       zero
.   insignificant
n.e.c. not elsewhere classified

ABBREVIATIONS

ABC	Agricultural Bank of China
ADB	Asian Development Bank
ADBC	Agricultural Development Bank of China
AEC	ASEAN Economic Community
AMC	asset management corporation
APEC	Asia Pacific Economic Cooperation
APPCDC	Asia Pacific Partnership for Clean Development and Climate
ASEAN	Association of Southeast Asian Nations
BOC	Bank of China
BVAR	Bayesian vector autoregression
CANET	China Academic Network
CAR	capital adequacy ratio
CBC	Construction Bank of China
CBRC	China Banking Regulatory Commission
CERD	Chinese economy with regional details
CES	constant elasticity of substitution
CESG	cultural, educational and sports goods
CET	constant elasticity of transformation
CGAP	Consultative Group to Aid the Poorest
CGE	Computable General Equilibrium (model)
CIA	Central Intelligence Agency
CNY	Chinese yuan
CNNIC	China Internet Network Information Centre
CPC	Communist Party of China
CPI	consumer price index
CRN	China Research Network
CSRC	China Securities Regulatory Commission
DFN	Deutsche Forschungsnetz [German Research Network]
DPP	Democratic Progressive Party
EEFSU	Eastern Europe and the former Soviet Union
EEM	electronic equipment and machinery
ETE	electronic and telecommunications equipment
FDI	foreign direct investment

FPC	Funding the Poor Cooperative
FTA	Free Trade Agreement
GATT	General Agreement on Tariffs and Trade
GDP	gross domestic product
GLS	generalised least squares
GTAP	Global Trade Analysis Project
HRS	Household Registration System
HSBC	Hong Kong and Shanghai Banking Corporation
ICBC	Industrial and Commercial Bank of China
LDC	less developed country
LIR	labour insurance regulations
MCA	Ministry of Civil Affairs
MFI	microfinance institution
MFN	most-favoured nation
MNE	multinational enterprise
MOF	Ministry of Finance
MOFERT	Ministry of Foreign Economic Relations and Trade
MOFTEC	Ministry of Foreign Trade and Economic Cooperation
NBS	National Bureau of Statistics
NGO	non-governmental organisation
NATO	North Atlantic Treaty Organisation
NIE	newly industrialised economy
NPC	National People's Congress
NPL	non-performing loans
NRC	net relative change
NSSF	National Social Security Fund
NYSE	New York Stock Exchange
OECD	Organisation for Economic Cooperation and Development
PAYGO	pay-as-you-go
PBOC	People's Bank of China
PECC	Pacific Economic Cooperation Council
PEO	Pacific Economic Outlook
PPI	producer price index
PRC	People's Republic of China
PRCGEM	People's Republic of China General Equilibrium Model

RCA	revealed comparative advantage
RCC	rural credit cooperative
RCCU	Rural Credit Cooperative Union
RCF	rural credit foundation
RFI	rural financial institution
RIETI	Research Institute of Economy, Trade and Industry
RMB	Renminbi
RPI	retail price index
RPS	rural postal savings
SARS	Severe Acute Respiratory Syndrome
SASAC	State-Owned Asset Supervision and Administration Commission
SCO	Shanghai Cooperative Organisation
SCORES	China Society for Research on Economic Systems
SITC	Standard International Trade Classification
SME	small and medium enterprise
SEZ	special economic zone
SOB	state-owned bank
SOCB	state-owned commercial bank
SOE	state-owned enterprise
SSF	social security fund
TFP	total factor productivity
TIFA	Trade and Investment Framework Agreement
TRQ	tariff-rate quota
TVE	township and village enterprise
UHIDS	Urban Household Income Distribution Survey
UMLS	Urban Minimum Living Security Program
UNCTAD	United Nations Conference on Trade and Development
VAR	vector autoregression
VAT	value added tax
VECM	vector error correction model
WHO	World Health Organization
WTO	World Trade Organization
XUAR	Xinjiang Uighur Autonomous Region

CONTRIBUTORS

Jennifer Chang is an economic analyst with the Australian Treasury, Canberra.

Chunlai Chen is Senior Lecturer, Crawford School of Economics and Government, The Australian National University, Canberra.

Ron Duncan is Professor Emeritus of the Crawford School of Economics and Government, The Australian National University, Canberra.

Jikun Huang is Professor and Director of the Center for Chinese Agricultural Policy, Chinese Academy of Sciences, Beijing.

Tingsong Jiang is Senior Economist, Centre for International Economics, Canberra.

Feng Lu is Professor, China Center for Economic Research, Peking University, Beijing.

Lucy Rees is a consultant with Bain and Company, Sydney.

Scott Rozelle is Professor, Freeman Spogli Institute for International Studies, Stanford University, California.

Rod Tyers is Professor of Economics, College of Business and Economics, The Australian National University, Canberra.

Xiaolu Wang is Professor, National Economic Research Institute, China Reform Foundation (NERI-China), Beijing.

Jun Yang is Associate Professor, Center for Chinese Agricultural Policy, Chinese Academy of Sciences, Beijing.

PREFACE

This book draws together research on the impact on its agricultural sector of China's accession to the World Trade Organization (WTO), and on the impact on agriculture in China and ASEAN countries of the implementation of the China-ASEAN Free Trade Agreement (FTA). This research was funded by the Australian Centre for International Agricultural Research (ACIAR) and was jointly carried out by researchers in the Crawford School of Economics and Government at The Australian National University (ANU), Canberra, and researchers in the China Center for Economic Research (CCER), Peking University, Beijing. During the project extension period, researchers in the Center for Chinese Agricultural Policy (CCAP) of the Chinese Academy of Sciences (CAS) joined the research team.

Not all of the research completed under the ACIAR project is published in the book. However, Chapter 1 by Chunlai Chen and Ron Duncan provides a summary of all the research carried out in the project and indicates where the research not included in the book is available.

An important part of the ACIAR project was the construction of a computable general equilibrium model of the Chinese economy that is disaggregated by province/region and has detailed disaggregation of the agricultural sector. The model also includes representative rural and urban households and therefore allows detailed analysis of regional impacts of changes in policies, particularly as they affect rural households. The full model is available on request.

The project researchers from the National Centre for Development Studies of the Asia Pacific School of Economics and Management at the Australian National University (ANU) (now the Crawford School of Economics and Government) were originally Ron Duncan, Yiping Huang and Yongzheng Yang. The staffing of the project was reorganised with the departure of Yiping Huang and Yongzheng Yang. Rod Tyers, Chunlai Chen, Xiaolu Wang and Tingsong Jiang, all of ANU, joined Ron Duncan to carry out the project. During the application and early implementation of the project extension, Christopher Findlay from the Australian National University (now with University of Adelaide), joined the research team.

The main collaborators in China were Justin Yifu Lin, Wen Hai, and Feng Lu at the China Centre for Economic Research (CCER) at Peking University. Other researchers in China included Shi Li and his colleagues at the Institute of Economic Research at the Chinese Academy of Social Sciences (CASS), and Weiming Tian and his team at the China Agricultural University. During the project extension period, Jikun Huang and Jun Yang from the Centre for Chinese Agricultural Policy (CCAP) of the Chinese Academy of Sciences (CAS) joined the research team.

The support of Ray Trewin and Trish Andrew from ACIAR is acknowledged. We also thank the members of the committee, Associate Professor Robert Scollay from the APEC Study Centre of the University of Auckland and Professor Fang Cai from the Institute of Population and Labour Economics of the Chinese Academy of Social Sciences who reviewed the project for ACIAR, for their constructive suggestions on method, output, marketing, and project extension.

Chunlai Chen and Ron Duncan
Canberra
May 2008

01 Achieving food security in China
Implications of WTO accession

Chunlai Chen and Ron Duncan

The impact of China's entry into the World Trade Organization (WTO) on its agricultural sector has been a major concern of the Chinese government and one of the hottest topics among policymakers and academics inside and outside China. In 1998, the Australian Centre for International Agricultural Research (ACIAR) funded a multi-year research project (ADP/1998/128) to examine the impacts of China's WTO accession on its agricultural sector. The project was a joint undertaking by researchers from Australia and China.

The project began before China's accession to the WTO, but it was anticipated that the accession application would be successful and that it was therefore important to understand fully the impact that the accession commitments could have within China. There was a concern to see that Chinese policymakers appreciated the benefits that would flow from trade liberalisation. At the same time, it was understood that the trade reforms would lead to structural adjustments that would involve the reduction of some activities, with the accompanying loss of employment and asset values. It was believed that it was desirable for policymakers to have an understanding of these consequences, as this would reduce the chances of the adoption of poor policies in response to any perceived or real adverse impacts. Another objective of the project was to analyse various policy options in order to offer input to China's policymaking process.

China made substantial commitments to freer trade in agriculture in its accession agreement. Underlying these commitments, therefore, is a substantial shift away from its previous basic agricultural policy position, which had an emphasis on food 'self-sufficiency' and restrictions on food imports. As might be expected, there remains strong support for food self-sufficiency policies and agricultural import restrictions, but these policies have very high economic costs for the country because they maintain resources in activities that do not use them efficiently. Therefore, one of the aims of the research project was to demonstrate the high costs of food self-sufficiency policies, in order to reduce the chances of any move back towards such policies.

The research project had a strong general equilibrium focus, which has been followed through the use of global and China-specific computable general equilibrium models and the analysis of the impact of all of China's WTO accession commitments, not only its agricultural commitments. The focus on general equilibrium research has two related justifications. First, when analysing the impacts of agricultural policies, we should look beyond the agricultural sector because agricultural policies will have economy-wide impacts. Analogously, trade policy initiatives in other sectors will have impacts on the agricultural sector. Further, the impact of macroeconomic policies on particular sectors can be as important, if not more important, than sector-specific policies. Second, it is clear from recent experience in developing countries that incomes of rural households are increased more by increases in off-farm income earned by household members than by increases in on-farm income. Therefore, in examining the welfare implications of the trade liberalisation for farm households, it is important to examine all the ways in which farm household incomes could be affected through the structural adjustments to the trade reforms.

This chapter summarises the results of the collaborative research project that has explored various implications of China's accession to the WTO, particularly those relating to agricultural policy and the agricultural sector, and to the issue of food security.

China's agricultural development and policies

Huang and Rozelle (Chapter 2) review China's agricultural development and its policy regime before and since its accession to the WTO. They

2

argue and demonstrate that China's WTO accession commitments are a continuation of policy changes that have been taking place since economic reforms began in 1979. Moreover, they see the accession commitments as verifying a fundamental shift by the government from direct participation in the economy to taking on a more indirect regulatory and fostering role. They see a trend in policies pushing agricultural activities in a direction that is more consistent with the country's resource endowments. The accession agreements will allow more land-intensive products into the domestic market from overseas and stimulate the export of labour-intensive crops.

Huang and Rozelle see the post-WTO accession policy changes taking two basic forms: those changes necessary to honour the obligations of the accession commitments, and those policies necessary to minimise any adverse impacts from the accession commitments. Some of the latter policies include further land reform to allow farm amalgamation, higher productivity and incomes, promotion of farmer organisations, abolition of agricultural taxes and investment in new technology. Overall, Huang and Rozelle see a strong government commitment to modernising the agricultural sector in China.

Food self-sufficiency and food security

Maintenance of grain self-sufficiency has long been a major plank of China's agricultural policy. As recently as 1996, the Vice-Minister of Agriculture, Wan Baorui (1996), announced that the grain self-sufficiency rate was to be maintained at above 95 per cent. Along with the widening of per capita income disparities between rural and urban areas, the rhetoric of food self-sufficiency is the most prominent weapon of China's protectionists. There has been concern that China will go the way of Japan, Korea and Taiwan and protect its agricultural sector as the sector shrinks in relative importance and national per capita incomes increase. China's accession to the WTO could have come just in time to make this possibility less likely. While WTO accession lessens the risk that the protectionists will succeed, the agriculture ministry has been assigned a prominent role in trade policy formation and negotiation, with the power to press for further agricultural protection on self-sufficiency as well as distributional grounds (Anderson et al. 2002; Tong 2003).

That there exists genuine concern for food self-sufficiency in China is understandable. Widespread famines have been experienced, although, in hindsight, these have probably been due more to bad policies than to bad weather. Concern about the possibility of food trade embargoes can also be understood—although, again, experience has shown that trade embargoes are difficult to implement (Lu 1997; Yang 2000). The difficulty of financing large volumes of food imports would also have been a legitimate concern in the past; however, this is no longer the case. In 2000, China's total export revenue was about US$250 billion. The importation of 22 million tonnes of grain (the WTO import quota commitment) would cost US$3-4 billion—only a small fraction of total export earnings.

Food self-sufficiency is not, however, the same as food security. Food security is a matter of whether households have sufficient income to maintain an adequate diet. The important question with respect to food self-sufficiency for China is the extent to which it is prepared to rely on the international market for the gap between its domestic production and its effective demand. China is such a large country that, inevitably, most of the goods and services consumed have to be produced domestically.

To illustrate the economic costs of adopting policies that aim to maintain grain self-sufficiency near the present level or to increase it, protectionist scenarios were modelled by Duncan et al. (see Chapter 8) using an adaptation of the GTAP model,[1] a global, multi-region, multi-product general equilibrium model. Following Yang and Tyers (2000), independent representations of governments' fiscal regimes were added to the standard GTAP base, including direct and indirect taxation, separate assets in each region (currency and bonds) and monetary policies with a range of alternative targets.

In earlier analysis, Yang and Tyers (1989) used a global agricultural sector model to examine the impact of rapid income growth in China on the composition of food consumption and the implications of this for food self-sufficiency. They found that the anticipated redistribution of consumption towards livestock products would raise import demand for feed grains and that this would make the maintenance of self-sufficiency through protection very costly. Because their analysis was restricted to the agricultural sector, however, they could not examine the redistribution and economy-wide effects of the protection needed to maintain self-sufficiency. The use of the GTAP model overcomes these limitations.

The modelling first projects the base case to 2010 under conservative output and productivity growth assumptions and then asks two questions. First, if China's present food self-sufficiency rates are to be held constant until 2010, will increases in protection be required? Second, what increases in protection would be required to achieve full food self-sufficiency by 2010 and what would be the economy-wide and distributional effects of this protection? Consistent with Yang and Tyers (1989), the base-case projection to 2010 shows substantial declines in Chinese food self-sufficiency (see Table 1.1), particularly for beverages, livestock products and feed grains (basically as the result of income growth), so that substantial increases in protection are needed to maintain the 2001 levels. To achieve self-sufficiency in all agricultural products by 2010, considerable further protection would be required. In both cases, this protection would be contractionary and redistributive, and it would retard growth in other sectors. The sensitivity analysis shows that the strength of the results rests quite heavily on some parameters, particularly the income elasticity of demand for livestock products.

The model employs the original GTAP constant difference of elasticities of substitution (CDE) system. Its non-homotheticity is an asset in that it permits a range of income elasticities to exist either side of unity. While this system is more general than the homothetic ones often used in such models, it is still restrictive in the width of the parameter range compared with still more general systems. The CDE system is employed here because of its parametric economy. Because of the restrictiveness of the CDE system, the lower bound for the income elasticity of rice cannot be set below 0.1, despite evidence suggesting that it is now negative (Ito et al. 1989; Peterson et al. 1991). As a result, the differences between the model's income elasticities of livestock products and processed foods—which are superior goods—and those of grains are likely to be smaller than they really are. One consequence of this is that the results likely underestimate the growth in demand for livestock products and processed foods and hence underestimate the associated derived demand for cereal feeds and other agricultural inputs. This means it is likely that there is a downward bias in the estimates of the cost of achieving and maintaining agricultural self-sufficiency.

Because the declines in self-sufficiency in the base-case projection to 2010 are significant, the tariffs necessary to retain 2001 self-sufficiency rates are substantial, particularly for the beverages, 'other crops' and livestock product groups (see Table 1.2). These taxes on imports are, effectively, taxes on all China's trade; thus, they also reduce China's exports, causing exporting industries to contract. Overall, the increased protection induces a 1 per cent contraction in gross domestic product (GDP) along with some restructuring across industrial sectors. The more heavily protected agricultural industries are favoured, mostly at the expense of manufacturing, particularly light manufacturing.

The additional tariffs required to achieve full food self-sufficiency by 2010 are very large, particularly on imports in the livestock products, processed food and 'other crops' groups (see Table 1.2). These tariff increases distort incentives in the economy substantially, shifting resources into agriculture and contracting the manufacturing and service sectors. Throughout the economy, this decline in allocative efficiency reduces returns to installed capital and therefore investment. The level of 2010 GDP is reduced by nearly 2 per cent. The tariffs that would achieve agricultural self-sufficiency in 2010 reduce exports from China's growth powerhouse—its light manufacturing industries—by half. Domestic resources are reallocated to the agricultural sector, raising costs in manufacturing and reducing the international competitiveness of China's manufacturing industries. The resulting misallocation of labour is particularly striking. The higher tariffs cause employment in agricultural and food-processing activities to be substantially greater, at the expense primarily of light manufacturing.

Higher agricultural tariffs raise land rents by a considerable margin but reduce real wages and capital returns. Real wages grow less in agriculture and in the modern sector. This is true for production and skilled workers, and it is also true for the owners of physical capital. The capital losses occur because the industries that are hurt by the tariffs are more capital intensive than agriculture. In the end, landholders are the only winners from the tariffs.

We might well ask: what is gained by self-sufficiency? Would food be more readily available in China? No! China's 2010 prices for imported foods would be increased by up to 60 per cent because of the increased tariffs;

even the prices of home-produced food products would increase by at least 10 per cent. The key consequence of political significance would be a reduction in interdependence with the global economy. This cuts two ways. Reduced reliance on food imports means curtailing the principal source of China's overall economic growth since the 1980s: access to foreign markets for its labour-intensive goods. Curtailed exports reduce its capital returns, thereby cutting incentives for investment and, ultimately, the growth rate of its economy.

Table 1.1   Food self-sufficiency in China, 2001-10 (per cent)

Commodity	2001	2010
Rice	100	100
Beverages	99	91
Other crops	94	89
Livestock	99	95
Processed food	88	83
Fish	99	99
Minerals	95	94
Energy	80	80

Source: Authors' simulations.

Table 1.2   Changes in tariffs required for food self-sufficiency by 2010 (per cent)

Commodities	Extra protection to hold self-sufficiency rates at 2001 levels	Extra protection to achieve full self-sufficiency
Rice	0.0	0.0
Beverages	35.4	50.6
Other crops	19.2	72.7
Livestock	39.2	78.7
Processed food	11.1	67.3
Fish	16.1	31.9

Source: Authors' simulations. The changes in protection are shown as proportional changes to nominal protection coefficients.

Commodity, regional and household impacts of WTO accession

The project examined the likely commodity/regional impacts of China's WTO agricultural commitments. This was done through the calculation of 'production concentration indices', which Lu (see Chapter 3) defines as the ratio of the sown area (or output) of the commodity per capita of the agricultural population of a region divided by the same ratio for the country. The interpretation of the index is that a region has a comparative advantage (disadvantage) in the commodity if the production concentration index (PCI) is greater (less) than one (analogous to the use of the export concentration index as a measure of 'revealed' comparative advantage).

The calculations show relatively high PCIs for the labour-intensive commodities (vegetables, fruits, meats and fish products) in the eastern region, relatively low PCIs for these commodities in the western region, and values in between for the central region. On average, the eastern region has comparative advantage in all four labour-intensive commodity groups. The central region has comparative advantage in vegetables and meat products but not in fruits and fish products. The western region does not have comparative advantage in any of the four labour-intensive product groups.

PCIs for the land-intensive products (grains, oil seeds, cotton and sugar) are relatively high for the western region, relatively low for the eastern region and in between for the central region. The western region has comparative advantage in all four land-intensive commodity groups, while the central region has comparative advantage in grains, oil seeds and sugar. In the eastern region, only sugar has comparative advantage.

In assessing the adjustment impacts of the reduction in protection for agricultural commodities, Lu argues that, in line with China's perceived comparative advantage, liberalisation will strengthen the tendency for imports of land-intensive commodities and encourage exports of labour-intensive commodities. It is expected that the export promotion effect will follow closely the regional distribution of comparative advantage of labour-intensive commodities. With regard to the negative import-substitution effect, the study postulates two possibilities: one is that imports of land-intensive products will substitute largely for domestic production in the regions with relatively high domestic production costs for these commodities. The second is that imports will substitute for domestic production in proportion to the existing regional concentration of the commodities.

The two possibilities will have different implications for regional adjustment costs. The first is to be preferred, as it allows the principle of comparative advantage to play a larger role in resource allocation. The second, less desirable, possibility is in line with the policy stance that has emphasised provincial grain self-sufficiency. Under the first scenario, those provinces with comparative advantage in labour and land-intensive commodities will have relatively small adjustment costs. Those provinces without comparative advantage in labour-intensive or land-intensive commodities are likely to be worst off as they will benefit least from export expansion and have to bear the largest adjustment costs. The eastern region is seen as likely to be the major beneficiary from export expansion of labour-intensive products, while the inland regions are expected to receive much smaller benefits. On the other hand, the coastal regions could experience a larger share of the adjustment costs from the import growth while inland provinces could have less adjustment.

Under the second scenario, the regional distribution of the import-substitution effects changes considerably. The central and western regions could incur a large share of the adjustment costs—that is, they are likely to have a combination of small export-promoting effects and large adjustment costs. This could result in an increase in the income gap between the coastal and inland regions.

To maximise the likelihood for scenario one to materialise, agricultural policies should be changed in a more market-oriented manner. There should be a move away from the emphasis on regional self-sufficiency. As well, the state monopolies in domestic marketing and distribution of bulk agricultural commodities as well as in transport should be removed to allow goods to flow more freely across provincial borders.

In another of the project studies, Jiang (2002) examined the growing income gap between the coastal and inland regions for the period 1978–2000. This study found that the gap was unchanged or had even declined from 1978 to 1990—the period in which China undertook major agricultural reforms and experienced high growth in the agricultural sector. This development favoured the poorer regions that had higher agricultural

9

shares in total output. In the period 1991–2000, however, which followed the major industrial reforms, the eastern region benefited most from growth of foreign investment and development of the private sector. During this time, the regional income gap between the coastal and inland regions widened considerably. Income disparity within the regions, however, declined.

The increasing gaps between rural and urban incomes and between incomes in coastal and inland regions are of considerable concern in China. The likelihood of WTO accession increasing these disparities was a major factor behind resistance to accession. Jiang (see Chapter 6) modelled how WTO accession was likely to affect these income disparities using the CERD model developed under this project (Chen and Duncan 2008).

In simulations of the WTO accession commitments, crops, food processing, motor vehicles and parts, and machinery sectors are affected adversely by the accession, particularly the motor vehicles and parts sector. Other sectors benefit, particularly the light manufacturing sector. Agricultural production declines most in the eastern region and least in the central region. The study results show the usefulness of general equilibrium analysis, which has taken account of there being a higher return to labour-intensive activities outside of agriculture in the eastern region. The eastern region does by far the best overall because it realises most of the gains in allocative efficiency. Hence, there is an increase in regional income disparity. Rural household incomes increase most in the eastern region because they have the best opportunities for earning off-farm income. Across the regions, however, rural household incomes increase less than urban household incomes.

In the project, Jiang (see Chapter 7) also undertook some simulations to examine the impacts of policies that could be adopted to mitigate these adverse effects on farm incomes. One policy option tested was the use of a production subsidy. It was found that if agriculture were to be subsidised to maintain the pre-accession grain self-sufficiency rate, the subsidy would be 7.2 billion yuan. If the target were to maintain the food self-sufficiency rate, the subsidy would be 180 billion yuan. Another policy option tested was an increase in agricultural research and development to improve agricultural productivity. It was estimated that China would have to almost double the level of agricultural research and development in order to maintain the food self-sufficiency rate.

Surplus agricultural labour

Another of the project studies (Wang, Chapter 5, and Wang 2002) argued that a major reason for the widening rural-urban income gap since 1991 was the excess supply of labour in agriculture. This excess supply results in low labour productivity—much lower than in the industrial and tertiary sectors—and slower growth in incomes. As a result of the slower productivity growth and the excess of labour in agriculture, the agricultural share of GDP fell from 51 per cent in 1952 to only 14 per cent by 2002, while the share of agricultural workers in the total work-force declined only from 84 per cent in 1952 to 50 per cent in 2002. The share of the rural population in the national total declined even more slowly—from 85 per cent in 1953 to 61 per cent in 2002. The excess supply of labour in agriculture has been attributed in large part to the restrictions on the movement of people from rural areas to cities, accompanied by discrimination in the form of denial of access to housing, education, job training and health facilities.

Because of the development of rural industries, the sources of rural household income have changed remarkably. In 1990, the agricultural share of rural household incomes was 74 per cent. By 2002, this share had fallen to 47 per cent. The development of the rural industrial sector—particularly the township and village enterprise (TVE) sector—was very important in providing opportunities for diversification of rural household incomes. Employment in the TVE sector increased from 28 million to 135 million in the period 1978-96, accounting for more than one-quarter of the rural labour force (Wang and Duncan, Chapter 4, and Wang and Duncan 2003). Despite this, the number of farmers has increased and there appears to be more surplus agricultural labour than ever. The diversification of rural household incomes varies greatly between the major geographical regions, with the share of off-farm income in rural households in the coastal region about 75 per cent but about only 25 per cent in the western region. Moreover, the rural-urban income disparity has widened most in the western region: between 1980 and 2000, the rural/urban income ratio fell from 54.4 per cent to 43.2 per cent in the eastern region, from 49.5 per cent to 40.1 per cent in the central region, and from 44.9 per cent to 30.4 per cent in the western region.

Development of rural industries slowed in the late 1990s, due mainly to greater market competition, the unfavourable location of rural enterprises, difficulties in accessing external finance and lack of infrastructure, technical inputs and human resources. Meanwhile, urbanisation accelerated and larger numbers of rural labourers moved to urban areas in search of jobs. In 2001, the urbanisation rate (the ratio of urban to total population) in China reached 38 per cent, compared with 26 per cent in 1990 and 19 per cent in 1980.

Restrictions on migration

In spite of the acceleration of urban development, the urbanisation rate in China is, on average, 10-20 percentage points lower than in other countries at a similar income level—even allowing for the so-called 'floating population' (Wang and Xia 1999). In particular, there are relatively few medium and large cities. In 2001, 121 million people—only 9.6 per cent of the population—lived in cities of more than 500,000 people. In the less-developed western region, only 5.6 per cent of the population lived in cities of that size. If China had an urbanisation rate similar to that of other countries of a similar income level, an additional 120-240 million people would be living in urban areas. This number can be thought of as the excess number of people in the rural economy (see Wang and Duncan, Chapter 4).

Wang and Duncan (Chapter 4) note that there are positive correlations between rural industrialisation (measured as the share of TVE employment in rural labour) and rural incomes and between the urbanisation rate and rural incomes. They therefore undertook a causality test of the relationship between urbanisation and regional economic growth. Because urbanisation and rural incomes could be a function of economic growth, the causality test was carried out within an endogenous growth model. The results from the modelling indicate that each percentage point increase in the urbanisation rate increases provincial economic growth by 0.37 percentage points above the already high 7-10 per cent growth rate—that is, urbanisation has a long-run impact on economic growth. When regional dummies are introduced, there are seen to be significant impacts from urbanisation on economic growth in the eastern and central regions, but the impact is insignificant in the western region. There could be two reasons for this last result: the rate of urbanisation in the western region has been low and the urban economy in the western provinces has not experienced much restructuring, and is therefore less market-oriented and less efficient.

Macroeconomic implications of WTO accession

Most other studies of China's WTO accession have focused on the medium and long-run impacts of accession. There are, however, important short-run issues, in particular those relating to the macroeconomic policy environment in which the reforms take place. The study by Tyers and Rees (see Chapter 9 and 2002) therefore examined the short-run impacts of the reforms under scenarios of capital controls, fixed and floating exchange rate regimes and alternative fiscal policies.

An adaptation of the GTAP model similar to that used in Duncan et al. (see Chapter 8) was used for the analysis. Because of the inclusion of independent representation of governments' fiscal regimes (with inclusion of direct and indirect taxation) and monetary policies with a range of targets, it is possible to study a range of policy regimes. To be representative of short-run conditions, the model also allows for labour market rigidity and departures from full employment.

In order to undertake the short-run analysis, a simulation of the long-run effects of the accession commitments was first carried out. The results allowed for a derivation of investors' expectations on the assumption that they took changes in long-run returns on installed capital into account in determining short-run changes in their investment behaviour. The results from the long-run simulation show the expected allocative efficiency gains from the trade reform, which are reflected in increased GDP and increased returns on installed physical capital, which induce greater investment and larger net inflows on the capital account. The increased average long-run return on installed capital is therefore part of investors' expectations in the short run and so tends to increase the level of investment in the short run—even if capital controls are maintained. The trade reforms also cause consumption to switch away from home-produced goods and the relative prices of such goods to fall, and hence there is a real depreciation. There is also an increase in export competitiveness, and exports expand.

Manufacturing, particularly light manufacturing, is the main beneficiary of the trade liberalisation, together with the transport sector. This result—which is contrary to intuition from the Heckscher-Ohlin-Samuelson (HOS) trade model—derives from the model's departure from the HOS model in two ways: first, there is extensive use of intermediate inputs from the same sector (intra-industry trade); and second, competing imports are differentiated from home products. Under these assumptions, the tariff reductions on imported intermediate inputs have a direct effect on the home industry's total cost. The indirect effect of the reductions in tariffs on competing but differentiated products depends on the elasticity of substitution between imports and home-produced goods. For manufacturing, the input-cost effect of tariff reductions is considerably greater than the impact from the loss of protection against competing imports. Cost reductions of similar origin are the reason for the gains accruing to the domestic transport sector.

As the reforms cause the most substantial reductions in protection in China's food-processing sector and therefore lead to long-run contractions in rice and 'other crops', there is substantial relocation of employment from agriculture to the manufacturing, energy and transport and other services sectors.

Simulations of short-run effects

For the short-run base-case simulation, China is assumed to maintain a fixed exchange rate against the US dollar and rigid capital controls, while nominal wages are 'sticky'. The other regions specified in the model have inflation and CPI targeting, no capital controls, full short-run nominal wage rigidity in the industrial countries and fully flexible nominal wages elsewhere. Government spending in all regions is assumed to absorb a fixed proportion of GDP and the rates of direct and indirect taxes are constant, so that government deficits vary in response to shocks.

Five different macroeconomic regimes were simulated to study the impact of the trade reforms.

1. Rigid capital controls and fixed tax rates; monetary policy targets the CPI, and the exchange rate floats.

2. Fixed exchange rate and fixed tax rates; there are no capital controls.

3. Fixed tax rates, monetary policy targets the CPI, the exchange rate floats and capital controls are removed.

4. Rigid capital controls and a fixed exchange rate; the direct tax rate adjusts to maintain the government deficit as a fixed proportion of GDP.

5. The closure is the same as number four, however, capital controls are removed.

The short-run effects of the trade reform vary considerably under the different macroeconomic regimes. When capital controls are in place and the exchange rate is fixed, the allocative gains from the tariff reductions are insufficient to offset the contractionary effects of the deflation that is due to the rise in the relative prices of foreign goods. When capital controls are weak, the trade liberalisation attracts increased inflows on the capital account and mitigates the real depreciation and associated domestic price deflation. The real volume of investment rises irrespective of the target of monetary policy, as does the level of GDP. The choice of monetary policy still matters, however, with CPI targeting leading to a smaller GDP price deflation, more modest gains in the real production wage and better short-run GDP gains.

As with monetary policy, the impact of the different fiscal policies depends on the strength of capital controls. Given tight capital controls, if tax rates are held constant and the fiscal deficit expands, domestic interest rates rise and private investment is crowded out. Where income tax increases to compensate for the tariff cut, there is less pressure on the domestic capital market and the interest rate increase is less, as is the fall in investment. In the absence of effective capital controls, the case of no increase in the tax rates performs better than the alternative policy. The increased government borrowing draws in international savings at international interest rates and does not crowd out private investment. Both fiscal policies, however, give superior results in the absence of capital controls.

The key determinant of the short-run structural adjustment resulting from the trade liberalisation is the size of the real depreciation; the real depreciation is larger when capital controls are in place. Traded-goods sectors, such as light manufacturing, are advantaged while non-traded sectors are not. When there are no capital controls, the manufacturing gains are smaller and the non-traded services sectors also benefit. Across the board, however, for the same reasons as in the long-run simulation, agriculture and food processing are disadvantaged by the reforms. It is important to note that some relaxation of the monetary policy regime can reduce the adverse effects on the agricultural sector from the trade reform. When capital controls are in place and the exchange rate is fixed, almost the entire agricultural sector is hurt. Where capital controls are in place but the exchange rate is floated, the 'other crops', livestock and fisheries sectors expand.

Employment in food processing falls regardless of the macroeconomic policy regime. Significant structural change is required in the short run with the movement of employment from agriculture to manufacturing; however, in the long run, the size of the employment shift is smaller. Under either fiscal regime, the greatest contraction in employment in food processing occurs when capital controls are tight and monetary policy targets the nominal exchange rate. Unlike in the long run, employment in the other agricultural sectors is not necessarily contractionary—the outcome is dependent on the macroeconomic policy regime.

In summary, if capital controls are too tight and the fixed nominal exchange rate is retained, the reforms are deflationary. If the labour market is slow to adjust, employment growth will slow and the reform package will be contractionary. To obviate this, the government has to allow sufficient net inflow on the capital account to at least maintain the level of domestic investment. If it does not do this, a small nominal depreciation would achieve the same result. The fiscal policy response to the loss of import revenue has comparatively little influence on China's economic performance in the short run. Regardless of whether government spending or the government deficit is held constant, the optimal macro-policy environment is a floating exchange rate with no capital controls.

Macroeconomic impacts on migration

In a follow-up study to Tyers and Rees (see Chapter 9 and 2002), Chang and Tyers (see Chapter 10 and 2003) analyse the slow-down in China's income growth since the East Asian financial crisis. In particular, they examine the slow-down in rural income growth and the widening rural-urban income gap and ask to what extent this is due to: 1) the remaining obstacles to rural-urban migration (as suggested by Ianchovichina and Martin 2002); 2) the WTO trade reforms (as suggested by Anderson et al. 2002); or 3) restrictive macroeconomic policies. Using the GTAP model adaptation of Tyers and Rees (see Chapter 9),[2] the researchers test the extent to which China's fixed exchange rate and capital controls—and its WTO accession commitments—have contributed to the relatively poor performance of the rural sector. The East Asian financial crisis was seen as leading to a large (largely illegal) outflow of capital. This capital flight and the trade reforms are hypothesised to have led to a real exchange rate depreciation. The pegging of the yuan to the US dollar has therefore necessitated a deflation. If wages are 'sticky' and fall more slowly than prices, employment declines. It is hypothesised that the resulting real wage increase in the modern sectors has reduced labour demand and hence 'bottled up' workers in the rural sector and reduced rural per capita incomes.

Analysis of the data shows that while restraints on rural-urban migration have been relaxed to some extent, the migration flow has decreased rather than increased. The simulation results support the hypothesis that the fixed exchange rate and the capital controls have restricted the flow of workers from the rural sector. The model shows the rate of migration into the manufacturing sector falling by at least one percentage point a year and into the services sector by at least two percentage points a year. In all sectors there is a stark contrast between employment growth under tight capital controls and a fixed exchange rate regime on the one hand and an open capital account and floating currency regime on the other. Indeed, with an expansionary macroeconomic policy and optimistic assumptions about the productivity effects associated with the WTO accession reforms, simulated worker relocation demands from the reforms exceed the average

17

of China's recent rural-urban performance. This suggests that there is ample scope for a more rapid rate of migration of labour out of the rural sector, which would help reduce the rural-urban income gap.

China's agricultural trade after WTO accession

Chen (see Chapter 11) examines how the pattern of China's agricultural trade has changed since the country's accession to the WTO. In that time, the economy has grown very rapidly. The average annual growth rate of China's real GDP was more than 9.8 per cent during 2002-05. China's foreign trade has been expanding even more rapidly—at an annual growth rate of 28.6 per cent, compared with 9.4 per cent during the 1990s. Undoubtedly, China's economy has benefited from its more open international trade regime.

In real terms, China's agricultural exports and imports hardly changed in the period 1992-2001; but since then both have increased rapidly, although imports have grown much more rapidly than exports—so much so that in 2004 and 2005 China experienced its first agricultural trade deficits since at least the early 1990s. Chen shows that the changes in China's trade since 2001 have been strongly consistent with the country's comparative advantage, which is in labour-intensive activities. While increasing very rapidly, agricultural exports have, however, become a considerably smaller share of total trade. Moreover, within agricultural trade, processed agricultural products are dominating exports and imports are being dominated increasingly by cereals, vegetable oils and oil seeds, and raw materials for textiles. Processed agricultural products are labour-intensive activities, while cereals, vegetable oils and oil seeds, and raw materials for use in textiles are land-intensive activities.

Implications of the ASEAN-China Free Trade Agreement

In parallel with WTO accession, China has also engaged actively in regional and bilateral free trade agreements (FTAs). At present, China has bilateral FTAs with Pakistan, Chile, Jordan, Thailand, the Association of South-East Asian Nations (ASEAN) (currently the Early Harvest Program, EHP, is in operation but the full FTA is not expected until 2010), Hong Kong and Macau (as a Closer Economic Partnership Arrangement, CEPA). China is currently negotiating free trade agreements with Australia and New Zealand, and has proposed to negotiate FTAs with South Korea and India.[3] Among these bilateral FTAs, the ASEAN–China Free Trade Agreement (ACFTA) could have the largest impact on the economies of China and ASEAN.

Agricultural trade between China and ASEAN: complementary or competitive?

Yang and Chen (see Chapter 12) show that bilateral agricultural trade between China and ASEAN has increased rapidly in recent years, especially since the negotiation and implementation of the ACFTA and the launch of the EHP. ASEAN's agricultural exports to China have increased rapidly, reaching US$5 billion in 2005. China's agricultural exports to ASEAN have also increased but at a slower pace, reaching US$2.2 billion in 2005. ASEAN has been enjoying an increasing surplus with China in agricultural trade.

The share of exports to China in ASEAN's total agricultural exports has increased rapidly from 4.8 per cent in 1999 to 10.2 per cent in 2005. China became the third largest export market for ASEAN's agricultural products in 2005. With the full implementation of the ACFTA, the share is expected to rise further. A similar trend has not, however, been witnessed with respect to China's exports to ASEAN.

Yang and Chen find that agricultural trade between China and ASEAN is more complementary than competitive. China can be expected to export more labour-intensive agricultural products to ASEAN and import more land-intensive agricultural products from ASEAN. As China's revealed comparative advantage (RCA) in labour-intensive agricultural products is higher than that of ASEAN, it should be possible for China to increase its exports to ASEAN in these kinds of agricultural products (that is, fruits and vegetables, processed products, animal products and fish). Compared with China, ASEAN has an overwhelming comparative advantage in certain land-intensive agricultural products (such as rubber and palm oil). Therefore, both sides will be better off if they exploit their comparative advantage in agricultural sectors by deeper integration of their economies.

The agricultural production structure in ASEAN has experienced some adjustments to match Chinese market demand. The trade complementary index (TCI) for ASEAN's exports and China's imports in food and living animals (SITC0) rose from 1.0 in 1998 to 1.23 in 2005. The TCI for ASEAN's exports and China's imports of all agricultural products also increased quickly, from 1.07 in 2001 to 1.48 in 2005. These trends demonstrate that the complementarity of ASEAN's exports and China's imports has been increasing in recent years. This implies that ASEAN might have undergone a structural adjustment in its agricultural sectors in response to China's rising status as an export destination for its agricultural products. Such an adjustment has not, however, been witnessed in China—at least, not one as significant as in ASEAN.

It should be relatively easy for ASEAN to gain access to the Chinese market during the integration between the two economies. The high and increasing value of the TCI for ASEAN's exports and China's imports reveals the strong market match between ASEAN and China: ASEAN is selling what China wants to buy. Therefore, the structural adjustment in agricultural production in ASEAN (shown by the rising TCI) should improve ASEAN's capacities to grasp the opportunities provided by China's huge market.

There is, however, also competition in some agricultural products and certain structural adjustments are inevitable. For example, as China's imports from ASEAN in tropical fruits increased quickly in recent years, many Chinese farmers producing tropical fruits in coastal areas found that they were losing profits and domestic market shares (Newspaper of Southern Agriculture 2006). As a result, many fruit trees have been destroyed. Therefore, certain policies should be taken to assist the transition to different farming activities or to help farmers move to non-agricultural sectors.

Economic effects of the ASEAN–China Free Trade Agreement

Yang and Chen (see Chapter 13) assessed the economic effects of the ACFTA in its two stages up to 2010. The analysis is based on an improved recursive GTAP model. The data are based on Version 6.0 of the GTAP database for 2001, together with data derived from other sources. There are two distinguishing characteristics of this study. The first is that, in addition to the commitments in the ACFTA, it incorporates trade liberalisation in China (China's WTO commitments) and trade liberalisation in ASEAN (ASEAN free trade commitments). The second is that it has separated and explored the different effects of the two-stage implementation of the ACFTA.

Yang and Chen find that all member countries will gain from the ACFTA: it will increase social welfare and promote real GDP growth in the EHP phase of 2004-06 and in the full implementation during 2006-10. As the EHP includes only a small package of agricultural commodities, the gains in the second stage of the full implementation of the ACFTA will be much larger in all member countries.

There is a large trade creation effect among ACFTA members; total exports of ACFTA members will increase. A trade diversion effect is, however, also apparent. Trade between ACFTA members and other regions can be expected to decline due to the creation of the ACFTA. Because the trade creation effect is much larger than the trade diversion effect, global trade will be increased by the ACFTA, especially in the second stage of its full implementation.

The ACFTA will bring about substantial structural changes in China and in ASEAN countries. Trade liberalisation will improve the exploitation of comparative advantages in ACFTA member states. The structural changes will take place in agricultural and industrial sectors. The results also show that the different policy arrangements stemming from the two-stage trade liberalisation will have different impacts on the shifts in economic structure during the process of implementation.

The rest of the world will have to face the challenges brought about by the creation of the ACFTA. Because the ACFTA will enhance the competitiveness of China and ASEAN in each other's markets, exports of non-member countries will be substituted. Social welfare and real GDP will decline in the non-member countries as a result of the creation of the ACFTA.

Summary and discussion

Across the wide range of analysis carried out in this project, it is shown consistently that the trade reforms China adopted in order to accede to the WTO will mean substantial structural changes within the agricultural sector. Looked at from a partial equilibrium agricultural perspective, the reforms can be seen to result in substantial negative impacts across the sector and a worsening of food security in the sense of reduced access to income. It cannot be stated too strongly, however, that the outcomes of the reform have to be analysed from an economy-wide perspective. In China, as in other rural-based countries, the main factors behind reductions in rural poverty will be the scope for rural households to earn off-farm income and for people to move from rural areas into industrial and services activities in urban centres. Therefore, to a very large extent, the success of the trade reforms will depend on policies beyond agriculture.

There was considerable internal resistance to China joining the WTO, with concern about food security (more commonly seen in China as food self-sufficiency) and the perceived adverse impacts on the agricultural sector. The analysis carried out in this project has confirmed that structural change driven by productivity growth—which the trade reforms will promote—will lead to agriculture becoming a smaller and smaller part of the economy. As incomes increase, consumption patterns change and the share of agriculture shrinks, China will become less self-sufficient in many commodities. Because it has such a large population, however, China will always have to produce most of the food that it consumes. As the modelling has shown, trying to hold this development at bay or reverse it would have exorbitant costs. No doubt, however, there will continue to be resistance to the reforms in agriculture. The large proportion of the population that is still supported by agriculture can be a significant political weapon. Unfortunately, resistance to reforms can prevent exploitation of the potential for their incomes to increase, thus providing further ammunition for arguments to provide government support for the sector. It is important, therefore, that the economic arguments against 'food self-sufficiency' be made over and over.

The gap between urban and rural incomes has widened in the past decade or so. Partly, this is an outcome of the very rapid growth of the urbanised industrial and services sectors, but it is also the result of a slowing of the growth of rural incomes. This is partly the result of the slow-down in the growth of the TVE sector and the lessened opportunities for rural households to earn off-farm incomes. Incomes in the agricultural sector have not grown as rapidly as they could because of poor agricultural policies, such as ineffective and costly price support policies, regional self-sufficiency policies and monopolistic marketing and distribution of bulk commodities, fertilisers and seeds. China's accession to the WTO should help to maintain pressure to bring about reforms in these areas.

The current government grain reserve system is inefficient in many respects and very costly. It is run by different government agencies at the central, provincial and municipal levels, each with different interests and not very clearly defined roles. These arrangements give rise to conflicting interests in operations dealing with market instability and cannot serve the goal of food security well. A smaller, well-managed grain reserve system with a clear, single objective would better serve this goal. It should be understood, however, that grain markets will never operate efficiently while there is a government-run storage system, as it will crowd out efficient private storage.

The restrictions on rural-urban migration and constraints on the development of urban centres have also restricted opportunities for the rural-urban income gap to be reduced. The research shows that urbanisation has significant positive impacts on rural incomes and regional economic growth. The development of urban centres, particularly in inland regions, appears to be a matter of high priority. The 'keep them down on the farm' policy of China (and many other developing countries) seems to be due partly to a concern about problems associated with urban development, such as congestion and pollution. Urban centres should, however, be seen in a positive light. They exist because they provide efficiencies of scale and scope. Problems associated with cities derive largely from the lack of good planning of infrastructure, inappropriate property rights to land and inappropriate taxes and subsidies. China should therefore persist with infrastructure development to promote inland cities as an offset to the geographical disadvantages of the inland regions. Otherwise, it will face continued pressures for migration to the coastal cities. Removal of the remaining restrictions on rural-urban migration and discrimination against rural migrants should accompany the promotion of the development of urban centres.

It is shown that rural households in the coastal region will do best from the trade reforms as they have the most opportunity for earning income in off-farm employment. These opportunities arise because the non-agricultural industries in the eastern region benefit most from the trade reforms. This is partly the result of the bulk of private-sector development having taken place in this region. The government should reduce its support of state-owned enterprises (SOEs) in the inland regions, which crowd out private enterprises. The private sector, however, also needs the government to provide infrastructure to overcome the geographical disadvantages of the inland regions.

The modelling has also shown that China's monetary policy regime of the fixed yuan and capital controls has increased the rural–urban income gap by raising real wages and reducing employment growth in the non-agricultural sectors. The results show that moving away from this monetary policy regime could lead to a much more rapid relocation of labour out of agriculture and thereby promote a reduction in the rural–urban income gap.

A monetary policy regime change could also reduce the adverse impacts of the WTO trade reforms on the agricultural sector. Modelling with capital controls in place and the fixed exchange rate results in almost the entire agricultural sector being adversely affected by the tariff reductions. Even partial relaxation of the monetary policy regime could reduce this adverse impact. For example, when capital controls are in place but the exchange rate is floated, the 'other crops', livestock and fisheries sectors expand.

The government is, however, right to undertake any change in its monetary policy regime gradually and cautiously, as shown by the East Asian financial crisis. Adoption of a floating exchange rate would be premature, considering the underdeveloped state of China's financial sector, its only partially reformed banking sector and its still-vulnerable SOEs. The priority at this stage should be to accelerate the reforms in each of these areas.

Notes

1   A detailed description of the original model is provided by Hertel (1997).
2   The key difference between the long-run analysis in this paper (Chapter 10) and that in Tyers and Rees (Chapter 9) is that in the model used for the simulations reported in this paper (Chapter 10) there is an assumed positive relationship between the trade reforms and productivity growth.
3   See http://en.wikipedia.org/wiki/List_of_free_trade_agreements

References

Anderson, K., Jikun, H. and Ianchovichina, E., 2002. 'Impact of China's WTO accession on rural-urban income inequality', presented at the Australian Agricultural and Resource Economics Society Pre-Conference Workshop on WTO: Issues for Developing Countries, Canberra, 12 February.

Chang, J. and Tyers, R., 2003. *Trade reform, macroeconomic policy and sectoral labour movement in China*, Working Papers in Economics and Econometrics, No.429, Faculty of Economics and Commerce, The Australian National University, Canberra.

Chen, C. and Duncan, R., 2008. A*chieving Food Security in China: implications of WTO accession*, ACIAR Technical Reports No.69, Australian Centre for International Agricultural Research, Canberra.

Hertel, T. (ed.), 1997. *Global Trade Analysis Using the GTAP Model*, Cambridge University Press, New York.

Ianchovichina, E. and Martin, W., 2002. *Economic impacts of China's accession to the WTO*, World Bank Policy Research Working Paper 3053, The World Bank, Washington, DC.

Ito, S., Peterson, E. and Grant, W., 1989. 'Rice in Asia: is it becoming an inferior good?', *American Journal of Agricultural Economics*.

Jiang, T., 2002. 'WTO Accession and Regional Incomes', in R. Garnaut and L. Song (eds), *China 2002: WTO entry and world recession*, Asia Pacific Press, Canberra:45-62.

Lu, F., 1997. *Food trade policy adjustment in China and an evaluation of the risk of a food embargo*, Working Paper C1997007, China Centre for Economic Research, Peking University, Beijing.

Newspaper of Southern Agriculture, 2006. 'Pressures by ASEAN fruit import', 18 April.

Peterson, E., Jin, L. and Ito, S., 1991. 'An econometric analysis of rice consumption in the People's Republic of China', *Agricultural Economics*, 6(1):67-78.

Tong, J., 2003. 'WTO commitment: further marketisation and trade liberalisation', in R. Garnaut and L. Song (eds), *China 2003: new engine of world growth*, Asia Pacific Press, Canberra:141-51.

Tyers, R. and Rees, L., 2002. Trade reform in the short run: China's WTO accession, Paper presented to ACIAR Project 98/128 Workshop at the China Centre for Economic Research, Peking University, Beijing, September.

Wan, B., 1996. Prospect and policy of China's agricultural development, Keynote speech by the Agricultural Vice-Minister, Wan Baorui, to the international symposium on China's Grain and Agriculture: Prospect and Policy, 7 October, Beijing.

Wang, X., 2002. 'The WTO challenge to agriculture', in R. Garnaut and L. Song (eds), *China 2002: WTO entry and world recession*, Asia Pacific Press, Canberra:81-95.

Wang, X. and Duncan, R., 2003. 'The impact of urbanisation on economic growth', in R. Garnaut and L. Song (eds), *China 2003: new engine of world growth*, Asia Pacific Press, Canberra:217-30.

Wang, X., and Xia, X., 1999. 'Optimum city size and economic growth', *Economic Research*, (9), Beijing.

Yang, Y., 2000. Are food embargoes a real threat to China?, Mimeo., National Centre for Development Studies, The Australian National University, Canberra.

Yang, Y. and Tyers, R., 1989. 'The economic cost of food self-sufficiency in China', *World Development*, 17(2):237-53.

——, 2000. *China's post-crisis policy dilemma: multi-sectoral comparative static macroeconomics*, Working Papers in Economics and Econometrics, No.384, Faculty of Economics and Commerce, The Australian National University, Canberra.

02 Agricultural development and policy before and after China's WTO accession

---

Jikun Huang and Scott Rozelle

China's economy has experienced remarkable growth and significant structural changes since economic reforms were initiated in 1979. The annual average growth rate of gross domestic product (GDP) has been about 9 per cent in the past three decades (NBSC 2006). This rapid growth has been accompanied by a significant structural shift in the economy from agriculture to industry and services.

Although agriculture's share of the economy has been falling, China has still enjoyed agricultural growth rates that have considerably outpaced the increase in population. Food security—one of the issues of central concern to policymakers in China—has also improved significantly since the late 1970s. Contrary to the predictions of many analysts that China would put pressure on global food security in the course of its rapid industrialisation and the liberalisation of its economy, large net food import growth has not occurred (Huang et. al 2007). At the micro level, China has made remarkable progress in improving household food security and reducing the incidence of malnutrition during the past two and half decades.

While past accomplishments are impressive, there are still great challenges ahead. For example, income disparity has risen along with rapid economic growth. There are significant income disparities among regions, between urban and rural regions and households, and among households within the same location (Cai et al. 2002; World Bank 2002). Within the agricultural sector, China is also facing increasing challenges due to the diminishing amount of cultivated land in eastern China and increasing water scarcity in the northern part of the country (Sonntag et al. 2005). Many have predicted that almost all gains in agricultural output will have to come from new technologies that significantly improve agricultural productivity (Fan and Pardey 1997; Huang et al. 2002, 2003).

Trade liberalisation might also challenge China's agricultural and rural economy. There have been serious debates about the impact of World Trade Organization (WTO) accession on China's agriculture. Some argue that there will be substantial adverse impacts from trade liberalisation, affecting hundreds of millions of farmers (Carter and Estrin 2001; Li et al. 1999). Others believe that although some impacts will be negative and even severe in particular areas, the overall effect of accession on agriculture will be modest (Anderson et al. 2004; van Tongeren and Huang 2004; Martin 2002). In part, the differences in opinion can be traced to a lack of understanding of the policy changes that occurred before China's accession.

The objective of this chapter is to review China's agricultural development and policy before and since WTO accession. The chapter is organised as follows: first, we briefly review China's agricultural performance and policies before accession to the WTO. Next, we examine the main features of the agreement that China must adhere to as it enters the WTO. Third, we review changes in China's agricultural policy since joining the WTO. Finally, we conclude by drawing out the implications of the policy changes.

Agricultural performance and policy before China's WTO accession

Agricultural performance

China's economic liberalisation and structural changes have been under way for several decades. Although the reforms have penetrated the whole economy since the early 1980s, most of the transformations began and in some way have depended on growth in the agricultural sector (Nyberg and Rozelle 1999). In the period from 1978 to 1984, de-collectivisation, price increases and the relaxation of local trade restrictions on most agricultural products accompanied a take-off of China's agricultural economy. Grain output has increased by 4.7 per cent per annum, and there has been even higher growth in horticultural, livestock and aquatic products (Table 2.1). Although agricultural growth decelerated after 1985—following the one-off efficiency gains from de-collectivisation—the country still enjoyed agricultural growth rates that outpaced the increase in population (Table 2.1).

Despite the healthy increase in agricultural output, the even faster growth of the industrial and services sectors during the reform era has begun to transform the rural economy—from agriculture to industry and from rural to urban. During this process, the share of agriculture in the national economy has declined significantly. Whereas agriculture contributed more than 30 per cent of GDP before 1980, its share fell to 16 per cent in 2000 and to less than 13 per cent in 2005. Meanwhile, the share of services rose from 13 per cent in 1970 to 40 per cent in 2005 (Table 2.2).

Rapid economic growth and urbanisation have boosted demand for meats, fruits and other non-staple foods—changes that have stimulated sharp shifts in the structure of agriculture (Huang and Bouis 1996). For example, the share of livestock's output value in agriculture more than doubled from 14 per cent in 1970 to 30 per cent in 2000 (Table 2.2). Aquatic products rose at an even more rapid rate. One of the most significant signs of structural change in the agricultural sector is that the share of crops in total agricultural output fell from 82 per cent in 1970 to 56 per cent in 2000, and it has continued to fall since China's WTO accession in 2001. Moreover, the most significant declines in crop growth rates have been experienced in the grain sector (Table 2.1).

Changes in the external economy for agricultural commodities have paralleled those in domestic markets. Whereas the share of primary (mainly agricultural) products in total exports was more than 50 per cent in 1980, it fell to less than 10 per cent in 2000 (NBSC 1980-2000). In the same period, the share of food exports in total exports fell from 17 per cent to 5 per cent. The share of food imports also fell sharply—from 15 per cent to 2 per cent.

Crop-specific trade trends show equally sharp shifts and suggest that exports are moving increasingly towards products in which China has a comparative advantage (Huang et al. 2007). The net exports of land-intensive bulk commodities such as grains, cotton, oil seeds and sugar crops have fallen; exports of higher-valued, more labour-intensive products, such as horticultural and animal (including aquaculture) products, have risen (Figure 2.1).

Table 2.1 Annual growth rates of China's economy, 1970-2004 (per cent)

| | Pre-reform | | Reform period | | |
	1970-78	1979-84	1985-95	1996-2000	2001-04
Gross domestic product	4.9	8.8	9.7	8.2	8.7
Agriculture	2.7	7.1	4.0	3.4	3.4
Industry	6.8	8.2	12.8	9.6	10.6
Services	n.a.	11.6	9.7	8.3	8.3
Population	1.80	1.40	1.37	0.91	0.63
Per capita GDP	3.1	7.4	8.3	7.2	8.1
Grain production	2.8	4.7	1.7	0.03	-0.2
Rice:					
Production	2.5	4.5	0.6	0.3	-0.9
Area	0.7	-0.6	-0.6	-0.5	-1.2
Yield	1.8	5.1	1.2	0.8	0.2
Wheat:					
Production	7.0	8.3	1.9	-0.4	-1.9
Area	1.7	0.0	0.1	-1.4	-5.1
Yield	5.2	8.3	1.8	1.0	3.3
Maize:					
Production	7.4	3.7	4.7	-0.1	5.5
Area	3.1	-1.6	1.7	0.8	2.5
Yield	4.2	5.4	2.9	-0.9	2.8
Other production					
Cotton	-0.4	19.3	-0.3	-1.9	6.5
Soybean	-2.3	5.2	2.8	2.6	2.4
Oil crops	2.1	14.9	4.4	5.6	0.6
Fruits	6.6	7.2	12.7	8.6	29.5
Meats (pork/beef/ poultry)	4.4	9.1	8.8	6.5	4.6
Fish	5.0	7.9	13.7	10.2	3.5
Planted area:					
Vegetables	2.4	5.4	6.8	6.8	3.8
Orchards (fruits)	8.1	4.5	10.4	1.5	2.2

Note: Figures for GDP for 1970-78 are the growth rates of national income in real terms. Growth rates were computed using regression analysis. Growth rates for individual commodities and groups of commodities are based on production data.
Sources: National Bureau of Statistics of China, various issues (1985-2005). *China Statistical Yearbook*, China Statistics Press, Beijing; Ministry of Agriculture, 1985-2005 (various issues). *China's Agricultural Yearbook*, China Agricultural Press, Beijing.

Table 2.2   Changes in the structure of China's economy, 1970-2005 (per cent)

	1970	1980	1985	1990	1995	2000	2005
Share of GDP							
Agriculture	40	30	28	27	20	16	13
Industry	46	49	43	42	49	51	47
Services	13	21	29	31	31	33	40
Share of agricultural output							
Crops	82	80	76	65	58	56	51
Livestock	14	18	22	26	30	30	35
Fisheries	2	2	3	5	8	11	10
Forestry	2	4	5	4	3	4	4

Source: National Bureau of Statistics of China, various issues. *China Statistical Yearbook*, China Statistics Press, Beijing; National Bureau of Statistics of China, various issues. *China Rural Statistical Yearbook*, China Statistics Press, Beijing.

Figure 2.1   Agricultural trade balance by factor intensity, 1985-97 (US$ million)

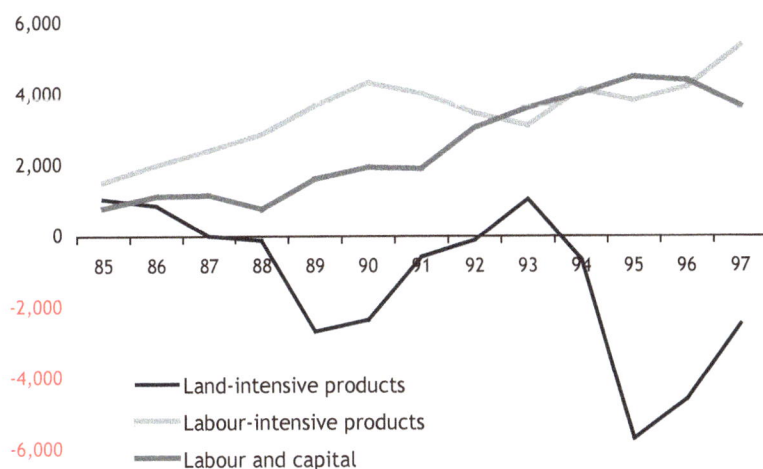

Source: Huang, J. and Chen, C., 1999. *Effects of Trade Liberalization on Agriculture in China: institutional and structural aspects*, United Nations ESCAP CGPRT Centre, Bogor.

We believe that, taken as a whole, the trends in China's economic structure and agricultural trade in the two decades before WTO accession reveal that the changes that are being experienced as a result of accession are not new. China was already moving in a direction that was more consistent with its resource endowments. To the extent that the new trade agreements reduce barriers to allow more land-intensive products into the domestic market and the reductions in import restrictions overseas stimulate the export of labour-intensive crops, the main impact of WTO accession is to push forward trends that were already under way. The commitments that China undertook in its WTO Protocol of Accession are largely consistent with the nation's long-term reform plans.

Agricultural development policies

Despite the continuity with the past, few can dispute that the terms of China's WTO accession agreement pose new challenges to the agricultural sector. In some cases, there could be large impacts on rural households and these will undoubtedly elicit a sharp response. While the nature and severity of the impacts will depend on how households respond, perhaps of even greater importance is how China's agricultural policymakers manage their sector as the new trade regime takes effect. To examine this set of issues more carefully, in this section we review agricultural policy before China's WTO accession and in the next section we examine how WTO measures change the policy environment in rural China.

Fiscal and financial policies. While government expenditure in most areas of agriculture increased gradually during the reform period, the ratio of agricultural investment to agricultural gross domestic product (AGDP) declined monotonically from the late 1970s to the mid 1990s. In 1978, the AGDP was 7.6 per cent; by 1995, it had fallen to 3.6 per cent (NBSC 2001). Moreover, a significant outflow of capital from the agricultural sector to the industrial sector and from the rural sector to the urban sector took place during the 1980s and 1990s through the financial system and government agricultural procurement (Huang et al. 2006; Nyberg and Rozelle 1999). After the mid 1990s, when China was preparing for WTO accession, investment in agriculture was increased significantly. This trend of increasing investment in the agricultural and rural sectors has continued since accession.

Foreign exchange and trade policies. China's policies governing the external economy have played a highly influential role in shaping the growth and structure of agriculture for many decades. During the entire Socialist Period (1950-78), the over-valuation of China's domestic currency destroyed incentives to export, effectively isolating China from international exporting opportunities (Lardy 1995). The government, however, allowed the real exchange rate to devalue by 400 per cent between 1978 and 1994. Except for recent years, when the exchange rate has experienced a slight appreciation, adjustments in the exchange rate throughout most of the reform period increased export competitiveness and contributed to China's export growth record. This, in turn, has assisted the expansion of the national economy. Perhaps more than anything, China's 'open-door' policy, including its exchange rate policy, has contributed to the rapid growth in the importance of the external economy.

Rural development and labour market policies. The shift of labour from the rural sector to the urban sector lies at the heart of a country's modernisation effort, and China has been experiencing this in primarily two ways: by the absorption of labour into rural firms and by massive movement of labour into cities. Rural industrialisation has played a vital role in generating employment for rural labour, raising agricultural labour productivity and farm incomes. The share of rural enterprises in GDP rose from less than 4 per cent in the 1970s to more than 30 per cent by 1999. Rural enterprises dominated the export sector throughout the 1990s (NBSC 2001). Perhaps most importantly, rural enterprises employ about 35 per cent of the rural labour that works off the farm. In addition to wage-earning jobs in rural areas, a large and increasing part of the rural labour force—up from 8 per cent in 1990 to 13 per cent in 2000 (de Brauw et al. 2002)—also works in the self-employed sector.

Although China's input markets still had a number of structural imperfections—such as employment priority for local workers, housing shortages and the urban household registration system—labour poured into cities during the 1980s and 1990s and the emergence of labour markets has been transforming the economy (Lin 1991; Lohmar 1999; de Brauw et al. 2002). According to a national survey of 1,200 households, more than 100 million rural workers found employment in the urban sector in the late 1990s (de Brauw et al. 2002). China's labour markets have allowed migrant work to become the dominant form of off-farm activity. Young and better-educated workers increasingly dominate this activity. It has expanded fastest in regions that are relatively well off; and has recently begun to draw workers from parts of the population, such as women, who earlier had been excluded. If China continues to change at the pace it did in the 1980s and 1990s, and other provinces experience the changes that have taken place in the richest provinces, China's economy should continue to follow a healthy development path and be well along the road to modernisation.

Improved incentives. China's rural economic reform, initiated in 1979, was founded on the household responsibility system (HRS). The HRS reforms dismantled the communes and contracted agricultural land to households, mostly on the basis of family size and the number of workers in the household. Most importantly, after the HRS reforms, control and income rights belonged to individuals. With the exception of the right to sell their land, farmers became the residual claimants of the outcome of their efforts.

There is little doubt that the changes in incentives resulting from these property rights reforms triggered strong growth in output and productivity. In the most definitive study on the subject, Lin (1992) estimated that China's HRS accounted for 42–46 per cent of the total increase in output during the early reform period (1978–84). Fan (1991) and Huang and Rozelle (1996) found that even after accounting for technological change, institutional change during the late 1970s and early 1980s contributed about 30 per cent of output growth. Other researchers have documented impacts that go beyond increases in output. For example, McMillan et al. (1989) found that the early reforms also raised total factor productivity, accounting for 90 per cent of the increase in output (23 per cent) between 1978 and 1984. Jin et al. (2002) also showed that the reforms had a large impact on productivity, accounting for growth in total factor productivity of more than 7 per cent annually.

Domestic output price and market liberalisation policies. Although early in the reforms China's leaders had no concrete plan to liberalise markets, they did take steps to change the incentives faced by producers that were embodied in the prices they received for their marketed surplus. The important contribution of China's pricing policy was in the timing and breadth of the policy change. The first major price rise occurred in 1979, almost at the time reformers were deciding to de-collectivise. Agricultural procurement prices have continued to increase gradually since the 1980s. Studies confirm the strong impact of these price changes on output during the early and late years of transition (Lin 1992; Fan 1991; Huang and Rozelle 1996; Fan and Pardey 1997).

In addition to pricing changes and de-collectivisation, another major task of reformers after the mid 1980s was to create more efficient institutions of exchange. Markets—whether classic competitive ones or some workable substitute—increase efficiency by facilitating transactions among agents to allow specialisation and trade and provide information through the pricing mechanism to producers and consumers about the relative scarcity of resources. After 1985, although the process proceeded in a stop-start manner, market liberalisation began in earnest. Changes in the procurement system, further reductions in restrictions on commodities trading, moves to commercialise the state grain trading system and calls for the expansion of markets in rural and urban areas led to a surge in market-oriented activity. For example, in 1980, there were only 241,000 private and semi-private trading enterprises registered with the State Markets Bureau; by 1990, there were more than 5.2 million (de Brauw et al. 2002).

Despite its stop-start nature, as the right to private trading was extended to include surplus output of all agricultural products after contractual obligations to the State were fulfilled, the foundations of the state marketing system began to be undermined (Rozelle et al. 2000). Reformers eliminated all planned procurement of agricultural products other than rice, wheat, maize and cotton; government commercial departments could buy and sell only through the market. For grain, improved producer incentives were introduced through reductions in the volume of the compulsory delivery quotas and increases in procurement prices. Even for grain, the share of the compulsory quota procurement reached 29 per cent of output in 1984 but declined to 18 per cent in 1985 and to 13 per cent in 1990. The share of the negotiated procurement at market prices increased from only 3 per cent in 1985 to 6 per cent in 1985 and to 12 per cent in 1990.

Technology and water infrastructure development. Agricultural research in China remains almost completely run by the government. Reflecting the urban bias of food policy, most crop breeding programs have emphasised fine grains (rice and wheat). For national food self-sufficiency considerations, higher yields have been the major target of China's research program, except in recent years when quality improvement was introduced into the nation's development plan. Although there have been several private domestic and joint-venture investments in agricultural research and development, policies still discriminate against them. Today, the record of reform of the agricultural technology system is mixed and its impact on technological developments and crop productivity is unclear. Empirical evidence demonstrates that while agricultural technology has played a critical role in China's agricultural productivity growth, the effectiveness of China's agricultural research capabilities has been declining (Jin et al. 2002).

Before the economic reforms, the State's agricultural investment was focused on building dams and canal networks, often with the input of corvée labour from farmers. After the 1970s, greater focus was given to increasing the use of China's massive groundwater resources (Wang et al. 2006). By 2005, China had more tube wells than any country in the world—except possibly India. Although initially this investment was made by local governments with aid from county and provincial water bureaux, by the 1990s the national government was encouraging the shift in ownership that was occurring as pump sets and wells and other irrigation equipment went largely into the hands of farm households (Wang 2000). At the same time, private water markets (whereby farmers pumped water from their own wells and sold it to other farmers in the village) were also encouraged. The main policy initiative in the surface-water sector after the mid 1990s was management reform (with the goal of trying to make water use more efficient).

Trade policy. In addition to important changes in foreign exchange policy (changes that saw the nation's currency depreciate steeply and trading rights become more accessible during the 1980s and 1990s), there have been a number of other fundamental reforms of China's international trading system. Lower tariffs and rising imports and exports of agricultural products began to affect domestic terms of trade in the 1980s. Initially, most of the decline in protection affected commodities that were controlled by single-desk, state traders (Huang and Chen 1999). For many products, competition among non-state foreign trade corporations began to stimulate imports and exports (Martin 2002). Although trade in many major agricultural commodities was not liberalised, the moves spurred the export of many agricultural goods. In addition, policy shifts in the 1980s and 1990s also changed the behaviour of state traders, who were allowed to increase imports.

Moves to relax the rights of access to import and export markets were matched by action to reduce the taxes that were being assessed at the border. A new effort began in the early 1990s to reduce the level of formal protection of agriculture. From 1992 to 1998, the simple average agricultural import tariff fell from 42.2 per cent to 23.6 per cent, and by 2001 it had reached 21 per cent (Rosen et al. 2004).

Overall, agricultural sector trade distortions declined in the 20 years before China joined the WTO (Huang et al. 2004). Much of the decline in protection came from decentralising the authority for imports and exports and from relaxing licensing procedures for some crops (for example, moving oil and oil seed imports away from state trading firms), as well as foreign exchange rate changes. Other trade policy changes have reduced the scope of non-tariff barriers (NTBs), relaxed the real tariff rates at the border and increased quotas (Huang and Chen 1999). Despite this real and, in some areas, rapid set of reforms, the control of a set of commodities that leaders consider to be of national strategic importance—including rice, wheat and maize—remains with policymakers to a large extent (Nyberg and Rozelle 1999).

Given the changes made before the country's accession to the WTO, it is not surprising that, while it was a major event in China (and will have an impact on many sectors), in its most basic terms it is a continuation of previous policies. Hence, the commitments embodied in China's WTO accession agreement with respect to the agricultural sector—market access, domestic support and export subsidies—are essentially a continuation of what China was doing in the 1990s.

Summary of China's transition-era agricultural policies. The scope of China's policy efforts during the transition era is impressive. Policy shifts were made in pricing, the organisation of production, marketing, investment, technology and trade. Although the rate of investment has risen during the reform period, China is still under-investing in agriculture compared with other countries. Taxes—those that are explicit and those that are implicit in pricing and trade policies—also have fallen. While China has not reached a point during the transition era where it heavily subsidises the agricultural economy in a way that characterises its neighbours in East Asia, it appears, however, to be heading in the direction in which developing countries at a certain point begin to turn from extraction from agriculture to net investment in the sector.

Outside of agriculture, many policies and other factors have affected the sector. Other rural policies—for example, fiscal reform, township and village enterprise emergence and privatisation, and rural governance—almost certainly have a large, albeit indirect, effect on agriculture. Urban employment policies, residency restrictions, exchange rate management and many other policy initiatives also affect agriculture by affecting relative prices in the economy, the access to jobs off the farm and the overall attractiveness of remaining on the farm.

Taken together, these policies have been shown to have a dramatic effect on China's agricultural sector. They have increased the output of food, driven agricultural prices down and improved supplies of non-grain food and raw materials for industry. The mix of policies—pricing, property rights, market liberalisation, investment and trade—has also made producers more efficient; it has freed up labour and other resources that are behind the structural transformation of the economy. The most convincing indicators that agriculture is beginning to play an effective role in the nation's development are that the importance of grain is shrinking inside the cropping sector, the importance of the cropping sector is shrinking inside the agricultural sector and the importance of agriculture is shrinking in the general economy. Rural incomes and productivity have increased; however, much of the increase in welfare is being generated by individuals (and there have been more than 200 million of them) who have been able to escape from grain production and move into high-valued crops, escape from cropping and move into livestock and fisheries production and, most importantly, escape from agriculture (the rural economy) and move into off-farm jobs (in the city).

China's commitments to WTO accession and provisions related to agriculture

In their most basic terms, the commitments in the agricultural sector can be classified into three categories: market access, domestic support and export subsidies (Martin 2002; Colby 2001; Rozelle and Huang 2001). The commitments on market accession have lowered tariffs on all agricultural products, increased access to China's markets for foreign producers of some commodities through tariff rate quotas (TRQs) and removed quantitative restrictions on others. In return, China is supposed to gain better access to foreign markets for its agricultural products, as well as a number of other indirect benefits.

Its substantial market-access commitments make China's accession unique among developing countries that have been admitted to the WTO. Overall, agricultural import tariffs (in terms of simple averages) declined from about 21 per cent in 2001 to 17 per cent by 2004. Previously, the simple average agricultural import tariff had been reduced from 42.2 per cent in 1992 to 23.6 per cent in 1998.

With a few exceptions—for example, in the case of several 'national strategic products'—most agricultural products have become part of a tariff-only regime. According to this part of the agreement, all non-tariff barriers and licensing and quota processes should be eliminated. For most commodities in this group, effective protection fell substantially after January 2002 (Table 2.3). To the extent that tariffs are binding for some of these commodities, the reductions in tariff rates should stimulate additional imports.

It is important to note, however, that although published tariff rates have fallen on all of these commodities, imports will not necessarily grow. Indeed, China has comparative advantage in many of the commodities in Table 2.3. For example, lower tariffs on horticultural and meat products might impact on only a small part of the domestic market—such as those parts that buy and sell only very high-quality products, such as meats for five-star hotels that cater to foreigners. Although tariffs have fallen for all products, since China exports many commodities at below world market prices, the reductions have not affected producers or traders.

The real challenge for agricultural products with tariff-only protection is, therefore, for commodities such as barley, wine and dairy products. To

Table 2.3   Import tariff rates on major agricultural products subject to tariff-only protection in China

	Real tariff rates in 2001	Effective as of 1 January 2002	Effective as of 1 January 2004
Barley	114 (3)[a]	3	3
Soybean	3[b]	3	3
Edible oils	114 (3)[a]	9	9
Citrus	40	20	12
Other fruits	30-40	13-20	10-13
Vegetables	30-50	13-29	10-15
Beef	45	23.2	12
Pork	20	18.4	12
Poultry meat	20	18.4	10
Dairy products	50	20-37	10-12
Wine	65	45	14
Tobacco	34	28	10

[a] Barley was subject to licence and import quotas; the tariff rate was 3 per cent for imports within the quota and no above-quota barley with the 114 per cent tariff was imported in 2001.
[b] The tariff rate was as high as 114 per cent before 2000 and lowered to 3 per cent in early 2000.
Source: World Trade Organization, 2001. *China's WTO Protocol of Accession*, November, World Trade Organization, Geneva.

understand what could happen to some of these products it is instructive to examine the case of soybeans, where producers clearly did not have a comparative advantage. Before 2000, the import tariff for soybeans was as high as 114 per cent, importers required licences and soybean demand was met mostly by local producers. In anticipation of WTO accession, however, tariffs were lowered to 3 per cent in 2000 and import quotas were later phased out. Subsequently, imports surged from 4.32 million metric tonnes (mmt) in 1999 to 14 mmt in 2001 and to more than 20 mmt after 2003. Prices also fell and the nominal protection rate for soybeans declined from 44 per cent in early 2000 to less than 15 per cent in October 2001 (Rozelle and Huang 2001). From this case, it can be seen that when protection rates are reduced from high levels and there is strong demand for the commodity, imports can increase sharply.

Such behaviour has, however, been constrained for a class of commodities called 'national strategic products'. China's WTO agreement allows the government to manage the trade of rice, wheat, maize, edible oils, sugar, cotton and wool with TRQs. These commodities are covered under special arrangements. As shown in Table 2.4, the in-quota tariff is only 1 per cent for rice, wheat, maize and wool (for sugar it is 20 per cent and for edible oils, 9 per cent). The amount brought in at these tariff levels is, however, strictly restricted. For example, in 2002, the first 8.45 mmt of wheat came in at a tariff rate of 1 per cent. The in-quota volumes were to grow over a three-year period (2002–04) at annual rates ranging from 4 per cent to 19 per cent. For example, maize TRQ volumes increased from 5.7 mmt in 2002 to 7.2 mmt in 2004. China does not have to bring in this quantity, but provisions are in place such that there is supposed to be competition in the import market so that if there is demand for the national strategic products at international prices, traders will be able to bring in the commodity up to the TRQ level.

At the same time, there are still ways, theoretically, to import these commodities after the TRQ is filled. Tariffs on out-of-quota sales (that is, more than 7.2 mmt in 2004 for maize) dropped substantially in the first year of accession and fell further between 2002 and 2005. If the international price of maize were to fall more than 65 per cent below China's price after 2004, traders would be allowed to import. During the transition period, however, the tariff rates were so high (for example, 65 per cent for grains and sugar in 2004 and edible oils in 2005) that in 2002–05 they were not binding.

After the first four to five years of accession, other changes will take place. For example, China agreed to phase out its TRQ for edible oils after 2006. State wool-trading monopolies could also be phased out and disappear gradually for most other agricultural products (Table 2.4). Although the China National Cereals, Oil and Foodstuffs Import and Export Company will continue to play an important role in rice, wheat and maize, it will face increasing competition in grains trading from private firms.

In its accession commitments, China made a number of other agreements, some of which are China-specific. First, China must phase out all export subsidies and not introduce any such subsidies on agricultural products in the future. Moreover, despite clearly being a developing country, China's *de minimis* exemption for product-specific support is equivalent to only 8.5

Table 2.4   China's market access commitments on farm products
subject to tariff rate quotas

	Import volume (mmt) (state trading share, %)			Quota growth (% p.a.)	In-quota tariff (%)	Out-of-quota tariff (% as of 1 January)		
	Real 2000	Quota 2002	Quota 2004			2002	2003	2004
Rice	0.24 (100)[a]	3.76 (50)	5.32 (50)	19	1	74	71	65
Wheat	0.87 (100)	8.45 (90)	9.64 (90)	8	1	71	68	65
Maize	0 (100)	5.70 (67)	7.20 (60)	13	1	71	68	65
Cotton	0.05 (100)	0.82 (33)	0.89 (33)	5	1	54.4	47.2	40
Wool[b]	0.30	0.34	0.37	5	1	38	38	38
Sugar[c]	0.64	1.68	1.95	8	20	90	72	50

[a] Figures in parentheses are the percentage of non-state trading in the import quota.
[b] Designated trading in 2002–04 and phased out thereafter.
[c] Phased out quota for state trade.
Source: World Trade Organization, 2001. *China's WTO Protocol of Accession*, November, World Trade Organization, Geneva; National Bureau of Statistics of China, 2001. *China Statistical Yearbook*, China Statistics Press, Beijing.

per cent of the total value of production of a basic agricultural product (compared with 10 per cent for other developing countries). Moreover, measures such as investment subsidies for farmers and input subsidies for the poor and other resource-scarce farmers, which are generally available for policymakers to use in other developing countries, are not allowed in China—that is, China must include any such support as part of its aggregate level of support, which should be less than 8.5 per cent of agricultural output value.

Because of its socialist background and the difficulty that the world has had in assessing the scope of the government's intervention in business dealings of all types, China agreed to a series of measures governing the way it deals with the rest of the world in cases of anti-dumping and countervailing duties. Put simply, special anti-dumping provisions will remain for 15 years after China's WTO accession. In cases of anti-dumping, China will be subject to a different set of rules that countries can use to prove their dumping allegations. In addition, the methods that countries can use against China to enforce anti-dumping claims will differ from most of the rest of the world. In essence, this set of measures makes it easier for countries to bring, prove and enforce dumping cases against China. It should be noted, however, that although the rules differ from those governing trade among other countries, China will have the same rights in its dealings with other countries—which could help it in some dumping matters when it concerns a partner's export behaviour.

China's WTO commitments and privileges associated with the measures in other parts of the agreement also affect its agriculture. For example, on agricultural chemicals, China has committed to replace quantitative import restrictions on three types of fertilisers (DAP, NPK and urea) by TRQs. Tariffs were cut on accession and further cuts have been phased in since 2005 on almost all industrial products (for example, tractors and pesticides). Furthermore, China has reduced significantly its non-tariff measures and eliminated all quotas, tendering and import licensing on non-farm merchandise since 2005. For textiles and clothing, however, the current 'voluntary' export restraints will not be phased out completely until the end of 2008, meaning that exports might not expand as fast as they would under a less restrictive regime. Substantial commitments to open up services markets in China also have been made.

Policy shifts since China's WTO accession

While the substantial institutional and marketing reforms implemented in agriculture since the late 1970s can facilitate the response of households to the changes that have arisen with the implementation of WTO accession—and more generally in China's transition to the post-accession regime—China still requires considerable reform to meet its WTO commitments (Martin 2002). In fact, the government has realised for a considerable time that it faces a real challenge. In many instances, officials have taken this challenge as an opportunity to accelerate continuing reforms in international and domestic policies.

Policy responses to WTO accession have taken two forms. First is the policy response required of China in order for it to keep its commitments and to adjust its domestic policies to be consistent with those promulgated by the WTO's rules. The other is the introduction of measures that are allowed under the new framework that could help to boost China's economy and minimise adverse shocks that arise as a result of the accession. Identifying the two kinds of policy changes is essential in studying how accession affects the ways that policymakers respond.

Legal and legislative changes

Many of the most important changes that have occurred because of the accession are in legal and legislative areas. China reserved the right to use a transitional period of one year from the date of accession to amend or repeal any institution, regulation, law or legal stipulation in its current economic policies in order to make them consistent with the WTO spirit of non-discrimination and transparency. The government had recognised the need for this and had already begun to rectify legal rules and legislation since the late 1990s.

To provide guidance to ministerial and local government authorities in amending or repealing relevant regulations, laws and policies, the State Council decreed two important regulations in January 2002: Regulations on Formulation Process of Laws and Regulations on Formulation Process of Administrative Laws. Essentially a guide for local governments and ministries, these new regulations were issued with the aim of handing over many government functions to the market and directing the government to take a more regulatory, indirect role in commerce and trade. They try to limit the role of government and emphasise that its role is primarily one of providing social and public services. The regulations also seek to simplify administrative processes and increase the transparency of regulations and policies.

The effort to create and implement this new regulatory framework is widespread. For example, during the last stage of WTO negotiations, each ministry formed a leading group or committee to work on all of the laws and regulations under its jurisdiction. These committees typically comprised decision makers and experts who had the mandate to clean up all regulations and prepare proposals for amending or repealing those laws and regulations that were not consistent with the WTO rules and China's commitments to its accession. Local governments had similar committees. Ministries and provincial governments are also working closely under the Standing Committee of the National People's Congress on those laws and regulations to be amended or repealed by the congress.

Several recent experiences involving amending laws and regulations and creating new institutions related to agriculture demonstrate the effectiveness of these committees and China's commitment to its WTO obligations. For example, China's Patent Law—which was issued originally in 1984 and amended in 1992—was amended again on 1 July 2001. Many of the associated regulations were also revised. Moreover, a new set of regulations on plant variety protection (PVP) was put into effect in 1999 when China became the thirty-ninth member country of the International Union for the Protection of New Varieties of Plants (UPOV). Soon after passage of the new regulations, government agencies proposed and implemented detailed regulations facilitating the implementation of PVP. The Ministry of Agriculture and the State Forest Bureau also created a new set of institutions: a series of New Plant Variety Protection Offices. Finally, China's Seed Law was issued in 2000. Hence, the PVP and the Seed Law now protect the rights of new plant varieties. To assist in the implementation of these laws, the government has set up an intellectual property rights (IPR) affairs centre under the Ministry of Science and Technology.

To fulfil its legal obligations related to agriculture, the Ministry of Agriculture has, since 2000, also repealed several regulations that sought to subsidise certain types of enterprises and apply different rules in agricultural input industries to different economic actors. Officials have eliminated the Regulations on the Development of Integrated Agricultural, Industrial and Commercial Enterprises under State Farms (issued in 1983 to assist in the development of state-owned farms) and the Regulations on the Development of Rural Township and Village Owned Enterprises (issued in 1979 to assist collectively owned enterprises). Seed Management Regulations that gave monopoly powers to local seed companies and Pesticide Field Trial Rules that discriminated against foreign companies were abolished.

Despite these substantial efforts, considerable institutional reforms are still needed. There are a number of laws and rules that treat domestic and foreign companies and individuals differently; these need to be changed to allow China to fulfil its legal obligations under its Protocol of Accession to the WTO. It could be an even greater challenge to build the nation's capacity for effective implementation of the amended and new laws.

Agricultural trade reforms

Changes in tariff policy are more straightforward and simpler than non-tariff policy reforms. China followed its tariff reduction schedule specified in the protocol on the first day of 2002. The average tariff rate was reduced from 15.3 per cent in 2001 to 12 per cent. For agricultural products, the tariff reduction was from 21 per cent to 15.8 per cent. China has also implemented its three years of transition for progressively liberalising the scope and availability of trading rights for agricultural products, as discussed in the previous section.

Compared with the trend in tariff reduction in the past decade, the tariff changes necessary under China's WTO accession should present relatively few problems. Significant reforms of non-tariff measures are, however, required—particularly state trading. China agreed to and did phase out restrictions on trading rights for all products except those under TRQ trade regimes; here, a more gradual approach to phasing out the state-trading regimes has been adopted (Table 2.4). Three years after accession, the private sector has come to dominate the trade of almost all agricultural products. There are, however, provisions to maintain state involvement in three commodities: wheat, maize and tobacco.

Technical barriers to trade (TBT) and sanitary and phytosanitary (SPS) measures, as well as institutional arrangements to fulfil the agreement on Trade-Related Intellectual Property Rights (TRIPS), are the other important issues that China has to deal with. The agreements on TBT and SPS focus on using internationally accepted standards to govern the use of standards as protectionist devices. This rules-based approach can be valuable in improving policy formulation, but is likely to require investment in strengthening standards-related institutions. Comprehensive adoption of these measures should lead to improved policies and, by basing policies on a scientific approach, should lead to a move away from the time-consuming and inefficient approach of resolving these issues on a political basis. China has undoubtedly struggled in its efforts to create a fully transparent and open trade regime with respect to non-tariff barriers.

Domestic market reform and infrastructure development

After more than 20 years of reform, China's agriculture has become much more market oriented (de Brauw et al. 2000). Traders move products around the country with increasing regularity and factor markets adjust more rapidly. By the late 1990s, only grain, cotton and, to some extent, silkworm cocoon and tobacco were subjected to price interventions. Even in these cases, markets—especially those for grain—have become increasingly competitive, integrated and efficient.

Despite the gains in market performance in recent years, WTO accession makes demands on China's domestic agricultural markets. The domestic marketing policy response to the nation's WTO accession was substantial and will continue. Major changes have been aimed at improving the efficiency of market performance and minimising the adverse shocks that could arise from external trade liberalisation. Perhaps more than in any other sector, the full liberalisation of cotton and grain markets in China clearly showed that its leaders have been using this opportunity to develop healthy domestic agricultural markets.

In response to WTO accession, the government has ambitious plans to increase investment in market infrastructure. Leaders see a need to establish an effective national marketing information network. Officials in the Ministry of Agriculture have been attempting to standardise agricultural product quality and promote farm marketing. Some have advocated the creation of agricultural technology associations. Generally, all of these moves are part of an effort by leaders to shift fiscal resources that were once used to support China's costly price subsidisation schemes (including domestic and international trade subsidies) to productivity-enhancing investments and marketing infrastructure improvements.

Land-use policy

The policy implications of WTO accession on land use and farm organisations are also being hotly debated. Many concerns have arisen about the ability of China's small farms to compete after trade liberalisation. Although every farm household holds title to land, the average farm size is small (about 0.5 hectares). Leaders are pleased with the equity effects of the nation's distribution of land, as it allays concerns about food security and poverty. Land fragmentation and the extremely small scale of farms will, however, almost certainly in some way constrain the growth of farm labour productivity and hold back farm income growth.

Although most policymakers appear to favour more secure land rights, they are still searching for complementary measures that will not cancel all of the pro-equity benefits of the current land management regime. By law, land in rural areas is owned collectively by the village (about 300 households) or small group (*cunmin xiaozu*, with 15–30 households) and is contracted to households (Brandt et al. 2002). One of the most important changes in recent years has been the extension of the duration of the use contract from 15 to 30 years. By 2000, about 98 per cent of villages had amended their contracts with farmers to reflect the longer use rights.

With the issue of use rights resolved, the government is now searching for a mechanism that will permit the remaining full-time farmers to gain access to additional cultivated land and increase their income and competitiveness. One of the main efforts revolves around the development of a new Rural Land Contract Law. The Standing Committee of the National People's Congress has drafted a law and the congress is expected to approve it in the near future. According to this law, although the ownership rights of land will remain with the collective, almost all other rights that would have been held under a private property system are given to the contract holder. In particular, the law clarifies the rights for transfer and exchange of the contracted land— something that could be taking effect already, as researchers are finding that increasing areas of land in China are rented. The new legislation also allows farmers to use contracted land for collateral to secure commercial loans and allows family members to inherit land during the contracted period. The goal of this new set of policies is to encourage farmers to increase their farm and household short and long-run productivity.

Farm organisations

The other major attempt to increase farm productivity and agricultural competitiveness under trade liberalisation is to promote the development of farmer organisations. At one time, the creation of such organisations was a politically sensitive issue. Leaders were concerned about the rise of any organisation outside the government's authority. Such restrictions, however, caused a dilemma in reforming the nation's agricultural and rural economies. Policymakers are also aware that given the small scale of China's farms there are many increases in economic efficiency that might be achieved by the creation of effective rural organisations and that if they are successful in raising incomes, there might be an increase in political stability. The government has now given its support to self-organised farmer groups that focus on agricultural technology and marketing. The Farmers' Professional Cooperative Law was issued in October 2006 to facilitate the development of farmers organisations.

Export subsidies and agricultural support policies

In one of its most fundamental concessions (since most countries are not required to do so on the basis of their own WTO protocols), China agreed to phase out its export subsidies in the first year of WTO accession. Such subsidies have played a considerable role in assisting the export of maize, cotton and other agricultural products, and in this way have indirectly supported domestic prices. In fact, after phasing out export subsidies, several of China's agricultural sectors (for example, maize and cotton) are likely to be subject to intensive competition from imports.

Besides the elimination of export subsidies, the WTO accession commitments place strict controls on the types and amounts of certain investments. In particular, domestic support to agriculture is divided into 'Green Box' and 'Amber Box' support. As is the case with other WTO members, China faces no limitations on the amount of support classified as Green Box that it can give, but it faces carefully circumscribed rules regarding the amount of support that can be given to activities classified as Amber Box. WTO commitments will, therefore, most likely force China to shift the level and composition of its agricultural support in the future.

49

On Amber Box policies, the accession protocol allows a *de minimis* level of support equal to 8.5 per cent of agricultural gross value product. After intense negotiations, the level was set somewhat below that enjoyed by other developing countries (10 per cent) but above that allowed to industrialised countries (5 per cent). Moreover, the list of items used in the computation of China's Amber Box support is wider than that applied to other countries (for example, certain agricultural output and input subsidies for poor farmers that are considered as Green Box in developing countries are included in the computation of Amber Box in China). On paper, therefore, China's hands appear to be quite firmly tied regarding the scope of the support it is able to provide. When one considers the amount of fiscal funds that China has historically devoted to these areas, however, it could be that the *de minimis* limit will not be binding. The biggest impact could be some time in the future after China has grown further and its budget constraint has been relaxed somewhat. China's commitment should, however, be thought of as fairly limiting as it closes off options for supporting its rural activities in ways that its neighbours in East Asia have done (Martin 2002).

New agricultural and rural development policies

Since the early 2000s, China's leaders have been considering solutions to the 'Three Nongs' issues, which are at the top of the economic development agenda: agriculture, rural areas and farmers. To facilitate economic development in these areas, the government has issued four Number One Documents since 2004. Each of these policy documents was decreed on the first day of the year (the order in which policy documents are issued normally reflects the importance of the policy).

Each of these documents had a specific focus on agricultural and rural development. The 2004 document was aimed primarily at raising farmers' incomes through reducing and eliminating agricultural taxes and fees paid by farmers, increasing income transfer to farmers—particularly those in the western regions—promoting agricultural structural changes and facilitating farmers' off-farm employment. The 2005 document called for substantial increases in agricultural production capacity through massive government investment in technology, land, water and agricultural and rural infrastructure. A new rural development strategy—a new socialist countryside development—was the focus of the 2006 Number One Document.

The highlight of the 2007 document was nation-wide abolition of all agricultural taxes, a plan for increasing investment in agricultural and rural development in the eleventh Five-Year Plan period and a commitment to poverty alleviation and other issues such as rural education, sanitation, healthcare, cultural development and community participation. The 2007 document emphasised the critical importance of modern agricultural techniques to the new socialist countryside development.

Conclusions

In this chapter, we have attempted to meet three objectives. First, we briefly reviewed China's existing agricultural policy, the past performance of China's agriculture and how it changed during the 20 years of reform before China's WTO accession. Next, we examined the main features of the agreement that China committed to in order to join the WTO. Finally, we briefly reviewed the policy changes since China joined the WTO.

We believe that one of the most important messages from this chapter is that, contrary to some opinions, China has begun to adjust to a post-accession environment. Tariffs have been reduced, many laws have been amended, public investment portfolios shifted and policy strategies changed to help China meet its commitments and assist its farmers to take on new roles. That is not to say that the job is over—far from it. The strides that the government has taken so far show that it understands the role it must play as a member of the WTO and that it is committed to living up to its obligations.

Most fundamentally, the government's response to WTO accession involves an entire paradigm shift: from direct participation in the economy to taking on a more indirect regulatory and fostering role. The government has to establish institutions that allow it to create and manage public goods. It needs to regulate markets to correct for market failures. It needs to do the things that the private sector is not willing to do and to do those things that will enhance the productivity of the nation's economic actors. It also needs to move away from an attitude that China's producers need to produce everything and establish an environment that foreign firms will be willing to invest in and into which they are willing to bring their best technology and management practices.

Another finding of importance is China's commitment to agricultural and rural development. China has made agricultural and rural development a top priority in the process of modernising its economy. Investment in agricultural and rural development has increased substantially since China joined the WTO in 2001 and can be expected to grow in coming years.

References

Anderson, K., Huang, J. and Ianchovichina, E., 2004. 'Will China's WTO accession worsen farm household income?', *China Economic Review*, 15:443-56.

Brandt, L., Huang, J. Li, G. and Rozelle, S., 2002. Land rights in China: facts, fictions and issues', *The China Journal*, 47:67-97.

Cai, F., Wang, D. and Du, Y., 2002. 'Regional disparity and economic growth in China: the impact of labor market distortions', *China Economic Review*, 11:197-212.

Carter, C.A. and Estrin, A., 2001. China's trade integration and impacts on factor markets, January, University of California, Davis (unpublished).

Colby, H., 2001. *Agricultural trade and investment liberalization after China's accession to WTO*, OECD Working Paper, Paris.

de Brauw, A., Huang, J. and Rozelle, S., 2000. 'Responsiveness, flexibility and market liberalization in China's agriculture', *American Journal of Agricultural Economics*, 82(December):1,133-39.

de Brauw, A., Huang, J., Rozelle, S., Zhang, L. and Zhang, Y., 2002. 'China's rural labor markets', *The China Business Review*, March-April 2002:2-8.

Fan, S., 1991. 'Effects of technological change and institutional reform on production growth in Chinese agriculture', *American Journal of Agricultural Economy*, 73:266-75.

Fan, S. and Pardey, P., 1997. 'Research productivity and output growth in Chinese agriculture', *Journal of Development Economics*, 53:115-37.

Huang, J. and Bouis, H., 1996. *Structural changes in demand for food in Asia*, Food, Agriculture and the Environment Discussion Paper, International Food Policy Research Institute, Washington, DC.

Huang, J. and Chen, C., 1999. *Effects of Trade Liberalization on Agriculture in China: institutional and structural aspects*, United Nations ESCAP CGPRT Centre, Bogor.

Huang, J. and Rozelle, S., 1996. 'Technological change: rediscovery of the engine of productivity growth in China's rural economy', *Journal of Development Economics*, 49(2):337-69.

Huang, J., Hu, R. and Rozelle, S., 2003. *Agricultural Research Investment in China: challenges and prospects*, China's Finance and Economy Press, Beijing.

Huang, J., Rozelle, S. and Chang, M., 2004. 'Tracking distortions in agriculture: China and its accession to the World Trade Organization', *World Bank Economic Review*, 18(1):59-84.

Huang, J., Rozelle, S. and Pray, C., 2002. 'Enhancing the crops to feed the poor', *Nature*, 418:678-84.

Huang, J., Rozelle, S. and Wang, H., 2006. 'Fostering or stripping rural China: modernizing agriculture and rural to urban capital flows', *The Developing Economies*, XLIV-1(March):1-26.

Huang, J., Yang, J. and Rozelle, S., 2007. 'When dragons and kangaroos trade: China's rapid economic growth and its implications for China and Australia', *Farm Policy Journal*, 4(1):35-49.

Jin, S., Huang, J., Hu, R. and Rozelle, S., 2002. 'The creation and spread of technology and total factor productivity in China's agriculture', *American Journal of Agricultural Economics*, 84(4):916-39.

Lardy, N.R., 1995. 'The role of foreign trade and investment in China's economic transition', *China Quarterly*, 144:1065-82.

Li, S., Zhai, F. and Wang, Z., 1999. *The global and domestic impact of China joining the World Trade Organization*, Project Report, Development Research Center, The State Council, Beijing.

Lin, J.Y., 1991. 'Prohibitions of factor market exchanges and technological choice in Chinese agriculture', *Journal of Development Studies*, 27(4):1-15.

——, 1992. 'Rural reforms and agricultural growth in China', *American Economic Review*, 82:34-51.

Lohmar, B., 1999. Rural institutions and labor movement in China, PhD thesis, Department of Agricultural Economics, University of California, Davis.

Martin, W., 2002. 'Implication of reform and WTO accession for China's agricultural policies', *Economies in Transition*, 9(3):717-42.

McMillan, J., Walley, J. and Zhu, L., 1989. 'The impact of China's economic reforms on agricultural productivity growth', *Journal of Political Economy*, 97:781–807.

Ministry of Agriculture, 1985-2005 (various issues). *China's Agricultural Yearbook*, China Agricultural Press, Beijing.

National Bureau of Statistics of China (NBSC), various issues (1985-2006). *China Statistical Yearbook*, China Statistics Press, Beijing.

——, various issues. *China Rural Statistical Yearbook*, China Statistics Press, Beijing.

Nyberg, A. and Rozelle, S., 1999. *Accelerating China's Rural Transformation*, World Bank, Washington, DC.

Rosen, D.H., Rozelle, S. and Huang, J., 2004. *Roots of Competitiveness: China's Evolving Agriculture Interests*, Institute for International Economics, Washington, DC.

Rozelle, S. and Huang, J., 2001. *Impacts of trade and investment liberalization policy on China's rural economy*, OECD Working Paper, Organisation for Economic Co-operation and Development, Paris.

Rozelle, S., Park, A., Huang, J. and Jin, H., 2000. 'Bureaucrat to entrepreneur: the changing role of the State in China's transitional commodity economy', *Economic Development and Cultural Change*, 48(2):227-52.

Sonntag, B.H., Huang, J., Rozelle, S. and Skerritt, J.H., 2005. *China's Agricultural and Rural Development in the Early 21st Century*, Australian Centre for International Agricultural Research, Canberra.

van Tongeren, F. and Huang, J., 2004. *China's food economy in the early 21st century*, Report No.6.04.04, Agricultural Economics Research Institute (LEI), The Hague.

Wang, J., 2000. Property right innovation, technical efficiency and groundwater management: case study of groundwater irrigation system in Hebei, China, PhD thesis, Chinese Academy of Agricultural Sciences, Beijing.

Wang, J., Huang, J., Huang, Q. and Rozelle, S., 2006. 'Privatization of tube wells in North China: determinants and impacts on irrigated area, productivity and the water table', *Hydrogeology Journal*, 14(3): 275-85.

World Bank, 2002. *World Development Indicators 2002*, World Bank, Washington, DC.

World Trade Organization (WTO), 2001. *China's WTO Protocol of Accession*, November, World Trade Organization, Geneva.

03 China's WTO accession
The impact on its agricultural sector and grain policy

Feng Lu

China's WTO accession was approved on 27 September 2001, and China finally joined the WTO at the annual Ministerial Meeting of the WTO in November 2001. The further integration of China into the international economy will undoubtedly bring profound economic and social changes to China. The issue has been widely debated in China in recent years. Of the implications of China's accession to the WTO, the impact on China's agricultural sector is of particular importance as it links to sensitive issues such as rural income and food security. As a result, the significance of the debate on the issue of the agricultural impact has been far beyond that suggested by the share of agriculture in China's GDP or its external trade.

China's WTO entry commitments in the agricultural sector

By 2001, China had concluded bilateral agreements with every nation except Mexico[1] but only the US-China Agreement was available to the public. In principle, China's commitments could differ from agreement to agreement. The WTO Secretariat's task was to combine the best commitments from each and aggregate them into a single combined text. The combined text would become the basis for the final documents for China's WTO accession, and all member countries would have recourse to the combined text. However, as for the agricultural commitments, it is widely believed that the US-China Agreement has addressed most of the important issues in this sector.

Following the signing of the bilateral agreement on 15 November 1999, the China Trade Relations Working Group under the US government released a document that specified the contents of China's commitments including for the agricultural sector. It specifically mentioned that the upper limit of China's agricultural subsidy would be determined in the phase of multilateral negotiation in Geneva. Agreement on this issue was finally reached in June 2001. On the basis of the information currently available, the agricultural commitments made by China with respect to WTO accession mainly cover three aspects: tariff binding and tariff reduction, a tariff rate quota system and reduction of market distortion measures.

Tariff binding and tariff reduction

China commits to establish a 'tariff-only' import regime; all non-tariff barriers will be eventually eliminated. Any other measures, such as inspection, testing, and domestic taxes must be applied in a manner consistent with WTO rules on a transparent and non-discriminatory bases and all health measures must be based on sound science. The tariff on agricultural products will be reduced from an overall average of 22 per cent to 17.5 per cent, while the average duty on agricultural products of priority interests to the US will fall from 31 per cent to 14 per cent. Table 3.1 details the specific tariff reduction commitments for major agricultural products and the time path over which they will be phased in. For example, the tariff on imported fresh cheese will be reduced from the current level of 50 per cent to 12 per cent. The tariff reduction will be phased in by 2004 (the final year for implementation of Uruguay Round agreements for development countries) in equal annual instalments.

Tariff rate quota system

China will replace its agricultural import quota and licensing system with a tariff rate quota (TRQ) system. The TRQ system has been widely adopted by WTO members in respect to some sensitive bulk agricultural products. The system is characterised by a two-tiered discontinued tariff rate divided by a benchmark quantity of quota. Imports of a given product within quota are subject to very low duties whereas imports above the quota face very high, usually prohibitive, duties.

Table 3.2 reports TRQs for five bulk agricultural products that are regarded as sensitive by China. For example, cotton will have an annual import quota of 734,000 tonnes upon China's WTO accession with an in-quota duty of 1 per cent. The cotton quota will increase to 894,000 tonnes over the transition period of implementation that ends by 2004. To ensure the state monopolistic power in cotton trading will not be used to impede commercial imports of cotton after China's WTO accession, 67 per cent of the quota will be allocated to non-state enterprises. It is further specified that if the TRQ share reserved for importation by a state trading company is not contracted for by October of any given year, it will be reallocated to non-state trading entities.

Due to food security considerations, wheat, rice and corn are subject to TRQ regulation. The total quota for grain will increase from 14.47 million tonnes upon WTO accession to 22.16 million tonnes by 2004. The in-quota duty for pure grain is 1 per cent and for processed grain products may be higher than 1 per cent but not exceed 10 per cent. Market shares for grain imports will be assigned to non-state enterprises.

Reduction of market distortion measures

China also made commitments to reduce market distortions in both domestic distribution and the foreign trade of various agricultural products and inputs.

First, China has, in principle, committed to allow foreign companies to have full trading rights, including rights in retail, wholesale, warehousing and transportation. These measures imply that the traditional state monopolies in foreign trade and domestic distribution for some agricultural products will be reformed. China agrees that any entity will be able to import most products into any part of China. This commitment is to be phased in over a three-year period with all entities being permitted to import and export at the end of the period. China will permit foreign enterprises to engage in the full range of distribution services for chemical fertilisers after a five-year transition period. China will gradually reform the long entrenched state monopoly in grain and other bulk agricultural products. Non-state enterprises will take different market shares in sensitive bulk agricultural products Table 3.2).

Table 3.1 China's agreed tariff reductions for selected agricultural products (per cent)

Products	Current tariff	Tariff in 2004*
Fresh cheese	50	12
Crated/powdered cheese	50	12
Processed cheese	50	12
Yogurt	50	10
Lactose	35	10
Ice Cream	45	19
Other food preparations	25	10
Beer	70	0 (by 2005)
Distilled spirits	65	10 (by 2005)
Wine	65	20
Fishery products	25.3	10.6 (Jan. 1, 2005)
Sorghum	3	2
Barley malt	30	10
Barley	(Quota)	9 (replace the old quota)
Soybean and meal	3	3-5 (binding tariff)**
Soybean oil	(Quota)	9 (2006)***
Vegetable oil	(Quota)	10 (replace quota immediately)
Frozen beef cuts	45	12
Frozen beef tongue and offal	20	12
Frozen pork cuts and offal	20	12
Frozen chicken and turkey	20	10
Lettuce	16	10
Cauliflower and other 5 vegetables	13	10
Canned sweet corn	25	10
Canned tomato paste	25	20

Tomato ketchup	30	15
Hazelnuts and other 4 nuts	30-35	10-13
Oranges and other 5 citrus	35-40	12-15
Apples and other 8 other fruits	30-40	10
Grapes	40	13
Grape juice and other 5 other fruits	30-65	20
Other water based drinks	50	35
Soup, cigarettes and other 8 products	20-65	10-25
Wood and wood products	10.6	3.8

Note: * Tariff reductions were to be phased in by 2004 (the year by which the obligations of Uruguay Round agreements would be completed) in equal annual instalments unless otherwise stated.
** China's imports of soybeans exceeded 3 million tonnes in 1998. China has committed to bind its tariff for soybeans at the current applied rate of 3 per cent and also foreclosed its ability to establish a quota in the future. China will bind its tariff for soybean meals at 5 per cent.
*** A tariff quota is currently applied to soybean imports with the in-quota duty at 9 per cent and the over-quota at 74 per cent. After joining the WTO, the over-quota duty will fall to 9 per cent in 2006 and the quota will be phased out by 2006.
Source: China Trade Relations Working Group, US Government, 15 February 2000.

Table 3.2   China's TRQ system for selected agricultural products ('000 metric tonnes)

Products	Initial TRQ	2004 TRQ	In-quota duty (per cent)	Private share (per cent)
Cotton	734	894	1	67
Wheat	7,300	9,636	1-10*	10
Corn	4,500	7,200	1-10*	From 25 to 40
Rice	2,670	5,320	1-10*	50 and 10**
Soybean oil	1,718	3,261***	9	From 50 to 90

Notes: * 1 per cent of duty for grain and no more than 10 per cent for partially processed grain products.
** 50 per cent of duty for short and medium grain rice; 10 per cent for long grain rice.
*** TRQ will be phased out by the year 2006.
Source: China Trade Relations Working Group, US Government, 15 February 2000.

Second, China agrees not to use export subsidies for agricultural products after it joins the WTO. China also commits to cap and reduce trade-distorting domestic subsidies. Although a serious dispute occurred over the upper-limit level of the subsidy, final agreement was reached in June 2001 that total trade-distorting domestic subsidies for agriculture under the title of 'Amber Box' would not exceed 8.5 per cent of China's agricultural GDP.

Third, China commits to abide fully by the terms of the WTO Agreement on Sanitary and Phytosanitary Measures, which requires that all animal, plant, and human health import requirements are based on sound science, not on the basis of a political agenda or protectionism concerns. On the basis of this commitment, China and the United States agreed bilaterally the terms for the removal of the restrictions on importation of US wheat, citrus, and meat.

Growth of China's economy and agriculture: the setting for the impact assessment

To give an assessment of the impact of further opening up of China's agricultural sector, it is necessary to have a look at the background to China's recent economic growth and structural changes, especially with respect to its agricultural and rural sector.

Growth and institutional reform

Thanks to the implementation of market-oriented reform and opening-up policies, China's economy achieved remarkable growth during the last 20 years or so. Nominal per capita GDP grew from 379 yuan to 6,534 yuan during the period of 1978–99. Real GDP per capita in 1999 was 5.22 times as high as that in 1978, recording an average annual growth rate of 8.2 per cent (China Statistical Yearbook 2000:53, 56). Rural residents' income per capita increased from 133.6 yuan in 1978 to 2,210.3 yuan in 1999 (China Rural Statistical Yearbook 2000:249). Deflated by the retail price index, the average annual income growth rate for rural population was 7.5 per cent. Both urban and rural residents benefited from the enormous economic growth; however, rural income has grown more slowly. As a result, the income gap between rural and urban households has widened. Within the rural population, the income gap between the east, middle and west regions has also increased. The income inequality between rural and urban areas and between the rural regions has raised concern among policymakers and the general public.

Production of major food and other agricultural products has increased at different rates over the last 20 years or so (Table 3.3). Bulk agricultural products such as grain and cotton grew at rates of less than 3 per cent per annum during the period, in part because the growth in demand for consumption of the products has been relatively low. Output of other products such as fruit, meats and aquatic products increased more quickly; again because of changes in diets towards these products as incomes increased. In contrast to the widespread concern about severe shortages in the supply of grain and other agricultural products in the late 1980s and the mid 1990s, China's agricultural sector has been facing problems of oversupply in recent years.

Of many factors behind the unprecedented growth in China's agricultural sector, two have been critically important. One is the technological progress that has fundamentally changed the agricultural production function. The other is the market oriented reform that has provided incentives for farmers to work harder and better. The past performance of the Chinese agricultural sector is suggestive in gauging the growth potential of Chinese agriculture. If it can be assumed that technological progress in agriculture will not halt and the market reform will not be reversed, China's agricultural system

Table 3.3   Production growth for selected agricultural products in
China (million tonnes)

Products	1978-79	1998-00	Growth rate (per cent)	
			Total	Annual
Grain	312.70	510.30	63.2	2.4
Oil-bearing crops	6.45	24.57	281.4	6.9
Cotton	2.44	4.16	70.6	2.7
Fruit	6.70	58.50	772.0	11.4
Pork, beef and mutton	10.30	58.40	466.7	9.7
Milk	0.98	7.81	696.4	10.9
Aquatic products	4.08	40.14	883.9	12.1

Source: State Statistical Bureau (SSB), 1995 and 2000. *Zhongguo Tongji Nianjian* [China
Statistical Yearbook], Zhongguo Tongji Chubanshe, Beijing.

should be able to provide sufficient food and agricultural products to meet
the growing demands of this huge and rapidly changing economy.

The institutional system in the Chinese rural sector has undergone
fundamental change since the first market-oriented reforms were
implemented in the late 1970s. The household contract responsibility system
in the initial reform period successfully transformed the agricultural economy
from the collectively-owned and managed system to a peasant household-
based system. Reform of the agricultural distribution system began in
the mid 1980s. Deregulation of residence and labour mobility has made it
possible for peasants to migrate to urban areas to explore opportunities
for work at higher incomes. However, the reform agenda is unfinished. For
example, reform of the distribution system for bulk agricultural products
such as grain and cotton has undergone several setbacks since the mid-
1980s. The policy reversals were not only detrimental to farmers' incomes
and long-term economic growth but also caused hundreds of billions of
yuan of losses through non-performing loans in the China Agricultural Policy
Bank. Urban-biased policies still exist in areas from public expenditure to
job market regulation. We need to bear these factors in mind in assessing
the impact of China's WTO accession on agricultural sectors.

Trade pattern and comparative advantage in China's agriculture

The direct impact of WTO accession on a given sector will show in changes in its trade flow. To help assess the trade impact, it is useful to look at China's agricultural trade pattern since the 1980s. The statistical code in the Chinese customs data collection changed from the SITC to the HS system in 1992, causing difficulties in the observation of trends in agricultural trade over the last 20 years or so. A cross-coding system has been produced to facilitate analysis of the data for Chinese agricultural trade in a consistent way throughout the 1980s (Lu 2000).

Figure 3.1 reports data on Chinese agricultural trade. There are clearly positive trends in exports and imports of agricultural products. Exports increased from about US$4 billion in the early 1980s to a peak level of US$14 billion in the mid 1990s. However, exports declined significantly as a result

Figure 3.1   China's agriculture trade, 1981-99 (US$ million)

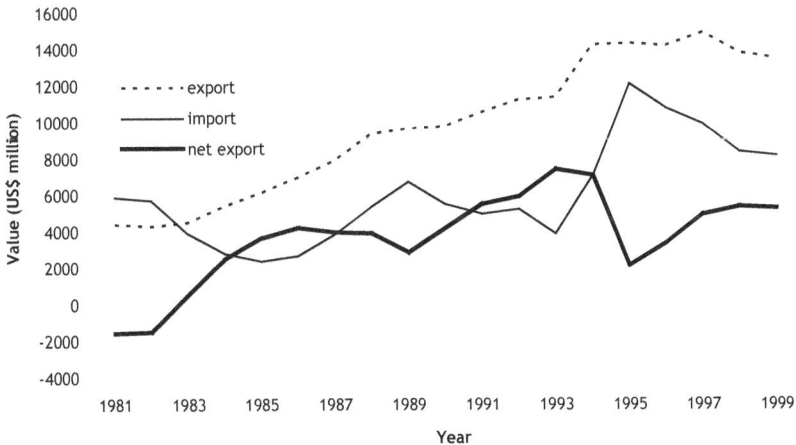

Source: China General Administration of Customs, various issues, *China Customs Statistical Yearbook*, Zhongguo Haiguan Chubanshe, Beijing; China General Administration of Customs, various issues, *China Customs Statistical Yearbook*, Zhongguo Haiguan Chubanshe, Beijing.

of shrinking markets in the East Asian economies due to the Asian economic crisis but surged again in 2000[2] with the economic recovery under way in the Asian economies. Imports declined substantially in the first half of the 1980s, followed by the first wave of growth in the second half of the 1980s. Imports declined again in the early years of the 1990s but surged from 1993 and peaked in 1995 when total imports reached US$12 billion. Total exports of agricultural products have exceeded imports in value terms for most years during the period. As a result, the value of net exports for the agricultural sector has been positive and growing for most of the period.

To investigate the structural changes in China's agricultural trade, agricultural products are divided into seven categories: bulk agricultural products; animal products as food; non-food animal products; fishery products; horticulture products; drinking and tobacco products; other miscellaneous products. Figures 3.2 and 3.3 report exports and imports, respectively, of the seven categories of agricultural products over the period 1981–99.

On the export side, fishery and horticultural products (vegetables and fruits in particular) showed remarkable growth momentum over the period. On the import side, bulk agricultural products such as grain and oil-bearing products held by far the largest share. From the perspective of the food trade, the basic structural features shown by the data present a noticeable 'food for food pattern'. As argued by Lu (1998), the most competitive Chinese agricultural products in the international market tend to derive from labour-intensive activities whereas those lacking competitiveness are usually land-intensive products. The evidence suggests that the evolution and structure of China's agricultural trade are consistent with the economic principle that links the structure of the trade flows with underlying comparative advantage and therefore to factor endowments among the different economies.

The setting for agricultural liberalisation

The combination of the factors briefly overviewed above present the setting for assessing the impact of China's WTO accession on its agricultural sector and its food policies. First, China is currently experiencing dynamic growth and has benefited enormously from market-oriented reform as well

Figure 3.2 Structure of China's agricultural exports, 1981-99 (US$ million)

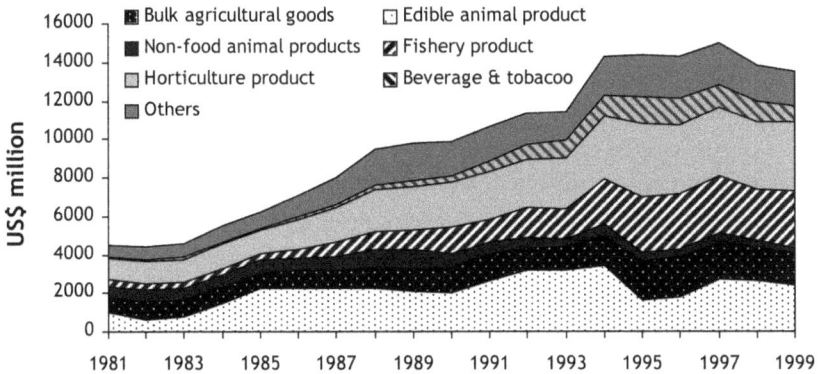

Legend:
- Bulk agricultural goods
- Non-food animal products
- Horticulture product
- Others
- Edible animal product
- Fishery product
- Beverage & tobacoo

Source: China General Administration of Customs, various issues, *China Customs Statistical Yearbook*, Zhongguo Haiguan Chubanshe, Beijing; China General Administration of Customs, various issues, *China Customs Statistical Yearbook*, Zhongguo Haiguan Chubanshe, Beijing.

Figure 3.3 Structure of China's agricultural imports, 1981-99 (US$ million)

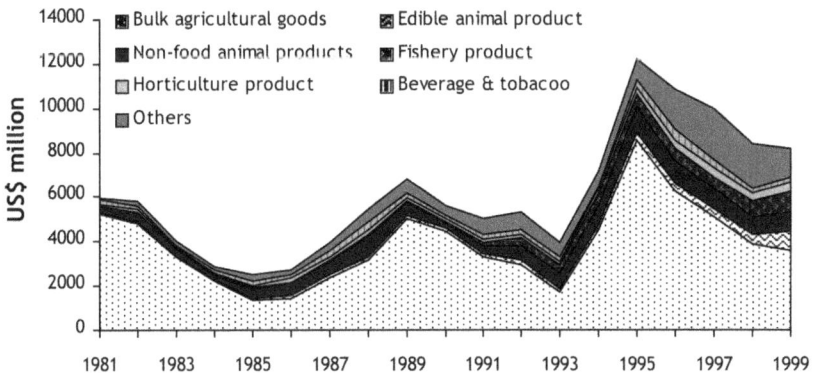

Legend:
- Bulk agricultural goods
- Non-food animal products
- Horticulture product
- Others
- Edible animal product
- Fishery product
- Beverage & tobacoo

Source: China General Administration of Customs, various issues, *China Customs Statistical Yearbook*, Zhongguo Haiguan Chubanshe, Beijing; China General Administration of Customs, various issues, *China Customs Statistical Yearbook*, Zhongguo Haiguan Chubanshe, Beijing.

as integration with the rest of the world over the last 20 years or so. During this period, the diet of the Chinese people has been substantially improved and food security has been more assured than at any time in its long history. Second, market reforms in agriculture and in other sectors are not complete. There are still large gains that may be tapped through completion of the reform agenda. Assessment of the agricultural impact needs to take into account the institutional impact. Third, income gaps between rural and urban areas as well as among rural residents in the different regions have become important issues for the long term economic growth and social stability of China. The income effects of agricultural liberalisation and their regional distribution deserve special attention. Finally, as indicated by China's agricultural trade pattern since the 1980s, China has comparative advantage in labour-intensive agricultural products and has comparative disadvantage in land-intensive products. It may be reasonably inferred that WTO accession will produce two kinds of trade impacts for the Chinese agricultural sector: growth of exports of labour-intensive agricultural products and increase in imports of land-intensive agricultural products.

Agricultural sector impacts of China's WTO accession

It has been widely acknowledged that WTO accession will be a milestone in the process of China's integration into the global economy. It will undoubtedly produce tremendous challenges and opportunities for the Chinese agricultural sector.

Opportunities and benefits of agricultural liberalisation

The potential benefits for the Chinese agricultural sector from WTO accession may be assessed through three effects. The first is its institutional reform impacts. Although market-oriented reform has made possible great achievements in the Chinese agricultural sector, the reform agenda is unfinished. Periodic reversals to the state monopolistic system in the domestic distribution of bulk agricultural products such as grain and cotton have led to huge claims against the Government budget over the past 15 years. The unstable policy environment has also been detrimental to agricultural growth and farmers' interests. The state monopoly in the distribution of agricultural inputs such as fertiliser and seeds also results in efficiency losses. It is widely expected that WTO accession will advance the reform agenda in these areas. Although it is difficult to forecast the institutional impacts, the strong interaction between institutional reform and economic performance in the Chinese economy over the past 20 years or so indicates the potential importance of these effects.

The second potential benefit of WTO accession is its export-promoting effect. WTO accession and future agricultural trade liberalisation will reduce trade barriers both in China and in the rest of the world. This will help China further expand exports of the agricultural products in which it has comparative advantage. On the basis of observation of the historical pattern of China's agricultural trade, exports of aquatic and meat products are likely to increase as a result of trade liberalisation. Abandonment of the grain self-sufficiency policy will be a crucial factor in achieving this export expansion because liberalisation of the grain trade will reduce feed grain costs and increase the competitiveness of animal production using feed grain inputs. Exports of horticultural products such as vegetables and fruits may increase, as their production is relatively more labour intensive than bulk agricultural products. Exports of traditional specialties such as honey and tea may also increase as foreign trade barriers are reduced. As the value of labour-intensive agricultural products usually has a relatively larger labour element, growth of these exports may produce favourable income effects in the rural economy.

The third potential benefit of accession is its welfare effect. Domestic prices for most agricultural products are likely to decline as a result WTO accession. A partial equilibrium model, China's Agricultural Policy Analysis and Simulation Model (CAPSiM), developed by the China Centre for Agricultural Policy (CCAP) at the China Academy of Sciences (CAS), compared a WTO accession scenario with a base scenario without WTO accession. The results indicate that the prices of bulk agricultural products such as corn, wheat and soybean will reduce by about 20 per cent (Huang 2000). Although the projected price reduction is hypothetical, it is generally agreed that trade liberalisation tends to reduce market prices for the products in question. With the price decline for food and other agricultural products, the purchasing power of a given income will increase and consumers' welfare will be improved.

Adjustment costs of WTO entry and their income implications

Agricultural liberalisation may also produce challenges and adjustment costs for the Chinese agricultural sector. The logic is simple. China's WTO commitments for the agricultural sector include tariff reductions, market access and reform of state monopolies in agricultural trade. Although these measures may increase the long-term competitiveness of China's agricultural sector and therefore contribute to income growth for the Chinese rural population, the structural adjustments necessary are likely to cause short-term difficulties for Chinese farmers. For example, as indicated by Table 3.1, tariffs for most dairy products will decline from 50 per cent to 10–12 per cent; tariffs for apples and eight other fruits will be reduced from 30–40 per cent to 10 per cent. The measures will reduce the Chinese renminbi (RMB) price of foreign imported goods and therefore increase competitiveness of these foreign products in Chinese markets. Other things being equal, the structural adjustments from the reduction of tariffs imply shrinking of the domestic production of these products and adjustment costs for farmers. The magnitude of the substitution effects from growing imports for domestic products will depend upon relative prices between the domestic market and the international market for the product after WTO entry. For example, the import-led substitution effects may be relatively small for fruits in which domestic production has an apparent comparative advantage. The external shock may be significantly larger for dairy products as dairy production is less competitive.

The TRQ system for sensitive products is a double-edged sword for domestic production. On the one hand, it provides a protective mechanism for domestic production, as imports are usually unable to exceed the quota threshold. On the other hand, the TRQ system makes it impossible to restrict imports of the products within the quota, should the import prices of these products be significantly lower than the domestic prices. As bulk agricultural products such as grain and oil-seed bearing products are usually land intensive, and China generally lacks comparative advantage in these products, imports of bulk agricultural products may increase. As a result, domestic production will face adjustment pressures. Possible import surges of grain, especially corn, are of most concern. If average annual imports of corn increase by 5 million tonnes as a result of WTO accession, domestic corn production has to be reduced by that amount.

Production adjustment costs have obvious implications for rural income. Although it may be reasonably believed that the resources released by reduction of production will be used more efficiently in other activities in the long run, in the short run farmers may not be able to find other profitable activities to employ the released resources, including their own labour. As a result, farmers' income may be negatively affected. This issue has been extensively analysed within the framework of simulation models (Huang 2000; Tian 1999; Li et al. 2000). On the basis of simulation of the above-mentioned CAPSiM model, bulk agricultural products such as grain, cotton, oil-bearing products and sugar products will decline by 2.5 per cent to 7.7 per cent. As a result, farmer income will be reduced by the amount equivalent to the total income of 3 million labourers in recent years (Huang 2000). It should be borne in mind, however, that the degree of accuracy of the estimated results depends upon the assumptions of the model, the interaction of various economic variables, and the reliability of the statistical data and the estimated parameters. The projections nevertheless highlight the importance of the short-term income effects of WTO accession for farmers. As income inequality between the rural and urban population has become a crucial issue for the overall economic and social development of China, the potential adverse income effects for the rural population deserve special attention.

Food security and WTO accession

In the debate over the effects of China's WTO accession, the impact on China's long-term food security has been frequently raised. There are concerns that WTO accession will place China's long-term food security at risk. For a huge country like China, food security has been and will continue to be a very important objective. There are no doubts whatsoever of the importance and legitimacy of the food security objective per se. However the assertion that agricultural liberalisation will harm food security is questionable.

Concerns about the detrimental impact of China's WTO accession on its food security are mainly based upon the possible surge in grain imports as a result of the implementation of the tariff quota system. Although the imports of grain into China may increase after China's entry into the WTO, they are unlikely to harm China's food security. We approach the issue from the following four perspectives.

First, the volume of the tariff quota for all grains increased from 14.5 million tonnes upon China's entry into the WTO to 22.2 million tonnes by 2004. China's imports of grain in 1995 were 20.81 million tonnes, only fractionally less than the peak level of the quota (China Statistical Yearbook 1996:592). These imports did no harm to China's food security; therefore, it does not appear convincing to argue that slightly higher imports will damage China's food security.

Second, the White Paper on grain security published by the Chinese Government in 1996 set 95 per cent of grain self-sufficiency as a reference line for assuring the objective of grain security for China. In principle, the desirable level of grain self-sufficiency for China may depend upon many changing factors in future, such as the level of food consumption, the international environment, and China's foreign exchange payment ability. It is arguable that China's food security objective may be consistent with a 90 per cent or even lower grain self-sufficiency to allow a larger role to be played by the international market. However, even with this official guideline that is cautious and conservative in nature, the tariff quota for grain may not exceed the upper limit of grain import ratios. China's total consumption of grain in recent years is about 500 million tonnes. As total grain consumption is likely to increase, a 22 million tonnes grain quota would be less than 4.4 per cent of total grain consumption by 2004. Taking into account other factors such as imports of other grains except corn, wheat and rice, imports are unlikely to exceed the 5 per cent of total grain consumption by a significant margin.

Third, the above discussion is based on the assumption that imports of grain will reach the quota level. However this assumption is quite uncertain. The tariff quota for grains in the package of market access arrangements for China's WTO accession only represents the import opportunity for foreigners not China's import obligations. How much grain will be imported during the period from the year of China's entry into WTO to 2004 will depend upon the relative price for the different grains between the Chinese domestic market and the international market. Although China's grain imports are likely to increase as grains are generally land-intensive products in which China lacks comparative advantage, it is also quite possible that future imports of grain will be significantly lower than the quota. If grain imports to the quota level are unlikely to harm China's food security, lower imports must make the gloomy predictions even more unrealistic.

Finally, the traditional argument asserting the detrimental impact of grain imports on food security usually mentions the constraint of the foreign exchange capacity to purchase grains. This argument made sense in the 1960s and 1970s during which time the grain import bills usually consumed more than 10 per cent of total export revenues. Around 1961 when China was hit by the severe famine brought about by the disastrous policy mistakes made by the Government, the proportion of grain import costs in total export revenue reached as high as 25 per cent.[3] Under these circumstances, foreign exchange was indeed a critical constraint on the growth of grain imports and therefore had serious implications for food security. However, the situation has fundamentally changed since the economic reforms were implemented. The past two decades or so have witnessed tremendous growth of China's export sector. Total export revenue increased from US$9.75 billion in 1978 to US$249.2 billion in 2000—an average annual growth rate of 15.1 per cent.[4] As a result of the growth of exports, future grain imports can at most consume only a small fraction of total export revenue. On the basis of projected grain prices, 22.2 million tonnes of grain may consume US$3-4 billion. In 2000, China's export revenue was more than US$200 billion; hence grain imports would be 1.2-1.6 per cent of total export revenue in 2000. It is likely that net grain imports will be an even lower percentage of total export revenue in the future.

The regional pattern of the agricultural impact of WTO accession

In light of the historical pattern of China's agricultural trade, the one-off trade flow effects of WTO accession may be negative for the agricultural products in which China lacks comparative advantage and positive for those products in which China has comparative advantage. As the provinces and regions of China are very diversified in terms of geographical settings and economic conditions, the trade impact of WTO accession on agriculture is likely to differ significantly among the different regions.

'Export promoting effects' and 'substitution effects from imports'

The logic and the analytical technique used for the analysis are simple. The historical performance of China's agricultural trade indicates that WTO accession is likely to strengthen the tendency for imports of land-intensive agricultural products on the one hand, and encourage exports of labour-intensive agricultural products on the other. In other words, WTO accession will likely produce two trade flow effects. One is the positive export-promoting effect and the other is the negative substitution effect from the growth in imports. As regards the pattern of export promoting effects, it may link closely to the domestic distribution of comparative advantage for labour-intensive agricultural products. Those regions with comparative advantage in labour-intensive agricultural products are likely to benefit more from the potential export expansion effects of the WTO accession, and vice versa.

The regional pattern of substitution effects will depend upon how domestic production of land-intensive products is substituted. From an analytical point of view there are at least two possibilities. One possibility is based on an assumption that growing imports of land-intensive agricultural products will largely substitute for domestic production of these products in the regions with relatively high production costs for these products. So those provinces with relative domestic advantage in land-intensive agricultural products may incur relatively small adjustment costs from the possible import surge effects, and vice versa. Alternatively, it may be assumed that growing imports of land-intensive agricultural products will substitute for domestic production of these products in proportion to the relative concentration of the production across the regions. The two assumptions have different implications for the regional distribution of the substitution effects and the adjustment costs.

Regional production concentration indices for major agricultural products

To examine the regional pattern of trade flow impacts on China's agricultural sector resulting from WTO accession, we need to know the regional distribution of comparative advantage of the two categories of agricultural products. On the basis of economic principles, a product or a production activity with comparative advantage in an economy or a region may be defined as one where the opportunity costs associated with the activity are relatively low. Alternatively, comparative advantage may be defined as the activities that utilise the abundantly endowed resource in an economy or region at a relatively high intensity. Providing a direct measure of comparative advantage for the various products across the regions of China would require data on opportunity costs or factor intensities, as well as production cost proportions. This would be highly demanding. Therefore, we use an alternative method to measure comparative advantage. Similar to the widely used method that takes the export concentration index for a commodity as an indicator of 'revealed' comparative advantage, we use the 'production concentration index' of a product for a region as an indicator of comparative advantage. The production concentration index of a product for a region is defined as the ratio of the sown area (or output) for the product per capita of the agricultural population for the region divided by the same ratio for the nation as a whole. The interpretation of the index is straightforward: a region has a comparative advantage if the measured value of the production concentration index is larger than one and has a comparative disadvantage if the value is less than one. The greater the margin by which the index exceeds (or is less than) one, the stronger the comparative advantage (or disadvantage) is for the given product in the region.

Table 3.4 presents the ratios of production concentration for major agricultural products for the provinces (as well as the autonomous regions and municipalities under the direct administration of the central government). The average indices for three broadly defined regions (the east, middle and west regions) are also reported. As indicated by the results, production concentration indexes for the labour-intensive products are relatively high for the east region and low for the west region, with the middle region in between. By comparison, production concentration indexes for the land-intensive products are relatively high for the west region and low for the east region, with the middle region again in between.

To facilitate the examination of the regional distribution of the agricultural impact of China's WTO entry, it is useful to present the concentration indices for the two categories of agricultural products in the framework of a plane-coordinate system (see Figure 3.4). The horizontal and

Table 3.4 Production concentration index for selected agricultural products for provinces and regions in China, 1997-99

Provinces and region	Land-intensive agricultural products					Labour-intensive agricultural products				
	Grains	Oil & oil seeds	Cotton	Sugar	Average	Vegetables	Fruits	Meat	Fishery products	Average
Beijing	0.93	0.19	0.12	0.00	0.91	1.87	1.78	2.01	0.50	1.54
Tianjin	0.92	0.34	0.31	0.00	0.89	1.86	0.91	0.90	1.29	1.24
Hebei	1.10	0.82	1.30	0.07	1.08	0.94	2.02	1.15	0.30	1.10
Liaoning	1.07	0.35	0.14	0.39	1.04	1.17	1.84	1.53	3.20	1.93
Shandong	0.93	0.85	1.21	0.01	0.94	1.44	1.26	1.10	2.20	1.50
Shanghai	0.75	1.06	0.19	0.16	0.78	2.06	0.32	1.95	1.76	1.52
Jiangsu	0.92	0.87	1.53	0.04	0.94	1.05	0.29	0.91	1.27	0.88
Zhejiang	0.64	0.56	0.33	0.16	0.62	0.77	0.72	0.48	2.75	1.18
Fujian	0.62	0.32	0.00	0.56	0.60	1.31	2.28	0.83	4.17	2.15
Guangdong	0.51	0.44	0.00	2.00	0.58	1.31	1.75	0.87	2.35	1.57
Guangxi	0.77	0.63	0.01	7.22	1.55	1.41	1.93	1.08	1.28	1.42
Hainan	0.96	0.78	0.00	7.90	1.70	2.21	1.57	1.05	2.97	1.95
East region	0.83	0.58	0.36	1.68	0.97	1.45	1.40	1.16	1.98	1.50
Shanxi	1.14	1.04	0.59	0.53	1.12	0.65	1.38	0.43	0.02	0.62
Inner Mongolia	2.88	2.92	0.01	3.60	2.90	0.70	0.58	1.54	0.10	0.73
Jilin	2.01	0.65	0.00	0.62	1.96	1.26	0.75	2.30	0.23	1.14
Heilongjiang	3.53	0.81	0.00	5.34	3.50	1.36	0.33	1.27	0.45	0.85
Anhui	0.97	1.73	1.58	0.06	1.12	0.68	0.19	0.92	0.69	0.62
Jiangxi	0.90	2.11	0.63	0.59	1.14	1.20	1.00	1.05	0.88	1.03
Henan	0.94	1.14	2.26	0.03	1.06	0.88	0.53	0.95	0.08	0.61
Hubei	0.97	2.03	2.21	0.26	1.24	1.53	0.62	1.08	1.29	1.13
Hunan	0.78	1.23	0.72	0.30	0.84	0.81	0.64	1.26	0.52	0.81
Middle region	1.57	1.52	0.89	1.26	1.65	1.01	0.67	1.20	0.47	0.84

Chongqing	0.96	0.56	0.01	0.04	0.93	0.85	0.37	0.95	0.17	0.58
Sichuan	0.85	0.85	0.39	0.22	0.84	0.73	0.39	1.15	0.14	0.60
Guizhou	0.82	1.07	0.02	0.22	0.85	0.77	0.21	0.61	0.04	0.41
Yunnan	0.93	0.30	0.01	4.05	1.11	0.56	0.61	0.85	0.10	0.53
Tibet	0.76	0.56	0.00	0.00	0.74	0.28	0.06	1.05	0.02	0.35
Shaanxi	1.16	0.76	0.27	0.04	1.13	0.54	2.59	0.50	0.04	0.92
Gansu	1.17	1.16	0.31	0.68	1.16	0.58	1.62	0.44	0.01	0.66
Qinghai	0.92	3.37	0.00	0.00	1.65	0.29	0.18	1.00	0.01	0.37
Ningxia	1.75	2.12	0.00	1.36	1.79	0.66	1.12	0.68	0.17	0.66
Xinjiang	1.45	1.85	23.46	5.17	8.90	0.73	1.83	1.34	0.14	1.01
West Region	1.08	1.26	2.45	1.18	1.91	0.60	0.90	0.86	0.08	0.61

Notes: The production concentration index for a given product for grain, oil, cotton, sugar, vegetables and fruits in a province is defined as the ratio of sown area of the product per capita for the rural population in the region divided by the national average for the same ratio. The index for meat and fishery products in a province is defined as the ratio of the output of the product per capita for the rural population in the region divided by the national average for the same ratio. The average index for the land-intensive agricultural products for a province is the weighted average of the indices for the four products using the distribution of land among the four products as the weights; the average index for the labour-intensive products for a province is the simple average. The average index of individual products and groups of land and labour-intensive products are simple averages of the indices of provinces covered in the large regions.

Sources: Data for rural populations are from *China Rural Statistical Yearbook* (1998 and 2000). Data on sown areas and output of meat and fishery products for provinces and regions are from *China Statistical Yearbook* (1998-2000).

Figure 3.4   Graphic presentation of the concentration ratios for land-intensive and labor-intensive agricultural products for provinces and regions, 1997–99

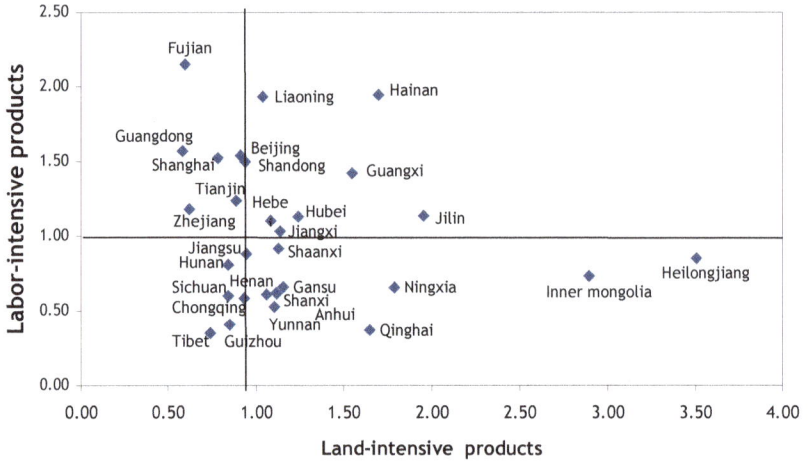

Notes: Xinjiang, with a concentration index for land intensive products of 8.9 and 1.01 is a remote outlier and therefore is not shown in the diagram.
Source: Data are from Table 3.4.

vertical axes of the system represent concentration ratios for labour-intensive and land-intensive products, respectively. Two additional lines representing the ratio of one for labour-intensive and land-intensive products divide the space of the plane system into four areas. All points located in the north east quadrant show indices for land and labour intensive products higher than one. The points in the south west quadrant represent the combinations of measurements for land and labour-intensive products lower than one. The points in the south east quadrant indicate land-intensive products with indices higher than one and indices for labour-intensive products lower than one; the north west quadrant shows the opposite combination of indices.

Regional pattern of the agricultural impact: Scenario one

Scenario one follows from the assumption that increased imports of land-intensive products resulting from WTO entry will mainly substitute for domestic production in the provinces that lack comparative advantage in these products. On the basis of this assumption, as well as discussion of the rationale for distribution of the export promoting effects, we may compare the agricultural trade impacts for provinces and regions in the different quadrants in Figure 3.5.

Seven provinces and regions (Hainan, Liaoning, Guangxi, Hebei, Hubei, Jiangxi and Jilin) located in the north-east quadrant are likely to have the best outcome from the agricultural liberalisation. As these provinces appear to have comparative advantage in labour-intensive agricultural products,

Figure 3.5   Regional pattern of the agricultural impact: scenario one

Adjustment costs from the possible import surge

		Large	Small
Benefits from the possible export expansion	Large	Fujian, Guangdong, Shanghai, Shandong Tianjin, Zhejiang Beijing	Hainan, Jilin, Guangxi Liaoning, Hubei, Hebei, Jiangxi
	Small	Jiangsu, Hunan Sichuan, Chongqing Guizhou, Xizang	Xinjiang, Heilongjiang Inner Mongolia, Ningxia, Qinghai, Shanxi, Shaanxi, Anhui, Yunnan, Gansu, Henan

they are likely to benefit more than the national average from export expansion of labour-intensive agricultural products. They may also incur relatively small adjustment costs from the import surge of land-intensive products as they also have comparative advantage in these activities.

The payoff for the group of six provinces and regions (Jiangsu, Hunan, Sichuan, Hunan, Chongqing and Guizhou) located in the south-west quadrant may be the most unfavourable as they lack comparative advantage in both labour-intensive and land-intensive agricultural products. As a result, they are likely to shoulder relatively large adjustment costs arising from the external shocks of import growth for land-intensive agricultural products and obtain a relatively smaller share of benefits from expansion in exports of labour-intensive agricultural products.

The seven provinces and municipalities directly under administration of the Central Government located in the north-west quadrant are likely to enjoy relatively large benefits from the export expansion of labour-intensive products. But they are also likely to face relatively large adjustment costs from the import growth of agricultural products that are land intensive in nature. Finally, there are 11 provinces and regions (Xinjiang, Heilongjiang, Inner Mongolia, Ningxia, Qinghai, Shanxi, Shaanxi, Anhui, Yunnan, Gansu and Henan) located in the south-east quadrant for which the benefits and costs are likely to be relatively small (Figure 3.5).

The coastal regions are likely to be the major beneficiaries from export expansion of labour-intensive products while the vast inland provinces and regions will have relatively small benefits. On the other hand, coastal regions may experience a relatively large share of the adjustment costs from the import growth while inland provinces and regions may experience smaller external shocks. It appears, therefore, that there could be a negative correlation between the benefits and costs across the broadly defined regions that is, the eastern, the middle and western regions.

Regional pattern of the agricultural impact: scenario two

In scenario two it is assumed that imports will substitute for domestic production in proportion to the domestic share of production across provinces and regions. Under this assumption, domestic adjustment costs resulting from the increasing imports will have to be mainly shouldered by provinces with a relatively high concentration index for the products. As for the export promoting effects, they are expected to be distributed in the same way as in scenario one.

Changes in the regional impact pattern under the new assumption may be simply captured by switching the position of the provinces in the left and right quadrants in Figure 3.6. For example, in line with scenario two, the seven provinces and regions located in the north-east quadrant switch to the north-west quadrant while those in the north-west quadrant move to the north-east quadrant. Similar shifts of position occur between those provinces in the south-east quadrant and the south-west quadrant.

In scenario two, the larger share of benefits from export expansion in labour-intensive products still goes to the coastal regions while the vast

---

Figure 3.6   Regional pattern of the agricultural impact: scenario two

---

Adjustment costs from the possible import surge

		Large	Small
Benefits from the possible export expansion	Large	Hainan, Jilin, Guangxi Liaoning, Hubei, Hebei, Jiangxi	Fujian, Guangdong, Shanghai, Shandong Tianjin, Zhejiang Beijing
	Small	Xinjiang, Heilongjiang Inner Mongolia, Ningxia, Qinghai, Shanxi, Shaanxi, Anhui, Yunnan, Gansu, Henan	Jiangsu, Hunan Sichuan, Chongqing Guizhou, Xizang

inland provinces and regions have a small share of the benefits. However, the regional distribution of the import-led substitution effects changes substantially. A large share of the adjustment costs from the possible growth of land-intensive agricultural imports may go to inland and western provinces while coastal provinces may only have to shoulder a small share. For example, 11 provinces, most of them either the major grain production bases (such as Heilongjiang, Henan and Anhui) or provinces located in western regions (such as Xinjiang, Shaanxi and Ningxia) will have the most unfavourable impacts from WTO accession, that is, a combination of small export-promoting effects and large adjustments costs. In general, under scenario two the agricultural impact of China's WTO accession will be more unbalanced in terms of the export-promoting effects and the import-led substitution effects than in scenario one. Scenario two is obviously less desirable than scenario one in terms of income distribution as it may see an increase in the income gap between the east and west regions.

The distribution of the adjustment costs from agricultural liberalisation is of course unlikely to accord exactly with either of the two scenarios. It may fall somewhere between the two cases. The above observations are nevertheless useful to an understanding of the mechanism affecting the regional distribution of the agricultural trade impact from China's WTO accession across provinces and regions in China. The regional impact may in part depend upon the policies adopted by the Chinese government in managing trade liberalisation. To maximise the likelihood for scenario one to materialise, the domestic agricultural polices will need to be adjusted in a more liberal and market-oriented direction. This will allow the principle of comparative advantage to play a bigger role in the allocation of domestic resources in the agricultural sector across provinces and regions. On the contrary, if the traditional policy stance with its emphasis on provincial grain self-sufficiency persists, it will be difficult for the coastal provinces to reduce grain and other bulk agricultural production in which they do not have a comparative advantage. As a result, the import-led adjustment costs will be distributed more proportionally across the provinces and the undesirable scenario two will be more likely.

Summary and policy implications

From a political economy point of view, WTO accession is undoubtedly one of the most complicated and challenging events in contemporary Chinese history. Of China's WTO commitments, the agricultural component is potentially the most contentious because of the number of people employed in agriculture, and its implications for food security as well as for social stability. The study has examined various aspects of the impact of WTO accession for China's agricultural sector. Several points emerge from the investigation.

The desirable impacts from WTO accession should be the institutional effects. WTO accession may help complete the unfinished agenda of institutional transformation in the agricultural and rural sectors through the dynamics from the interactions between the market-oriented reform and further opening up to the outside world. Implementation of WTO commitments is likely to reduce unnecessary government intervention in the distribution of agricultural products, and break up the state monopoly in foreign trade in bulk agricultural products. Improvements in the institutional framework will make important contributions to the long-term growth of Chinese agricultural and rural income. Without its institutional effects, the desirable impacts of China's WTO accession on its agricultural sector as well as on the whole economy would be substantially diminished.

Second, consistent with the structure of factor endowments in the agricultural sector, China has comparative advantage in labour-intensive agricultural products such as vegetables and fruits, fishery products and meat, but lacks comparative advantage in land-intensive bulk agricultural products such as grain and oil-bearing products. WTO accession is likely to help China expand exports of the products in which it has comparative advantage. On the other hand, it may also give rise to substantial adjustment costs for those products for which domestic production costs are relatively high and where imports may increase. Although the long-term impacts are generally expected to be positive, the short-term adjustment costs may be substantial. The income implications of the import-led substitution effects deserve special attention.

Third, it is interesting to gauge the regional pattern of the export-promoting benefits and the import-led adjustment costs that WTO accession may give rise to in the agricultural sector. The larger share of benefits from the export expansion of labour-intensive products is likely to go to coastal regions while the vast inland province and regions may have smaller benefits. As for the adjustment costs resulting from the potential growth of land-intensive agricultural imports, the regional distribution pattern will depend upon how extensively the imported products substitute for domestic production. Under the assumption that the increased imports will mainly substitute for domestic production in the provinces that lack comparative advantage in their production, the larger share of adjustment costs will be borne by the coastal regions while inland province and regions will have to shoulder only a small share. However, under an alternative assumption that the increased imports will substitute for domestic production in proportion to the relative density of current production across provinces, inland and western provinces will face the largest adjustment costs.

Finally, WTO accession and agricultural liberalisation may have significant implications for food security in China. As a direct impact of WTO accession, tariff quota arrangements for the three major grain products are likely to increase China's grain imports, with the gradual integration of the Chinese food economy with the world food market. As the evolving pattern of 'food for food exchange' will likely develop further, the traditional mechanism for achieving food security based on the concept of grain self-sufficiency needs to be modified. China needs to assess how best to assure its food security as domestic and external environments change. There are nevertheless no grounds for believing that the market access commitments made by China in its WTO accession package will harm its food security objective.

Bearing in mind the complexity of the subject and the incompleteness of this research, policy implications of this study are noteworthy. First, the Chinese government may need to minimise administrative interventions in the agricultural sector that were quite often proposed and implemented on the basis of the food security argument. Agricultural policy adjustments in line with domestic regional comparative advantage are necessary for the Chinese agricultural system to respond better to the potential export opportunities for those agricultural products in which China enjoys comparative advantage. Such policy adjustments will also be helpful in shaping the regional distribution of the adjustment costs of

WTO accession in a more desirable way. On the other hand, much needs to be done to improve the information system, quality control procedures and marketing skills of Chinese farmers and firms producing agricultural processed products to benefit fully from the potential opportunities in the international market.

Second, the negative effects from the possible surge of agricultural imports must be taken seriously. Particular attention should be given to the inland and western provinces and regions that will benefit little from the WTO accession, which may result in an even larger income gap between these economically less developed regions and the economically advanced coastal provinces. To help these regions, Chinese governments at central and provincial levels may need to take more responsibility for financing rural education rather than providing direct subsides to agricultural production activities. To strengthen the competitiveness of China's agricultural and rural sectors, the government may need to invest more in agricultural technology extension and rural infrastructure such as transportation and communication facilities.

Notes

1   It is reported that the Mexican Government promised not to be in the way of China's WTO accession even if the bilateral agreement with Mexico was not reached later in the year.
2   The estimated annual data on Chinese agricultural trade in 2000 are not reported in this figure.
3   The figures were calculated using the data reported in Department of Planning, Ministry of Agriculture, Husbandry and Fishery of China: 'Materials of Agricultural Economy (1940-1983)'(pp. 434-5), internal publications.
4   The figure for export revenues in 1978 is from *China Statistical Yearbook* (2000) and that for 2000 is from Summary of China Statistics (2001).

References

China General Administration of Customs, various issues, *China Customs Statistical Annual Report*, Zhongguo Haiguan Chubanshe, Beijing.

China General Administration of Customs, various issues, *China Customs Statistical Yearbook*, Zhongguo Haiguan Chubanshe, Beijing.

Huang Jikun, 2000. 'Impact on the Chinese agriculture from trade liberalisation and policy response (Maoyi ziyouhua dui woguo nongye deyingxiang yu duice)',*China's Direction (Zhongguo Zouxiang)*, Zhejiang People's Publishing House, Hangzhou:332-42.

Li Shangtong, Wangzhi, Zhaifan, Xulin, 2000. *WTO: China and World (WTO: Zhongguo yu Shijie)*, China Development Publishing House, Beijing.

Lu Feng, 1998. 'Grain versus food: a hidden issue in China's food policy debate', *World Development*, 26(9):1,641-52.

Lu Feng, 2000. China's bamboo product trade: performance and prospects, internal report for a commissioned study by International Network for Bamboo and Rattan, Beijing (unpublished).

State Statistical Bureau (SSB), various issues, *Zhongguo Tongji Nianjian* [China Statistical Yearbook], Zhongguo Tongji Chubanshe, Beijing.

State Statistical Bureau (SSB), various issues, *Zhongguo Nongcun Tongji Nianjian* [China Rural Statistical Yearbook], Zhongguo Tongji Chubanshe, Beijing.

Tian Weiming, 1999. Reform on the world trade system and China's choice on the strategy for agricultural trade (Shijie maoyi tixi gaige yu woguo nongchanpin maoyi zhanlue xuanze), internal report, School of Economics and Management, China Agricultural University, Beijing (unpublished).

04 Rural-urban income disparity and WTO impact on China's agricultural sector
Policy considerations

Xiaolu Wang and Ron Duncan

Since the beginning of agricultural reform in the late 1970s and early 1980s, grain output in China has increased significantly. The long-standing problem of grain supply shortages has basically been solved. However, grain production and pricing are still not fully liberalised and large fluctuations in grain prices, together with short-term shortages and surpluses of grain, have occurred several times. These events have seriously affected farmers' incomes. Partly as a result, rural-urban income disparities have grown, particularly in the 1990s.

In 2001, China entered the World Trade Organization (WTO). Because of the commitment made to open the domestic market for agricultural products, grain imports are increasing rapidly. Domestic grain production and farmers' incomes are facing new challenges. This chapter examines recent developments in rural-urban income differentials, especially in light of WTO commitments, the appropriateness of related domestic policies such as grain pricing and urbanisation policies, and the effects of adjustments of these policies in response to the new challenges.

Agriculture and rural-urban income disparity in China

After half a century of rapid industrialisation, the dominant role of China's agricultural sector (including farming, forestry, animal husbandry and fisheries) in the economy has been replaced by the industry and service sectors. In spite of significant growth in agricultural production, the share of the agricultural sector in GDP declined from 51 per cent to only 15 per cent from 1952 to 2001. However, because of the large size and continued growth of the rural population, the employment structure of the agricultural and non-agricultural sectors changed far more slowly than the change in the output structure. Agricultural employment accounted for 84 per cent of total employment in 1952, and declined to 50 per cent in 2001. At the same time, agricultural employment increased from 173 million to 365 million.[1]

The share of the rural population changed even more slowly. The rural population still accounts for 62 per cent of China's total population, as compared to 85 per cent in 1953 (Table 4.1).

Farming remains the most important component of Chinese agriculture; although reduced, it still accounted for 55 per cent of the gross value of agricultural output in year 2001. Surveys show that, in year 2000, 51 per cent of rural household income came from the agricultural sector as a whole and 39 per cent from farming (Figure 4.1). Grain production has been the major farming activity, especially in less developed regions.

Movement of the agricultural population to non-agricultural sectors was hampered by the restrictive central policies during the pre-reform period from the 1950s to the late 1970s. Rural industrialisation and urbanisation has been speeded up since the market reforms. Such developments have provided close to 200 million additional non-agricultural jobs to farmers in rural and urban areas over the past 24 years. Still, the number of farmers has increased and there appears to be more surplus agricultural labour than ever before. Rural income grew slowly, and differences between rural and urban incomes became larger.

Table 4.2 shows changes in income per capita in rural and urban areas of China and its three major regions from 1980 to 2000. It shows that rural income per capita in rural areas was 47 per cent of urban income in 1980; this ratio had declined to 35 per cent by 2000. Rural–urban income disparity increased in all three regions, but more seriously in the least-developed west region where the rural–urban income ratio fell from 45 to 30 per cent during the same period. In year 2000, the average annual rural income per capita in the west region was 1,713 yuan (or US$207), only half of the average rural income in the east region.

Table 4.1   The importance of the agricultural and rural sectors in China, 1952 and 2001

	1952	2001
Agricultural value-added		
Billion yuan	34	1,461
per cent of GDP	51	15
Agricultural workers		
Million persons	173	365
per cent of total employment	84	50
Rural population		
Million persons	505*	796
per cent of total population	85	62

* 1953 data.
Source: Calculated from National Bureau of Statistics, 2002, 1999b. *Statistical Yearbook of China*, China Statistics Press, Beijing.

Figure 4.1   The structure of rural household income, 2000

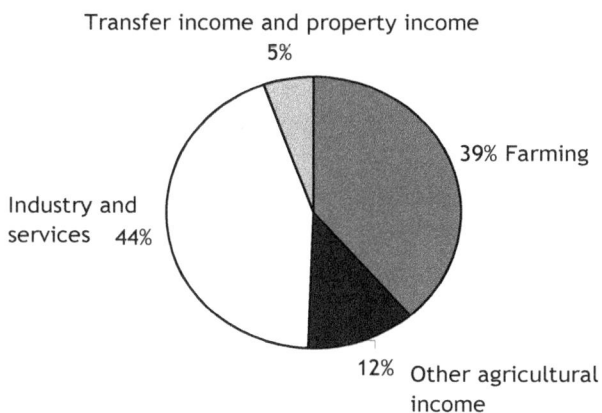

Transfer income and property income
5%

39% Farming

Industry and services   44%

12%   Other agricultural income

Source: National Bureau of Statistics, 2001. *Statistical Yearbook of China*, China Statistics Press, Beijing.

Table 4.2   Rural-urban income disparity, 1980 and 2000 (yuan in current prices)

	1980			2000		
	Urban	Rural	Rural/urban (per cent)	Urban	Rural	Rural/urban (per cent)
East	452	246	54.4	7,940	3,429	43.2
Central	386	191	49.5	5,217	2,091	40.1
West	383	172	44.9	5,642	1,713	30.4
China	413	193	46.7	6,635	2,301	34.7

Note: incomes are calculated by the authors as weighted averages from urban disposable income and rural pure income at the provincial level.
Source: National Bureau of Statistics, 2000a, 2001. *Statistical Yearbook of China*, China Statistics Press, Beijing.

The impact of WTO accession on agriculture and rural incomes

In 2001, China acceded to the WTO and committed to open its domestic food markets to imports to a much larger extent. There were several immediate changes in agricultural product trade policies following China's accession. The most important policies can be summarised as

- China accepted a tariff rate quota (TRQ) for major grain imports (wheat, corn and rice) at a token tariff rate of 1 per cent. The quota was 18.31 million tonnes for 2002, 20.2 million tonnes for 2003, and 22.16 million tonnes for 2004 (WTO, 2001). The quota is shared between the state and private trading enterprises, and all unused state quotas are to be transferred to private enterprises.

- The above-quota tariff rate for the major grains is 65 per cent.[2] There is no import quota for soybeans. The tariff rate for soybeans is 3 per cent

- The average tariff rate for all agricultural products is to be reduced from 22 to 17.5 per cent.

- All export subsidies are to be eliminated.

- Other non-tariff restrictions on imports of agricultural products, such as licensing, are to be eliminated. This includes restrictions on imports of wheat from the northwest areas of North America which may have TCK disease (a fungal disease).

The import quota of 22 million tonnes is large compared with the historical level of imports. During the 1990s, the average annual imports of grain were 9.0 million tonnes, and there were net exports of 0.8 million tonnes annually. There was only one year, 1995, when grain imports exceeded 20 million tonnes, and this resulted in a serious surplus of grain on the domestic market. The 2004 import quota is equal to 5.5 per cent of the total output in year 2000—not a very large proportion—however, the size of the domestic grain market is far smaller than total output, because half of the grain output is consumed by farmers and does not enter the market. Calculations show that the 2004 TRQ accounts for 11.9 per cent of the domestic grain market. It is estimated that full utilisation of the TRQ in the short run, without major structural adjustment, could mean the loss of 9 million farming jobs (Wang 2002).

China does not have a comparative advantage in the production of major grains such as wheat, corn and soybeans. A comparison shows that the average prices of these three products in rural markets in the period between 1995 and 2000 were 7 per cent, 30 per cent and 31 per cent higher than their 2000 c.i.f. prices, respectively. It is not surprising, therefore, that imports of soybeans and soybean oil are increasing rapidly. The domestic price of rice is lower than c.i.f. prices, although its average quality is also lower than that of imports. Thus, in the face of increasing imports, domestic production of these grains will decline significantly.

In 2000 and 2001, cereal imports were at moderate levels of 3.15 and 3.44 million tonnes, respectively. Due to surpluses and the low prices of grain in the domestic market,[3] as a result of the over-supply since 1998, it is unlikely that the TRQ will be fully used in the near future. However, the TRQ is preventing domestic grain prices from recovering to their recent levels. Meanwhile, soybean imports rose to the historically high level of 13.9 million tonnes in 2001, nearly as much as the domestic output (National Bureau of Statistics 2002). This has led to a large surplus in the soybean market.

It is clear that without major adjustments in the production structure, the growing grain imports will significantly affect farmers' incomes and further widen the rural-urban income gap.

However, the WTO accession provides great possibilities for reforming China's agricultural sector, leading to positive long-run effects on farmers' incomes, as it pushes agriculture towards its comparative advantages and higher efficiencies. In 2002, the area sown to grain in China fell by 2 per cent, whereas the total grain output increased by 1 per cent from the previous year. New breeds of soybean with significantly higher quality and higher oil content were introduced to north-east China, the main region of soybean production (China Central Television News 2003). These are signs indicating that further increases in efficiency in agricultural production are possible in response to the WTO challenge.

However, to increase agricultural efficiency significantly and reduce rural-urban income disparities, not only structural adjustments of agricultural production but also broad changes in China's domestic grain trade policy are needed.

The grain pricing system in China and its inconsistencies with WTO accession

The current domestic grain trade policy—which provides grain support prices to farmers and almost monopolises the domestic grain trade system—is not only inconsistent with the trade liberalisation but is hardly helpful to farmers.

Agricultural production and domestic trade in agricultural products was partially liberalised in the late 1970s and early 1980s. These reforms changed the long-term situation of grain supply shortages into surpluses. Grain output increased from 283 million tonnes to 407 million tonnes during the 1978-84 period. Grain production was first stimulated by the rise in state purchasing prices, and subsequently fuelled by the abolition of the people's commune system and introduction of the household responsibility system, which converted the collective-based production system into a private system. Grain production was affected by three prices at that time—the state quota price, the above-quota price, and the market price—but none of them were linked to world market prices.

In nominal terms, quota prices rose slightly in the 1980s at a time of high inflation. From 1985 to 1990, the real price of the major grains (as a weighted average of the prices of rice, wheat and corn) fell by 22 per cent. Market prices played a role at the margin, whereas government prices had a larger impact on grain production because the volume of state purchases was so large (Wang 2001). As the result of low government prices, grain output remained at the 400 million tonnes level in the middle and late 1980s, while demand for grain increased significantly. The above-quota prices have converged to a level close to market prices since the early 1990s and fluctuated in line with market prices because of less government intervention. This seems to be the main reason for grain production reaching 440 million tonnes in the early and middle 1990s.

Stimulated by the improvement in grain supply, the government decided to liberalise the quota control system in 1993. However, most of the state-owned grain dealers have become partially profit-oriented and expected to make profits from holding grain stocks. This resulted in supply shortages and sudden increases in market prices in the short run, which led farmers to hold their products for expected higher prices. Facing sharp increases in market prices and shortages, the government decided to give up the planned reform and, instead, to increase the quota prices substantially in 1994. Quota prices were further increased in 1995-97 to encourage grain production. In 1997, the level of quota prices in real terms (as a weighted average of the prices of rice, wheat and corn) was 51 per cent higher than in 1993 and 20 per cent higher than in 1985 (Ministry of Agriculture, various years). Domestic prices of several grain products significantly exceeded the world market prices. At the same time, the central government introduced a 'provincial governor responsibility system' to insure local self-sufficiency of grain supplies.

Responding to the higher prices, the area sown to grain increased by 4 per cent and grain yield per hectare increased by 10 per cent between 1994 and 1998. Total grain output reached 504 million tonnes in 1996 and 512 million tonnes in 1998. This output far exceeded domestic demand and resulted in decreases in market prices since 1997.

From 1996 to 2000, market prices of rice, wheat, corn and soybean dropped by 39 per cent, 35 per cent, 44 per cent and 44 per cent, respectively (authors' calculations on the basis of data, Information Centre, Ministry of Agriculture). A model simulation shows that the rigid government pricing system has caused serious price and output instability in grain markets (Wang 2001).

The rigidity of the state-monopolised grain trade system and mistakes of the grain trading companies added fuel to the flames of supply surpluses. While domestic supply was increasing, there were net imports of 18.67 and 10.25 million tonnes of grain in 1995 and 1996, respectively, which worsened the situation of domestic grain surpluses.

Faced with the low market prices, the government announced three policies in 1998: all grain products were to be purchased by the state grain companies at support prices that are higher than market prices; the state grain companies would have to sell grain at prices not lower than the support prices, so that the state subsidy could be reduced; and bank loans extended to these companies for the purpose of grain purchases were not to be used for other purposes. In addition, in order to make the support prices work, the grain market was monopolised by the state, and private businesses were prohibited from purchasing grain directly from farmers.

These policies did not achieve the goal of protecting farmers' incomes for several reasons. First, after two decades of market-oriented reform, it was extremely difficult to entirely monopolise grain purchases. Second, the role of policy executers given to the state grain companies conflicted with their role as profit makers. To make profits or to avoid losses, they tended to under-grade the quality of grain they were purchasing from farmers, or make extra deductions for 'wet' and 'impure' grain, so that they could pay no more than market prices. In some cases, they refused to purchase from farmers, only from private dealers, so that under-the-table deals could be made to share the price margin. Third, even if consumer prices could be monopolised at a higher level than would clear the market, the demand for grain was not under state control. As a result, the over-supplied grain could not be sold off and the losses had to be borne by either the state or farmers. In addition, while farmers benefited little from the support prices, the state had to bear the huge cost of storing a large amount of grain and to invest in many new storehouses.

Under these pressures, a few changes were made to partially liberalise the grain markets. First, some low quality grains were excluded from the support list in 2001. Second, up to 2002, the local grain market was liberalised in eight east coast provinces that had a grain trade deficit. Third, decision making on support prices was transferred from the central to the provincial governments, resulting in support prices moving closer to market prices.

In spite of these changes, the support price system still conflicted with the market mechanism, and generated more conflict following WTO accession, because it encouraged domestic production of grain which did not have a comparative advantage, and therefore resulted in surpluses and lower market prices. Further liberalisation of the grain market was needed.

Policy considerations in liberalising the grain market

There are several fears about liberalisation of the grain market.

- Without government protection, farmers may be hurt when the market price of grain is low.

- Without the government purchase and supply of grain, food security may not be assured when there are poor harvests or war threats.

- Liberalisation of the grain market and withdrawal of government subsidies would result in job losses in the state grain trade sector of up to two million.

- The accumulation of huge financial losses and non-performing loans in the state grain sector has been a hot political potato; once the market is liberalised, the non-performing loans in the state sector would need to be liquidated.

Contrary to these opinions, past experience shows that government intervention has resulted in instability of grain supply, larger price fluctuations, inefficient operations, and huge losses in the state grain trade sector. The earlier the state grain trade sector is reformed, the more quickly the losses can be avoided.

In those provinces in which the grain market has been liberalised, the situation is satisfactory, that is price levels, demand and supply are basically stable; the formal state grain companies have been either privatised or restructured; and at least some of these businesses have become profitable.

Liberalisation of the grain market will have at least some benefits. The grain surplus will soon be absorbed, which will help to stabilise market prices. Market prices formed on the basis of demand and supply will give farmers the best available information on which to base their production decisions. Other agricultural products in which China has comparative advantage, for example, some vegetables, fruits, animal products, and herbal medicines, will replace grains, and this will help to increase farmers' incomes. The accumulated huge losses and non-performing loans in the state grain sector have become a heavy burden on the economy. This burden will be soon removed after liberalisation of the grain market.

However, as well as deregulation, the government needs to be pro-active in certain areas.

- Anti-monopoly regulations should be enacted and a market supervision system established to reduce the likelihood of monopolistic practices, whether by government or private traders.

- Due to difficulties in farmers gaining access to market information, the government should accept responsibility for establishing a broad network of information services to provide farmers with supply and demand information and forecasts of the domestic and international food markets.

- In some areas, the agricultural technology support and training systems run by government have played an excellent role in helping farmers to adopt new technology. However, there are still many farmers, particularly in remote areas, who do not enjoy these services. These systems should be expanded. New products, breeds, fertilisers, and other technologies and production methods should be more widely and more quickly introduced and demonstrated to farmers.

- For emergency food supply situations, an effective nation-wide grain reserve system is needed. However, its sole function should be as emergency food supplies. Non-government stocks should not be allowed to perform that role. The current grain reserve system is too large and too complicated and inefficient, run as it is by different government agencies at the central, provincial, and municipal levels. Because of the different interests of the various agencies and the conflicts within and between the agencies, this arrangement can hardly serve a national goal. Given a shortage of supply, some agencies may support the market using their reserves, while others may buy and hoard grains to make a profit. Therefore, it is necessary to reform the grain reserve system to give it a single objective and to have it operate under clear guidelines and unified control.

Relationship between rural industrialisation and urbanisation and rural incomes

Grain market deregulation and structural adjustment in the agricultural sector are important. However, due to the huge surplus of agricultural labour, these measures will not be sufficient to absorb the shock of agricultural internationalisation. Policy adjustments are also needed to accelerate China's rural industrialisation and urbanisation.

During the economic reform period, particularly up to the mid 1990s, rural industry developed very quickly and made a major contribution to China's rapid economic growth (see the World Bank 1996; Cai 2000; and Wang 2000). Employment in the Township and Village Enterprise (TVE) sector increased from 28 million to 135 million during the period 1978-96, accounting for more than one-fourth of the total rural labour force. TVEs produce at least one-quarter of the total industrial output. Development of the TVE sector has also made a great contribution to increases in farmers' incomes. Nearly one-half of rural household income now comes from non-agricultural sources, mainly from the TVE sector. Figure 4.2 shows that the level of rural income per capita in China's 31 provinces is closely associated with provincial achievements in rural industrialisation.

Rural industries experienced substantial development in the 1980s, partially because the rural reform created a better market environment while the urban economy was still heavily subject to central control. Government policies also had an important influence, encouraging development of TVEs in rural areas and discouraging rural-urban migration.

Development of rural industries slowed in the late 1990s, mainly due to sharper market competition, unfavourable location of the rural enterprises, difficulties in external finance, and lack of infrastructure facilities, technical inputs, and human resources. Meanwhile, urbanisation accelerated. Large numbers of rural labourers migrated to urban areas to find jobs. In year

Figure 4.2   Relationship between rural industrialisation and rural
income, 2000

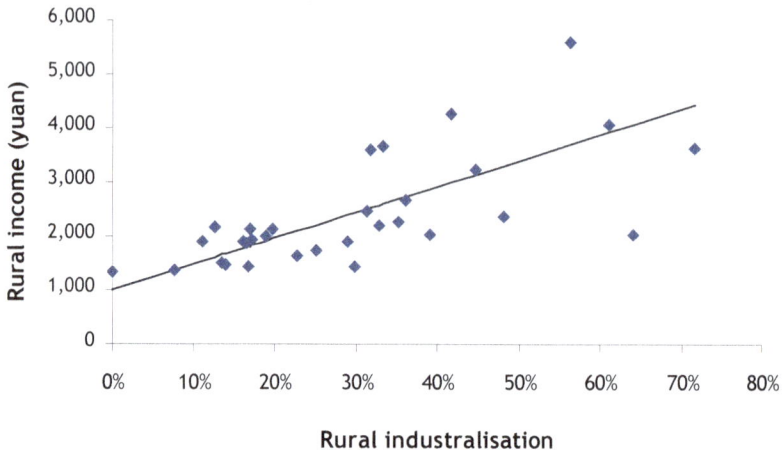

Rural industralisation

Note: Industrialisation is indicated by the share of TVE employment in rural labour. Rural income is per capita rural pure income (yuan).
Source: Calculated from National Bureau of Statistics, 2000a, 2001. *Statistical Yearbook of China*, China Statistics Press, Beijing.

2001, the urbanisation rate (the ratio of urban to total population) in China reached 38 per cent, whereas it was only 26 per cent in 1990 and 19 per cent in 1980. Improvement in the availability of rural finance will certainly help in the development of rural industries. Nevertheless, the importance of rural industrialisation in the economy is likely to be replaced by urbanisation.

In spite of the acceleration in urban development, the rate of urbanisation in China is still 10-20 percentage points lower than the average of other countries at a similar income level (Wang and Xia 1999). In particular, medium and large cities are in short supply compared with China's large population. In 2000, there were 121 million people living in the cities of 0.5 million population and above, accounting for only 9.6 per cent of the national population. In the less-developed west region, people living in cities of these sizes only account for 5.6 per cent of the region's population. If China had a similar urbanisation rate to the average of other countries at a similar income level, there would be an additional 120-240 million people living in urban areas. This number may also be thought of as the excess supply of people in the rural economy.

Figure 4.3 plots the urbanisation rate and rural income of China's 31 provinces in 2000. It indicates a positive relationship between urbanisation

Figure 4.3   Relationship between urbanisation and rural income, 2000

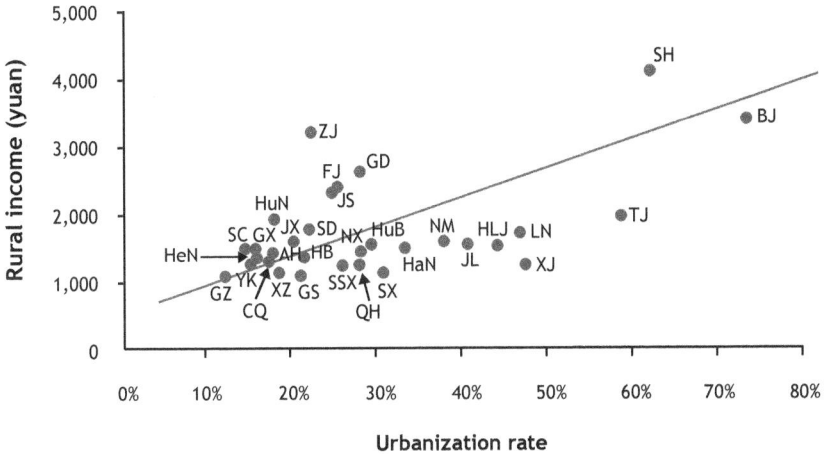

Urbanization rate

Symbols of provinces:

BJ: Beijing	TJ: Tianjin	HB: Hebei	SX: Shanxi	NM: Neimenggu
LN: Liaoning	JL: Jilin	HLJ: Heilongjiang	SH: Shanghai	JS: Jiangsu
ZJ: Zhejiang	AH: Anhui	FJ: Fujian	JX: Jiangxi	SD: Shandong
Hen: Henan	Hub: Hubei	Hun: Hunan	GD: Guangdong	GX: Guangxi
HaN: Nainan	CQ: Chongqing	SC: Sichuan	GZ: Quizhou	YN: Yunnan
XZ: Xizang	SnX: Shannxi	GS: Gansu	QH: Qinghai	NX: Ningxia
XJ: Xinjiang				

Note: The urbanisation rate used is the proportion of urban to total employment in each province.
Source: National Bureau of Statistics, 2001. *Statistical Yearbook of China*, China Statistics Press, Beijing.

and rural income: provinces having a higher urbanisation rate have a higher rural income. This relationship may imply that urbanisation helps to liberate redundant rural labour from arable land and therefore increases agricultural productivity. Considering that one-half of the Chinese labour force is still agricultural labour, urbanisation could have a significant impact on agricultural productivity.

A causality test on the impact of urbanisation and policy considerations

However, this urbanisation effect does not have to be the only explanation for the relationship shown in Figure 4.3, because both urbanisation and higher rural income can be an outcome of economic growth. In the following, we introduce a growth model to test the causality between urbanisation and regional economic growth. This model has its origins in neoclassical growth models and endogenous growth models (see, for example, Solow 1956; Lucas 1988) but includes variables representing urbanisation:

$$Y_{it} = A + a^1 K_{it} + a^2 H_{it} + a^3 D_{it} + a^4 L + a^5 U_{it-1} + a^6 DU_{i(t-1)} + \varepsilon_{it} \qquad (1)$$

where Y_{it}, K_{it}, H_{it} D_{it} and L_{it} are growth rates of GDP, capital stock, human resources (indicated by workers' average year of schooling), cultivated land area and total employment, respectively, of the ith province in year t. U_{it-1} is the urbanisation rate with a one-year lag (the urbanisation rate used is the ratio of urban employment to provincial total employment). $DU_{i(t-1)}$ is the difference of U_i between years t-1 and t-2. ε is the error term.

Lagged variables of U and DU are used for the causality test between urbanisation and economic growth. This specification identifies the effect of urbanisation on growth if U or DU is significant. The reason for including both U and DU in the model is to distinguish the possible growth effect (long-run effect) and level effect (short-run effect) of urbanisation. A significant estimate of U indicates a continuing effect of urbanisation on growth (meaning urbanisation brings about higher productivity growth), whereas a significant estimate of DU indicates an impact from changes in the urbanisation rate on growth, which is a short-run effect.

To impose the restriction of constant returns to scale ($a^3 = 1 - a^1 - a^2$), both sides of Equation 1 were divided by L_{it}

$$y_{it} = A + a^1 k_{it} + a^2 h_{it} + a^3 d_{it} + a^5 U_{i(t-1)} + a^6 DU_{i(t-1)} + \varepsilon_{it} \tag{1'}$$

where y, k, h and d are Y, K, H and D divided by L, respectively.

To see the possible differences in the urbanisation effect at the different levels of economic development, $U_{i(t-1)}$ is replaced by three urbanisation variables for the east, central and west regions. The east region is the most developed while the west region is the least developed of the three regions. Similarly, replacements are made to $DU_{i(t-1)}$.

$$y_{it} = A + a^1 k_{it} + a^2 h_{it} + a^3 d_{it} + a^7 U^1_{i(t-1)} + a^8 U^2_{i(t-1)} + a^9 U^3_{i(t-1)}$$
$$+ a^{10} DU_{i(t-1)} + a^{11} DU_{i(t-1)} + a^{12} DU_{i(t-1)} + \varepsilon_{it} \tag{2}$$

Panel data for 25 provinces and covering the 20 years from 1979 to 1998 have been used. The other six provinces were excluded due to incomplete data. Data were provided by Fang Cai and Dewen Wang (2002) and from National Bureau of Statistics (various years and 1999b). Capital stock was calculated from the historical data for capital formation in each province.

Both fixed effects and random effects models were estimated. Hausmann's test rejects the hypothesis of the appropriateness of the random effects model; therefore the results of the fixed effects model are reported in Table 4.3.

The two versions of the model produce similar results. Most coefficients are statistically significant at the 1 per cent level. According to version (1'), the elasticities of capital, human resources, employment and land with respect to economic growth are 0.348, 0.431, 0.205 and 0.016, respectively, all in reasonable ranges. The elasticity of land is minor and insignificant, which is not surprising because the farming sector only contributes a small proportion to the economy. The estimates of U are significant, which indicates that every one percentage point increase in the urbanisation rate accelerates provincial economic growth by 0.37 percentage points over the 7-10 per cent high growth rate. DU is omitted from the model due to insignificant coefficient estimates in previous regressions. These results suggest that urbanisation has a long-run impact on economic growth.

The major difference in the results of the second version of the model is the insignificant effect of urbanisation on economic growth in the less-developed west region. The effect is significant in the two other regions. There are two possible reasons for this result: the effect in the west region is insignificant because the achievement of urbanisation in the west has been relatively low or, contrary to the other regions, the urban economy in the west provinces has not experienced much restructuring, and is therefore less market-oriented and less efficient.

In general, the results identify a contribution from urbanisation to economic growth via productivity changes. A reasonable explanation for the higher productivity growth is improvement in resource allocation between the rural and urban sectors. This implies that accelerating urbanisation could be an important measure to counteract the short-run side effects of WTO accession on the agricultural sector, and to reduce the rural-urban income gap.

Table 4.3   Modeling results: urbanisation and economic growth

Variable	Version 1'		Version 2	
	Coefficient	t-ratio	Coefficient	t-ratio
K	0.3481	(8.891**)	0.3552	(8.931**)
H	0.4315	(5.122**)	0.4261	(5.045**)
D	0.0157	(0.903)	0.0159	(0.911)
U_{t-1}	0.3753	(3.619**)		
U1			0.3560	(2.698**)
U2			0.5525	(2.835**)
U3			0.0097	(0.031)
Constant	-0.0837	(-2.383*)	-0.2119	(-2.265*)
Dependent var.	y		y	
R^2	0.3697		0.3726	

Note: t ratios with * and ** are significant at the 5 per cent and 1 per cent levels, respectively. For simplicity, coefficients of provincial dummy variables are not reported in this table.
Source: Estimation results. Original data from National Bureau of Statistics, 1999b, 2001. *Statistical Yearbook of China*, China Statistics Press, Beijing.

Urbanisation in China being behind other countries is mainly a result of the previous central government policies restricting rural-urban migration and growth of large cities. Some of these restrictions have been removed over the past decade, however, rural-urban migration is still partially restricted by the urban household registration system, job entry barriers, non-access or harder access to health care benefits and schooling, and public security protection, for example, against rural migrants. The low level, or lack, of rural education, job training, and employment information services are also barriers against rural-urban migration. Policy adjustments to remove these restrictions, to improve rural education, and to provide government services on job training and employment information services are essential for accelerating urbanisation.

The current urban economy provides limited opportunity for rural migrants because the number and size of cities is limited. Expansion of the urban economy will provide employment opportunities, especially in China's under-developed services sector, which has a significantly higher proportion in the existing urban economy than in the rural economy. It is the government's responsibility to improve urban planning, urban infrastructure, and public utilities in order to promote urban development. With these changes, many small cities and towns can be expected to become large or medium sized cities.

Conclusion

The agricultural sector in China is seen as being labour-redundant, largely as the result of policies restricting the development of urban areas. These policies are also seen as a major reason for China's large and growing rural-urban income gap. The situation is worsening after the opening of China's grain market resulting from its WTO accession. In particular, the over-regulated domestic grain market and government pricing conflict with the new situation of openness, since these result in over-supply of grains. It is suggested that the domestic grain market should be liberalised and the grain pricing system should be deregulated.

For further solutions to the import shock to the agricultural sector as the result of WTO accession, the roles of rural industrialisation and urbanisation are examined. We find that urbanisation has positive effects on rural incomes and regional economic growth. Since the urbanisation rate in China is significantly lower than in other countries at a similar level of GDP per capita, we recommend that urbanisation should be accelerated by deregulating rural-urban migration, promotion of urban infrastructure, improvement in rural education, job training and employment information services.

Notes

1   Data used in this paper without acknowledgment are from the National Bureau of Statistics.
2   However, this tariff rate is unlikely to be used in normal cases since the quota is large compared with the historical record of grain imports.
3   In real terms, the compound market price of major grains, that is, rice, wheat, corn and soybean, in 2000 was only 57 per cent of the 1995 level and 78 per cent of the 1997 level (calculations by the authors). Ministry of Agriculture Information Centre (various years), Wu 2001 and National Bureau of Statistics (various years).

References

Cai, Fang, Dewen Wang and Juwei Zhang, 2000. 'Economic growth in China: labour, human capital and employment structure', in Xiaolu Wang and Gang Fan (eds), *The Sustainability of China's Economic Growth*, Economic Science Press, Beijing.

Cai, Fang and Dewen Wang, 2002. 'Regional comparative advantage in China: differences, changes and its impact on regional disparity', in Xiaolu Wang and Gang Fan (eds), *Changing Trend and Influential Factors of Regional Disparity in China,* Economic Science Press, Beijing.

China Central Television News, 8 January, 2003.

Ministry of Agriculture, China, Survey data, Information Center, Beijing.

Lucas, R.E., 1988. 'On the mechanics of economic development', *Journal of Monetary Economics*, 22:3–42.

Ministry of Agriculture, China, various years, *China Agricultural Development Report*, Beijing.

National Bureau of Statistics, various years and specifically 1999b, 2000a, 2001, 2002. *Statistical Yearbook of China*, China Statistics Press, Beijing.

Solow, R.M., 1956. 'A contribution to the theory of economic growth', *Quarterly Journal of Economics*, 70, February, 65–94.

Wang, Xiaolu and Xiaolin Xia, 1999. 'Optimum city size and economic growth', *Economic Research*, No. 9, Beijing.

Wang, Xiaolu, 2000. 'The role of rural industrialisation in China's economic growth', in Xiaolu Wang and Gang Fan (eds), *The Sustainability of China's Economic Growth*, Economic Science Press, Beijing.

Wang, Xiaolu, 2001. 'Grain market fluctuations and government intervention in China', research report for the ACIAR project China's Grain Market Policy Reform, The Australian National University, Canberra.

Wang, Xiaolu, 2002. 'The WTO challenge to agriculture', in R. Garnaut and L. Song (eds), *China 2002: WTO entry and world recession*, Asia Pacific Press, The Australian National University, Canberra:81–96.

World Bank, 1996. *The Chinese Economy: Controlling Inflation, Deepening Reform*, World Bank, Washington, DC.

World Trade Organization, 2001. *Accession of the People's Republic of China, decision of 10 November 2001*, World Trade Organization, Geneva.

Wu, Laping, 2001. Price comparison between world and domestic grain markets, prepared for the ACIAR project on China's grain market, The Australian National University, Canberra (unpublished).

05 The impact of WTO accession on China's agricultural sector

Xiaolu Wang

Agriculture is still very important in China, not only because it provides the major source of food for the people but also because it is still the major source of income for half of the 1.26 billion people in China. In the past half a century, from 1952 to 2000, China experienced rapid industrialisation, and the share of agriculture in GDP fell from 51 per cent to 16 per cent. However, the differentiation of the population changed far more slowly. During the same period, the proportion of rural people in the total population only declined from 85 per cent to 64 per cent, and the proportion of agricultural workers in China's total employment fell from 84 per cent to 50 per cent. The decline in the share of agricultural workers was faster than the decline in the share of the rural population due to rapid rural industrialisation (that is, the development of the Township and Village Enterprises sector) during the past two decades. Still, there were 334 million workers in the agricultural sector in the year 2000.[1]

From the large disparity between the agricultural share in GDP and the share of agricultural workers in total employment, one can see how low the labour productivity of the agricultural sector is compared with other sectors. In year 2000, rural per capita net income was only 2253 yuan (or US$272), which was equivalent to only 36 per cent of the average urban per capita disposable income.[2] Of the total rural per capita net income, agricultural income was 1136 yuan, of which 868 yuan was from farming, that is, 50 per cent and 39 per cent of rural net income, respectively. Thanks to the rapid development of the rural non-agricultural sectors over the past 20 years, non-agricultural activities provided 44 per cent of farmers' incomes (off-farm wage income plus income from non-agricultural household businesses).

Grain outputs, yields and inputs

Farming remains the main agricultural activity. In 2000, farming accounted for 56 per cent of the gross value of agricultural output. Grain production has been the most important component of farming. A rough calculation using output and average prices of major grain products shows that in gross value terms total grain output accounted for 65 per cent of farming and 38 per cent of agricultural output in 1998. China is the largest grain producer in the world with output ranging in recent years between 460 and 510 million tonnes (unprocessed grain). This output accounted for over one-fifth of total world grain output (see Table A5.1).

Grain output has increased significantly since the beginning of agricultural reform in 1978. It was 300 million tonnes in 1978, and reached its highest level to date of 512 million tonnes in 1998 (grain output declined to 462 million tonnes in 2000). The large increase in grain output can be mainly attributed to four effects: the market-oriented reforms and the adoption of the household responsibility system in the early 1980s; increases in domestic grain prices; technical progress (for example. development of hybrid rice); and continued increases in inputs.[3]

Grain output growth was due entirely to increases in yields, as arable land area has been slowly declining. The institutional reforms and technical progress made the major contributions to increases in yields up until the mid 1980s, and then these effects gradually diminished. Inputs such as fertilisers, insecticides and irrigation grew faster than output growth during the pre reform period, but rose more slowly than output during the reform period (Wang 2000a). Still, the intensity of material inputs has reached a high level relative to other major grain producers (see Table A5.2). The average grain yields for China as a whole reached 4.95 tonnes per hectare (measured by sown areas) in 1998-99, which is significantly higher than in most developing countries and is close to that of industrialised countries.[4]

Yield would be much higher measured in terms of cultivated area, because the average multiple-cropping ratio is usually 1.5 to 1.6. The level of labour intensity has been one of the highest among the world's major grain producers (see Table A5.2). Due to the rapid increases in marginal costs, further increases in yields, although possible, are not likely to increase farmers' incomes.

Agricultural labour intensity in China, measured by the number of agricultural workers per hectare of arable land, is 59 per cent higher than the world average, twice as high as for Japan, 20 times higher than the UK, Germany, and France, 50 times higher than the US, and 100 times higher than Australia and Canada (Table A5.2). The cultivated land area in China was only 0.39 hectares per farmer in year 2000. Because of the limitation of land area, the marginal product of labour diminished rapidly. The low labour productivity of the agricultural sector is mainly the result of natural limitations in the arable land areas and the huge agricultural population.

Transfer of the agricultural labour force

A major challenge to China's agricultural sector is how to shift a substantial part of its huge labour force to non agricultural sectors. Because of rapid rural industrialisation, led by the market-oriented reforms of the past 20 years, 100 million rural labourers moved from the agricultural sector to the township and village enterprise (TVE) sector. Total TVE employment was 128 million in 2000. Rural–urban migration was strictly restricted during the pre reform and early reform periods, but has increased rapidly since the late 1980s, partly as a result of the relaxation of the policy and partly as a result of employment pressures due to the slower growth of TVEs. Most rural workers migrating to urban areas have done so without officially changing their household registration status to urban residents. The number of these people, so-called 'floating labour', is statistically unavailable. Based on data from surveys, however, the author estimates that they were 14 million in 1990 and they totalled around 47 million in year 2000.[5]

However, during the period between 1980 and 2000, the total rural labour force increased from 318 to 499 million (they were re-estimated by the author as 347 million and 519 million for 1980 and 2000, respectively),[6] as a result of natural population growth. Therefore, in spite of the rapid rural industrialisation and continued urbanisation in the past two decades, the agricultural labour force has increased from 291 million to 334 million (the author's estimations are 317 million and 344 million for the two years, respectively). The growth rate of rural labour has decreased significantly due to the family planning policy and urbanisation, although it still remained positive at 0.9 per cent in the 1990s.[7] The author predicts that the rural labour force will grow at a low rate of 0.2 per cent per year, to reach a total of 530 million in 2010.

While the natural growth of rural labour is slowing, rural industrialisation has also stagnated since the mid 1990s. Total employment in the TVE sector grew at an annual rate of 12 per cent in the 1980s, but at only 3 per cent in the 1990s. TVE employment reached its highest level of 135 million in 1996, fell in 1996 and 1997, and slightly recovered to 128 million in 2000. There seems no reason to believe that employment growth in the TVE sector will recover in the medium term. We therefore assume an average 1.5 per cent growth rate of TVE employment from 2001 to 2010, to reach a total of 149 million by 2010.

Rural–urban migration is unlikely to grow faster in the near future, due to the weaker demand for, and oversupply of, unskilled labour in urban areas in recent years. In particular, a large number of urban workers in the state-owned enterprise sector have been laid off (SOE employment fell by 30 million from 1996 to 2000), and growth of export and domestic production has slowed.

After the deduction of those who have been employed by TVEs and those who migrated to urban areas, nearly all the remaining rural labourers are in the agricultural sector. This is because they are entitled to a small parcel of arable land under the current 'household responsibility system', which provides them with a form of minimum insurance. Therefore China's agricultural sector is in reality an enormous reservoir of underemployed labour.

Table 5.1 shows the growth of the rural labour force, TVE employment and the estimated rural–urban migration during the 1980s and 1990s, and the expected growth of these variables in the 2000s. If we assume that the speed of the net transfer of agricultural labour to the urban sector between 2001 and 2010 will be slightly slower than in the 1990s, that is, 3 million per year instead of 3.3 million per year, then the agricultural labour force in 2010 will be 304 million, a 40 million reduction from the 344 million in year 2000. In this case, and without further reduction in crop land, the labour intensity in the agricultural sector will fall by 12 per cent. This will mean only a minor improvement in labour productivity and in per capita farming incomes.

More optimistically, if urbanisation can be significantly accelerated, with the rural-urban migrants doubling from 2001 to 2010, that is, an increase from 3.0 to 6.0 million per year, the agricultural labour force will be reduced to 274 million by 2010. Labour intensity in agriculture will decline by 20 per cent—still only a minor improvement in labour productivity.

Table 5.1   Rural labour, TVE employment and rural-urban migration, 1980-2010 (million persons)

Year	Rural labour (statistics)	Rural labour (adjusted)	TVE employment	Rural-urban migration	Agricultural labour
1980	318	347	30	0	317
1990	473	473	93	14	366
2000	499	519	128	47	344
2010 (assumption I)		530	149	77	304
2010 (assumption 2)		530	149	107	274

Note: Figures in the third column are adjusted from the labour statistics according to the information from the 1962, 1984 and 1990 national census. See Wang (2000b) Agricultural labour is calculated as the balance of total rural labour minus TVE employment minus rural-urban migration.
The number of rural-urban migrants was estimated by the author on the basis of survey by Ministry of Labour and Social Security and the National Bureau of Statistics (1999). In the first assumption for 2001-2010, the growth rates assumed for rural labour and TVE employment are 0.2 per cent and 1.5 per cent, respectively, and rural-urban migration is assumed be 3.0 million per year. In the second assumption, rural-urban migration is 6.0 million per year, while the other assumptions remain the same.
Sources: Department of Training and Employment of the Ministry of Labour and Social Security, PRC, and Rural Social and Economic Survey Team of the National Statistical Bureau, 1999. The situation of rural labourers' employment and flow in China, 1997-1998, printed report, Beijing; National Bureau of Statistics, 1999 and 2001. *Statistical Yearbook of China*, China Statistics Press, Beijing; Wang, Xiaolu, 200b. 'Sustainability of China's economic growth and institutional changes', in Wang, Xiaolu and Fan, Gang (eds), *The Sustainability of China's Economic Growth*, Economic Science Press, Beijiing; author's estimations.

Imports, exports and prices of grain

Grain trade

In the 20 years from 1981 to 2000, there were net grain imports in 11 years and net exports in nine years. Total net imports were small compared to total output. From 1995 to 2000, average annual imports were 4.82 million tonnes but net imports were only 0.79 million tonnes per year. Annual grain imports have never reached the WTO tariff quota of 22.2 million tonnes during the past half century. Only in 1995 were imports close to the quota (20.8 million tonnes), although as discussed below this level of imports caused a serious oversupply of grain. Table A5.3 lists total imports and exports of grains and of some major grain products.

Grain prices

Before comparing domestic grain prices with those in world markets, a brief review of the structure of the domestic market and changes in the setting of domestic prices since 1985 is in order.

At the beginning of this period, domestic prices of grain (farm gate prices) took three forms: the state quota price, the state above-quota price, and the rural market price. The state purchase prices, especially the quota prices, were 7 per cent to 25 per cent lower than the rural market prices in 1985 and 19 per cent to 53 per cent lower in 1993 because the nominal state purchase prices increased more slowly than market prices (see Tables A5.4–A5.7). The state quota price in real terms, as a weighted average, decreased by 20.3 per cent from 1985 to 1993 (Table 5.2). Due to supply shortages, the state progressively increased its quota or above-quota prices from 1994 to 1996. Market prices also increased in 1994 and 1995 because market supplies were squeezed by the increased state purchases and increased state grain stocks. The highest domestic prices in real terms, as a combination of the state and market prices, were reached in 1996. According to an internal report, major grain prices exceeded world market prices by 38 per cent to 45 per cent at that time, except for rice which was 8 per cent lower.

Table 5.2   Real price changes: state and market, 1985-2000
             (1985 price=1.00)

Year	State	Market	Average
1985	1.000	1.000	1.000
1990	0.784	1.158	0.909
1993	0.797	0.965	0.853
1994	0.929	1.140	1.000
1995	0.969	1.399	1.112
1996	1.090	1.290	1.157
1997	1.202	1.028	1.144
1998	1.140	0.964	1.081
1999	1.001	0.912	0.971
2000	0.975	0.802	0.917

Note: Both the state and market price indexes are derived as a weighted average of the price index of rice, wheat, corn and soybean. The year 2000 shares in the sum of their output are used as the weights. The rural consumer price index is used as the deflator. For the average index, the weights are two-thirds and one-third for the state and market, respectively.
Source: Calculated from Tables 5.A4–A7 in Appendix.

In 1997, quota prices increased further even though market prices had started declining in 1996. This led to quota prices exceeding market prices in 1997 and 1998. In 1997, real quota prices were 50.8 per cent higher than in 1993, and 20.2 per cent higher than in 1985. However, real market prices declined from a level 29 per cent higher than the 1985 prices to only 3 per cent higher than the 1985 prices between 1996 and 1997 (Table 5.2).

The increases in grain prices over the 1994-97 period resulted in historically high output levels above 500 million tonnes between 1996 and 1999. Together with the large grain imports in 1995 and 1996 (20.8 and 12.2 million tonnes, respectively), this resulted in large domestic surpluses. As a result, the state purchase prices fell by 19 per cent from their highest levels in 1997 to the recent low of 2000, and the market price dropped by 43 per cent from its highest level in 1995 to a 15-year low in 2000.

Table 5.3 shows that in 2000 the real grain price, expressed as a weighted index of the state purchase prices and market prices, was 8 per cent lower than the 1985 level. By 2000 domestic prices of major grains had fallen to levels similar to, or even significantly lower than, world market prices.

Since 1998, the state quota and above-quota prices have been combined to form a support price, which was higher than market prices at the beginning but by 2000 was similar to market prices. This was because the decision making about support prices has been decentralised to the provincial or lower-level governments, and these governments cannot afford to subsidise grain prices.

Detailed discussion on the reasons for such large fluctuations in domestic market prices can be seen in Wang (2001). Table 5.3 provides a comparison of prices of major grains in rural markets between 1994 and 2000. It shows that the real price of rice and wheat in 2000 were at levels equal to only 70 per cent and 66 per cent, respectively, of their 1994 levels. The 1994 prices were low compared with those in 1995 and 1996.

There was a recovery in domestic grain prices in 2001. Up to September 2001, grain prices were 7.5 per cent above the levels of September 2000, and 5.7 per cent above the levels at the end of 2000 (Center for News

Table 5.3   Comparison of real grain prices in rural markets between 2000 and 1994 (yuan per tonne)

Items	1994	1996	1998	2000	2000/1994 (per cent)
Rice	2,057	2,818	2,175	1,721	70.0
Wheat	1,141	1,741	1,298	944	66.2
Corn	1,009	1,487	1,579	1,225	97.2
Soybean	2,451	3,212	2,928	2,534	82.7
RCPI (per cent)	100	102.7	104.2	102.6	
IPIA (per cent)	100	113.5	106.8	101.4	

Notes: RCPI and IPIA are the Rural Consumer Price Index and Input Price Index in Agriculture, respectively.
Real grain prices between 2000 and 1996 were derived using a compound deflator from RCPI and IPIA at 0.5:0.5.
Source: Calculated from National Bureau of Statistics, 2001. *Statistical Yearbook of China*, China Statistics Press, Beijing.

Gathering and Editing, China Agricultural Web 2002). The recovery in prices has been mild since the surplus of grain stock has not been fully absorbed. Total grain stocks were still large at the end of 2001.[8] Therefore, a larger recovery in grain prices could be expected in the coming years—without a major increase in imports—before normal storage levels are reached.

One may conclude from the above description that either the 1995-96 domestic market prices or the 1997 quota price significantly exceeded an intermediate internal equilibrium level, whereas the 2000 domestic prices were far below that equilibrium. To compare with world prices, we may use the average domestic market price level (in constant terms) during the period of 1985-2000 to represent an intermediate internal equilibrium price level. This average level is equal to 79 per cent of 1995 market prices and 137 per cent of 2000 market prices. It is 10 per cent higher than 1985 market prices. By accident, it is equal to the mean of market prices between 1995 and 2000.

We should note that the farm gate prices are not comparable with world market prices for two reasons: the price of imported grain will be higher than world market prices after including transport, insurance and other costs; and to compare domestic prices with prices of imported grain, domestic purchase, transport and wholesale costs should be added.

It is reasonable to compare domestic wholesale prices and the import CIF prices. Laping Wu (2001) has found that the CIF prices of imported rice and wheat were significantly higher than the domestic wholesale prices from 1996 to 2000. In this comparison, however, there are a few important things we should note. First, the quality of exported, imported, and domestically wholesale grain may not be fully comparable. For example, imported rice and wheat is usually of a higher quality than the domestic grains. Second, if we compare the domestic wholesale prices of grain with rural market prices or state purchase prices, the former were usually similar to, or even lower than, the latter. This was either a result of state subsidy in the earlier years or due to surpluses in the domestic market in recent years.

Historical data on quota, above-quota, rural market, wholesale, FOB and CIF prices for rice, wheat, corn and soybeans are presented in Tables A5.4-A5.7. They show that the rural market prices of wheat, corn and soybean were not significantly lower than the CIF or FOB prices, even in

Table 5.4   Comparing rural market prices with CIF prices (yuan per tonne)

Items	Rural market price	2000 CIF
Rice	2,452	3,859
Wheat	1,477	1,375
Corn	1,265	976*
Soybean	2,341	1,785

Note: Rural market prices are calculated as the mean of 1995 and 2000 rural market prices (1995 prices have been converted to 2000 constant prices). CIF prices are derived from the imported volumes and values in 2000, converted from US$ to RMB. The CIF price is for 1999 since data for 2000 are unavailable.
Source: Calculated from Table A5.4-A5.7 and National Bureau of Statistics, 2001. *Statistical Yearbook of China*, China Statistics Press, Beijing.

year 2000. Only the rice price is significantly lower than the CIF price. Given consideration of the quality differences and the domestic costs for purchase, transport, storage and wholesaling, domestic prices of the major grains, except for rice, would still be higher than import prices.

In Table 5.4, the means of real market prices between 1995 and 2000 are compared with CIF prices. The comparison shows that, except for rice, between 1995 and 2000 the means of the rural market prices for major grains are all higher than the 2000 CIF prices. If the domestic costs of purchasing, transport and wholesale activities are added, the domestic prices would be even higher.

In general, it is reasonable to accept the points of many Chinese experts who believe that only rice production has comparative advantage, albeit weak, in China, whereas all the other major grain products have either weak or strong comparative disadvantage. Even for rice, due to the quality differences between domestic and imported products, large imports are still possible in the intermediate term. Therefore, the WTO tariff quota will have major impacts on the domestic grain production.

The direct impact of the WTO quota for grain imports

Major changes after WTO accession

There will be several important changes in China's agricultural policy following from its WTO accession commitments.

- China adopts a tariff quota for total grain imports at a 1 per cent token tariff rate. The quota will be 18.31 million tonnes in 2002, 20.2 million tonnes in 2003, and increasing to 22.16 million tonnes in 2004 (Table 5.5). According to the accession agreement, the 22.2 million tonnes of tariff quota is to be shared between the state and private trading enterprises, and all unused state quotas are to be transferred to private enterprises.

- The above-quota tariff rates for the major grains will be 65 per cent, and for soybeans the tariff rate will be 3 per cent.

- The average tariff rate for all agricultural products to be reduced from 22 per cent to 17.5 per cent.

- The average rate of domestic support for agricultural products will be zero. There will be no export subsidies.

- Other non-tariff restrictions on imports of agricultural products, such as licensing, are to be eliminated. This includes restrictions on imports of wheat from the north-west areas of North America, which may have TCK disease.

The economic impact of the latter changes is not clear and the issues are strongly debated. More detailed information and analysis are needed. The above-quota tariff rate is not likely to be important because experiences with grain imports show little likelihood of grain imports exceeding the quota. In the following, the focus is on the impacts of the tariff quota on agricultural production, farmers' incomes, employment, and consumers. These impacts appear to be the most important. The impact of the elimination of domestic support for agricultural products is also discussed.

Table 5.5   Tariff quotas for grains, 2002-2004 (million tonnes)

Year	2002	2003	2004
Wheat	8.468	9.052	9.636
Corn	5.850	6.525	7.200
Rice	3.990	4.655	5.320
Total	18.308	20.232	22.156

Source: World Trade Organization, 2001. *'Accession of the People's Republic of China, decision of 10 November 2001'*, cited from http://www.moftec.gov.cn/

Table 5.6   Estimation of the volume of non-traded grain (unprocessed), 2000

	Total consumption (million tonnes)	Self-consumption ratio	Self-consumption (million tonnes)
Rural food grain	201.9	0.8	161.5
Feed grain			83.4
Meat	85.7	0.5	51.4
Poultry and eggs	22.4	0.5	13.4
Cultivated fish, etc.	13.0	0.3	3.9
Milk	9.2	0.3	0.8
Draught animals	17.4	0.8	13.9
Seed grain	16.3	0.6	9.8
Total			254.7

Note: Rural food consumption of grain is derived from household survey data. Feed grain is derived from livestock production data. The quantity of meat (pork, beef and lamb) is converted from the gross weight (with bones) at a conversion ratio of 0.5. The weight conversion ratios between meat and feed grain are assumed to be 2.8 for meat production, 1.0 for poultry, eggs, cultured fish and shrimp, etc., and 0.3 for milk production.
The feed consumption of draught animals was assumed to be 0.5 kg per animal per day. The consumption of seed grain was derived from the sown area of grain in year 2000 at an estimated average 0.15 tonnes per hectare.
Self-consumption ratios are the proportions of farmers' living or production consumption of grain that is provided by themselves as a percentage of their total consumption. They are based on the author's personal experience in various rural surveys.
Source: Calculated from statistical data on grain production, rural household grain consumption, livestock production, draught animals, grain sown area, rural population and household size in year 2000. National Bureau of Statistics, 2001. *Statistical Yearbook of China*, China Statistics Press, Beijing.

The size of the domestic grain market

To assess the impact of the 22 million tonnes grain import quota on the domestic grain market we need to calculate the size of the domestic market. Total grain output was between 435 and 512 million tonnes (unprocessed grain) during the 1990s (462 million tonnes in 2000). However, the domestic market is far smaller than the total output because a substantial part of the grain output is consumed by farmers and does not enter the market. In the 1990s, the state quota and above-quota purchases of grain were around 180 to 200 million tonnes per year (possibly higher in 1998-99); however, there are no statistics on the quantity of grain sold in the free market. The size of China's domestic grain market (grain traded both by the state dealers and in the free market) may be approximated from a calculation of the volume of non-market grain consumed by farmers in 2000 (Tables 5.6 and 5.7).

In Table 5.6, the estimated volume of non-traded grain (self-consumption) is shown as 255 million tonnes for the year 2000. The volume of traded

Table 5.7   The estimated size of the domestic grain market in 2000 (million tonnes)

	Unprocessed weight	Trade weight
Total production	462	400
Non-traded consumption	255	221
Change in stocks	-20	-17
Net exports	12	10
Domestically traded	215	186
Total domestic demand	470	407
Tariff quota (2004 and later)		22.2
TQ as per cent of traded grain		11.9 per cent
TQ as per cent of total demand		5.5 per cent

Note: Trade weights are derived from the unprocessed weight. Rice is converted from paddy at a conversion ratio of 0.68; other grains are unprocessed. The change in grain stock is assumed.
Source: Calculated from Table 5.6 and National Bureau of Statistics, 2001. *Statistical Yearbook of China*, China Statistics Press, Beijing.

grain is derived in Table 5.7 under an assumption of a negative change in grain stocks by 20 million tonnes. Under these calculations, the tariff quota (2004) for imported grain at the 1 per cent tariff rate accounts for about 12 per cent of the domestic market.

The impact on domestic grain prices

In estimating the impact of the WTO import quota on the domestic prices, we may consider two possible situations. First we may assume that the 22 million tonnes of grain imports is an external shock at a time when the domestic market is in equilibrium. The domestic equilibrium prices are based on the calculation in Table 5.4 with a 15 per cent increase to include the domestic purchase, transport and wholesale costs. In this situation, the wholesale price of rice before the shock will be 27 per cent lower than for imported rice (but the average quality of domestic rice is also lower). Prices of wheat, corn and soybean before the shock will be higher than the imported prices by 24 per cent, 49 per cent and 51 per cent, respectively. In this situation, we may assume that the entire wheat and corn quotas, and 50 per cent of the rice quota, will be used. Altogether, these sum to 19.5 million tonnes. If we include other grains (for example, soybean imports may increase dramatically), total grain imports in 2004 could be above 21 million tonnes, that is, an increase of 18 million tonnes from an average 3.47 million tonnes between 1998 and 2000. This equals 9.6 per cent of the market demand and 4.4 per cent of total demand. (By the same calculation, imports will be around 17 million tonnes in 2002 and 19 million tonnes in 2003.)

Grain consumption in China is price inelastic. Based on estimations of China's grain demand elasticity in the literature, the author uses 0.37 as the weighted average elasticity of total demand for grain (alternatively, we could use 0.81 as the elasticity of market demand, which can be derived from the ratio between total demand and market demand).[9] The price effect of the 18 million tonnes of imports will lead to a 12 per cent decline of prices in the domestic market.

As an alternative situation (but more likely to be true), the external shock comes when the domestic market is still in surplus. In this case, the domestic prices of rice and wheat before the shock will still be lower than the imported prices, but that of corn and soybean is higher. Imports will be significantly smaller than in the first case. However, because the remaining quota can be used at any time when the domestic price is going up, the effects of imports will be similar to that in the first case, that is, they prevent the domestic price from recovering instead of pushing the price down. Therefore, for simplicity, only the first case is discussed in the following.

The impact on domestic production

How much grain production declines due to the price reductions will depend on the price elasticity of supply. According to the estimate by Wang (2001), the total price elasticity of grain supply is 0.52, and the effect of price changes on output will be fully realised within two years. Based on this elasticity, grain output will decline by 6.2 per cent in two years. There is likely to be a 'cobweb effect' because the price elasticity of demand is smaller than that of supply. In a standard 'cobweb effect' case, both production and price fluctuate to a larger and larger extent and can never converge to the equilibrium. However, this effect can be reduced by government operation of a price stabilisation scheme and improvement in information services to grain producers. For simplicity, we may consider the case where, on average, the price level declines by 6 per cent and output falls by 3.1 per cent as an equilibrium result.[10]

Producer losses

The producer losses can be approximately derived from the following formula

(Price before shock) x (percentage price reduction) x (volume of market grain before shock - net imports) + (Output reduction) x (domestic price after shock) - (material input cost) + (fixed costs that cannot be reduced) - (net incomes from new jobs after shock).

Using the 2000 production and consumption data, we can calculate

[1,815 x 6 per cent x 215 + (462 x 3.1 per cent) x 2006 x (1- 6 per cent - 40 per cent + 10 per cent)] x (1 - 30 per cent) =28.0 (billion yuan)

where 1,815 is the mean of 1995 and 2000 real grain price levels as a weighted average (constant yuan of 2000 per tonne), which is the assumed internal equilibrium price before shock, 6 per cent is the calculated price reduction due to the import shock, 215 is the calculated volume of market grain that is produced domestically before the shock, 3.1 per cent is the calculated reduction in total grain output, 40 per cent is the ratio of material input costs derived from the ratio between gross output value and value-added of agriculture in 1998, 10 per cent is the assumed ratio of fixed costs to the reduced output, which cannot be proportionally reduced, and 30 per cent is the assumed proportion of farmers who can move to the non-grain sector in the short run and earn the same income as before.

In this calculation, farmers' self-consumed grain is excluded because it is offset by their own production.

The derived 28.0 billion yuan of net losses is equal to 0.31 per cent of GDP, or 2.0 per cent of the agricultural value-added in 2000, and 4.2 per cent of farmers' net income from farming. Those grain farmers who have no other employment opportunities would incur a 9.1 per cent net income loss on average.

Producer losses could be much smaller if grain farmers can efficiently shift to non-grain production with the same resources available. However, the current situation is that the number of grain farmers (those fully or mainly engaged in grain production) is still very large, at least accounting for 60 per cent of the total 350 million farmers. Grain farmers are concentrated in the less developed central and west regions. Most of them have very low productivity, annually producing only 2 tonnes of grain per capita, and receive very low incomes. However, the transfer of these farmers to other sectors has been slow. The major obstacles are

- limited employment opportunities for unskilled labour in other sectors and the low level of education and skills
- accessing domestic or international markets of non-grain products is difficult for many farmers due to poor information, telecommunication and transport services in the remote rural areas
- the government has imposed grain production quotas and provided price protection, and thus reduced farmers' incentives to move out of the grain sector. These policies have been gradually relaxed in recent years.

Consumer gains

Consumer gains from increased imports can be approximately calculated from the difference in grain prices before and after the increase in imports times the market consumption of grain after the shock, as follows

1,815 x 6 per cent x 215 x (1 + 9.6 per cent – 6.8 per cent) = 24.1 (billion yuan)

where 9.6 per cent is the ratio of imported grain to domestic grain before the import shock, and 6.8 per cent is the ratio of output reduction to the market grain before the shock. Again, the farmers' consumer gain from the price changes is excluded for the reason mentioned earlier.

The 24.1 billion yuan of consumer surplus is equal to 0.27 per cent of GDP, 0.52 per cent of total consumers' average income, or 0.83 per cent of urban residents' income.

The impact on employment

If we assume that the imported grain will replace domestic production and crowd out farmers without causing any decline in other farmers' income, then 4.4 per cent of the grain farmers, equal to 9.2 million, will lose jobs; although the increase in imports can only marginally increase employment opportunities in the grain export countries due to their far higher labour productivity. Assuming that all these grain imports would come from the United States, for example, this would create only 40,000 new jobs in its farming sector according to its average productivity per capita.

The direct impact of grain imports

The results show that the major side effect of grain imports in the short term is not losses in value but the unbalanced distribution of losses and gains (Table 5.8). The costs directly impact low-income farmers, resulting in a relatively large percentage decline of their incomes; whereas the benefits mainly goes to urban consumers who have a much higher income, and therefore accounts for only a small proportion of their income. In addition, due to the inelastic adjustment, job losses will exert pressure

Table 5.8   The direct impact of grain imports

Costs and benefits	Producer losses	Consumer gains
Value (billion yuan)	28.0	24.1
Per cent of GDP	0.31	0.27
Per cent of farmers' incomes	4.2	
Per cent of pure grain farmers' incomes	9.1	
Per cent of all consumers' incomes		0.52
Per cent of urban consumers' incomes		0.83
Rural employment opportunities	Domestic losses	Foreign gains
'000 persons	9,240	40

Note: Foreign gains in employment opportunities are calculated using the labour productivity of the US farming sector.
Source: Author's calculations.

on the economy in the years ahead. Given that this situation cannot be avoided, the key issue is how to accelerate the structural adjustment.

Structural adjustment

Adjustment of the agricultural structure

To deal with the grain import shock, an adjustment that might be made is to change the proportions of grains according to China's comparative advantage. Rice production may be expanded to replace other grains. Some adjustment is already in process. Compared with the year 1995, the total area sown to grain had fallen by 4.5 per cent by 2000. Of this, the wheat area was down by 7.6 per cent whereas the rice area was only reduced by 2.5 per cent. Further adjustment can be expected. However, the capacity for further adjustment is limited because rice normally requires irrigation, whereas most of China's north areas, which mainly produce wheat, corn, soybean and other grains and not rice, are dry. Only the north-east provinces, that is, Liaoning, Jinlin and Heilongjiang, may have the potential to expand rice production.

A second way of adjusting is to replace grain with other agricultural products, for example, vegetables, cotton, oil-bearing crops, tea, fruits, etc.

The proportion of land area sown to grain to total sown area has fluctuated between 70 per cent and 80 per cent over the past 20 years, and fell to its lowest level of 69 per cent in 2000 due to low grain prices. In addition, the government has abolished the grain production quotas on the major grain importing provinces in the coastal areas to provide more market for those provinces that have higher comparative advantage. Quotas to the major grain producing provinces have also been dropped because there is still a surplus of grain supplies. The formal abolition of the government quota system or price protection will have a positive effect on the market-oriented structural adjustment.

Export-oriented agriculture has had limited development in China, mainly in the coastal areas such as Guangdong. In 2000, exports of food and food animals were US$12.3 billion, accounting for only 4 per cent of the total value of output of China's agricultural products. There are possibilities for further development of export-oriented crops and animal products. Future adjustment should be directed towards increased exports of labour-intensive products with low land intensity. However, a major effort is needed, especially for remote inland areas, to develop business connections with world markets, and also to develop related human resources, infrastructure and other facilities. These are long-run tasks, not only as a reaction to the import shock but also a way towards modernisation of agriculture.

Removing price protection

Past experience indicates that government protection of grain prices has had a negative impact on farmers' incomes because it has distorted market prices and led to supply adjustment in the wrong direction. Government grain prices set at levels higher than market prices encourage farmers to produce grain in excess of market demand, eventually resulting in declines in market prices and surpluses of grain products. This has resulted in large fluctuations in grain prices and production in the past (see Wang 2001), and adversely affected farmers' incomes. Although currently price protection is not in place, under the WTO commitments of the government, price support is ruled out.

To replace the price protection and assist low-income farmers, more effective measures may be to improve information, technical, and training services to farmers to help them to adjust their production and to shift to the more efficient production areas. In particular, more effort can be made to help farmers to find employment opportunities in non-grain and non-agricultural sectors.

Industrialisation and urbanisation

A major structural adjustment that can be expected is the further transfer of agricultural labour and other resources to the industrial and service sectors. The transfer of agricultural labour to the rural TVE sector and the urban sectors during the 1990s was 68 million; but this only reduced the agricultural labour force from 366 million to 344 million (see Table 5.1). In the process there were only minor increases in labour productivity and farmers' incomes, far slower than in the 1980s. Employment growth in TVEs stagnated in the late 1990s and rural-urban migration has also faced more resistance because urban unemployment has increased rapidly. According to past experience and the current situation, it is assumed that rural industrialisation and urbanisation will together absorb 5.1 million rural labourers per year on average from 2001 to 2010, and on net reduce the agricultural labour force by 4 million per year (Table 5.1).

To absorb the grain import shock fully, at least an additional 6 million farmers should be employed by the TVE and urban sectors between 2002 and 2004 (assuming that 3.2 million will move to non-grain agricultural production during this period). This is obviously impossible. However, over a longer period, urbanisation could be accelerated via policy adjustments (including removing the policy bias against medium and large sized cities) and improvements in urban infrastructure (Wang and Xia 1999). Assuming that the speed of rural-urban migration can be doubled, an additional 30 million agricultural workers would move to the urban sector between 2001 and 2010. In this case, the grain shock would be absorbed and there would be a larger improvement in agricultural productivity. Greater absorption of the surplus agricultural labour by the industrial and service sectors may lead to a net reduction of the number of farmers by another 100 million or more over the coming 20 years.

123

Notes

1   Data are from the National Bureau of Statistics, various years. *Statistical Yearbook of China*, China Statistics Press, Beijing. The same source is used below unless specified otherwise.

2   The definition of 'net income' is similar to 'disposable income'.

3   For references see for example, Sicular (1988), Lin (1992, 1996), Rozelle and Boisvert (1993), Huang, Rosegrant and Rozelle (1998), and Wang (2000).

4   This is according to official statistics. However, according to the 1996 national agricultural census data (and supported by satellite imaging data), in the past the cultivated land area was understated by 27 per cent in the official statistics (calculated from NBS 1999, 2001). The same ratio should also apply to sown areas. Adjustments are therefore made by the author in Tables A5.1 and A5.2. After this adjustment, grain yield in China is still significantly higher than in most developing countries.

5   Major sources for the estimation were a recent survey by the Department of Training and Employment of the Ministry of Labour and Social Security, and the Rural Social and Economic Survey team of the National Bureau of Statistical (MLSS & NBS 1999), and National Bureau of Statistics (1991). The samples of the former survey cover 179,450 rural labourers in all provinces except Tibet. The author assumes that all the 'floating workers' who were working outside their home county and half of those who were working outside their home town but within their home county were in urban areas and were excluded from the TVE employment statistics. The changes from 1998 to 2000 were estimated according to the average growth rate of 'floating workers' from 1990 to 1998.

6   There were serious discrepancies between the 1990 data for the population and labour force, which were from the 1990 national census, and the data before and after 1990, which were from the regular statistical reports. The National Statistical Bureau made adjustments to the population statistics before and after 1990 according to the census, but did not adjust the labour statistics. The author adjusted the labour data from 1972 to 2000 according to the information from the 1953, 1964, 1982, and 1990 national census and rural and urban birth rates, as well as the labour participation rates for each year since the 1950s. For reference see Wang (2000b).

7   Estimated by the author (see Wang 2000b). Official statistics show a lower growth rate of 0.5 per cent during the same period.

8   From a speech at the symposium 'Agriculture and Private Enterprises in China' in Guangzhou, by Du Runsheng, the former director of the State Committee of Agriculture and the former director of the Rural Development Research Center of the State Council, 18 December 2001.

9   Lin, Liu and Wu (2001) estimated the price elasticity of rural demand for wheat, corn, paddy rice and beans as -0.857, -0.044, -0.155 and -0.549, respectively. The elasticities of wheat and beans are far higher than those of corn and rice, apparently because the former can be substituted by cheaper grains. This is not the case for rice because rice

is the only major grain consumed in the whole of south China. Their weighted average is 0.34. In this study, it is adjusted to 0.37 to include urban demand.

10 The impact of price fluctuations on production and farmers' incomes would be far more serious if the government incorrectly responds to the price changes. Lessons can be drawn from past experience (see Wang 2001).

References

Center for News Gathering and Editing, 2002. 'Grain Prices in China Continue to Increase in the Third Quarter of 2001', *China Agriculture Web*, http://www.aweb.com.cn.

Department of Training and Employment of the Ministry of Labour and Social Security, PRC, and Rural Social and Economic Survey Team of the National Statistical Bureau, PRC, 1999. *The situation of rural labourers' employment and flow in China, 1997-1998*, Beijing.

Huang J., Rosegrant, M., and Rozelle, S 1998. 'Public investment, technological change and reform: a comprehensive accounting of Chinese agricultural growth', *Working Paper series*, Department of Agricultural and Resource Economics, University of California, Davis.

Information Center of the Ministry of Agriculture, China, survey data and database.

International Labour Office, 1998. *Yearbook of Labour Statistics*, International Labour Organization, Geneva.

Lin, J. Y., 1992. 'Rural reform and agricultural growth in China', *American Economic Review*, 82:34-51.

——, 1996. Dual track price and supply response: theory and empirical evidence from Chinese agriculture' unpublished manuscript.

Lin Wanlong, Liu Guicai and Wu Laping, 2001. 'A research on food elasticity of China's rural residents', research report for the research project Chinese Food Supply and Demand Analysis, Ministry of Science and Technology and Ministry of Agriculture, Beijing.

Ministry of Agriculture, China, various years. *China Agricultural Development Report*, Beijing.

National Bureau of Statistics, various years. *Statistical Yearbook of China*, China Statistics Press, Beijing.

Rozelle S. and Boisvert, R. 1993. 'Grain policy in Chinese villages: yield response to pricing, procurement, and loan policies', *American Journal of Agricultural Economies*, 75:339-49.

Sicular, T. 1988. 'Plan and market in China's agricultural commerce', *Journal of Political Economy*, 96(2):283-305.

Wang Xiaolu and Xia Xiaolin, 1999. 'Optimum city size and economic growth', *Economic Research*, No. 9, Beijing.

Wang, Xiaolu, 2000a. 'China's rural economy: sustainable development and population holding capacity', F.C. Lo (ed.), *Sustainable Development Framework for Developing Countries: the case of China*, United Nations University, Tokyo.

——, 2000b. 'Sustainability of China's economic growth and institutional changes', in Xiaolu Wang and Gang Fan (eds), *The Sustainability of China's Economic Growth*, Economic Science Press, Beijing.

——, 2001. 'Grain market fluctuations and government intervention in China', research report for the ACIAR project China's Grain Market Policy Reform, The Australian National University, Canberra.

World Bank, 1997. *World Development Indicators,* World Bank, Washington, DC.

World Trade Organization, 2001. *Accession of the People's Republic of China, decision of 10 November 2001*, World Trade Organization, Geneva.

Wu, Laping, 2001. 'Price comparison between world and domestic grain markets', prepared for the ACIAR project on China's grain market, The Australian National University, Canberra.

Appendix

Table A5.1   World cereal yields and production

	Yield 1980 kg/ha	Yield 1995 kg/ha	Cereal production 1995 million tonnes	Areas under cereal production 95 million ha.
UK	4,944	6,978	22.0	3.15
France	4,854	6,458	53.6	8.30
Germany	4,228	6,051	39.9	6.59
Japan	4,843	5,737	13.4	2.34
China (unadjusted)	2,948	4,664	416.8	89.36
US	3,771	4,647	277.0	59.61
Indonesia	2,866	3,840	58.1	15.13
Vietnam	2,016	3,523	25.2	7.15
China (adjusted)	2,153	3,406	416.8	122.36
Poland	2,337	2,940	25.1	8.54
Canada	2,141	2,705	49.7	18.37
Ukraine		2,522	32.4	12.86
Brazil	1,576	2,504	49.6	19.83
Mexico	2,189	2,463	25.3	10.29
Bangladesh	2,006	2,424	25.9	10.70
Thailand	1,911	2,386	25.4	10.63
India	1,350	2,134	214.9	100.68
Turkey	1,855	1,977	28.2	14.24
Australia	1,052	1,770	26.6	15.01
Russia		1,165	61.8	53.05
World	2,309	2,730	1,896.4	694.52

Note: The table includes all the countries in which cereal output exceeded 25 million tonnes in 1995, plus UK and Japan. Countries are ranked according to their 1995 yields. Yields are by sown area.
Source: World Bank, 1997. *World Development Indicators*, World Bank, Washington, DC. Data are from the Food and Agriculture Organisation. China Data are adjusted according to satellite imagery data for cultivated land area.

Table A5.2 Agricultural inputs in China and other major producing countries

	Per cent irrigation of arable land 1994	Fertiliser kg/ha 1994/95	Tractors /1000 ha 1994	Share of labour in agriculture per cent 1990	Labour intensity person/ha 1990-95
Bangladesh	33.9	108	0.5	64	8.00
Vietnam	26.6	175	0.5	72	7.20
China	51.5	309	7.9 (100.1)	54	4.09
India	28.3	80	12.5	64	3.37
China (adjusted)	51.5	226	5.8 (73.1)	54	2.98
Indonesia	15.2	85	3.7	57	2.73
Thailand	23.1	62	11.4	64	1.78
Japan	62.9	403	876.1	7	1.75
Turkey	15.1	54	53.7	53	1.15
Mexico	24.7	62	16.7	28	1.04
Brazil	5.9	93	37.1	23	0.72
Poland	0.7	98	153.5	27	0.46
Ukraine	7.5	35	34.0	20	0.30
Germany	4.0	413	197.3	4	0.20
UK	1.8	384	158.7	2	0.20
Russia	4.1	12	21.6	14	0.20
France	7.6	297	173.5	5	0.15
US	11.4	103	80.5	3	0.06
Canada	1.6	49	40.3	3	0.03
Australia	4.5	15	21.0	5	0.03
World	17.3	85	37.4	49	1.88

Note: Countries are ranked according to their labour intensity per hectare in 1995. Fertiliser is measured by plant nutrient.
Tractors per hectare exclude garden tractors. Data in parenthesis include garden tractors. Labour intensity is the agricultural population divided by the area of arable land. The statistical base for the calculations is arable land per capita and the share of labour in agriculture.
Source: Calculated from World Bank, 1997. *World Development Indicators*, Washington, DC; International Labour Organization, 1998. *Yearbook of Labour Statistics*, International Labour Organization, Geneva.

Table A5.3   China's grain trade, 1953-2000 (10,000 tonnes)

Year	Grain imports Total [1]	Wheat [2]	Import [3]=[2]/[1]	Total [4]	Grain exports Rice [5]	Soybean [6]	Maize [7]	Net grain Import [8]
1953	1.5	1.4	93.33	182.6	56.1	92.0	:	-181.1
1954	3.0	2.7	90.00	171.1	54.0	90.7	:	-168.1
1955	18.2	2.2	12.09	223.3	70.0	105.8	:	-205.1
1956	14.9	2.3	15.44	265.1	107.7	112.4	:	-250.2
1957	16.7	5.0	29.94	209.3	52.9	114.1	:	-192.6
1958	22.4	14.8	66.07	288.3	139.7	122.4	:	-265.9
1959	0.2	...	...	415.8	177.4	172.7	:	-415.6
1960	6.6	3.9	59.09	272.0	107.2	111.1	:	-265.4
1961	581.0	388.2	66.82	135.5	42.8	40.9	:	445.5
1962	492.3	353.6	71.83	103.9	45.8	25.9	:	388.4
1963	595.2	558.8	93.88	149.0	68.5	40.9	:	446.2
1964	657.0	536.9	81.72	182.1	76.2	59.0	:	474.9
1965	640.5	607.3	94.82	241.6	98.5	65.3	:	398.9
1966	643.8	621.4	96.52	285.5	148.7	65.1	:	358.3
1967	470.2	439.5	93.47	299.4	157.7	67.0	:	170.8
1968	459.6	445.1	96.85	260.1	129.9	68.8	:	199.5
1969	378.6	374.0	98.78	223.8	117.9	59.5	:	154.8
1970	536.0	530.2	98.92	211.9	128.0	47.0	:	324.1
1971	317.3	302.2	95.24	264.8	129.2	58.8	:	52.5
1972	457.6	433.4	94.71	292.6	142.6	41.2	:	165.0
1973	812.8	629.9	77.50	389.3	263.1	40.0	:	423.5
1974	812.1	538.3	66.28	364.4	206.1	47.1	:	447.7
1975	375.5	349.1	92.97	280.6	163.0	40.5	:	94.9

Year								
1976	236.7	202.2	85.42	176.5	87.6	20.0	..	60.2
1977	734.5	687.6	93.61	165.7	103.3	13.0	..	568.8
1978	883.3	766.7	86.80	187.7	143.5	11.3	..	695.6
1979	1,235.5	871.0	70.50	165.1	105.3	30.6	..	1,070.4
1980	1,342.9	1,097.2	81.70	161.8	111.6	11.3	..	1,181.1
1981	1,481.2	1,307.1	88.25	126.1	58.3	13.6	..	1,355.1
1982	1,611.7	1,353.4	83.97	125.1	45.7	12.7	..	1,486.6
1983	1,343.5	1,101.9	82.02	196.3	56.6	33.4	..	1,147.2
1984	1,064.5	1,000.0	93.94	344.0	118.9	83.4	91.1	720.5
1985	617.1	563.2	91.27	888.0	101.9	115.1	595.7	-270.9
1986	728.2	575.4	79.02	909.5	95.7	130.1	570.6	-181.3
1987	1,627.8	1,334.1	81.96	718.7	98.9	171.4	384.7	909.1
1988	1,478.8	1,391.0	94.06	654.2	70.5	145.9	352.2	824.6
1989	1,640.3	1,470.3	89.64	622.1	33.9	117.1	349.7	1,018.2
1990	1,356.4	1,233.5	90.94	543.4	30.3	91.0	288.7	813.0
1991	1,398.3	1,282.5	91.72	1,066.0	69.2	106.5	748.7	332.3
1992	1,156.9	1,034.0	89.38	1,445.1	120.4	84.5	1043.5	-288.2
1993	733.0	642.4	87.64	1611.9	170.9	34.5	1178.6	-878.9
1994	920.0	730.0	79.35	1,346.0	152.0	83.0	205.0	-426.0
1995	2,081.0	1,159.0	55.69	214.0	5.0	38.0	12.3	1,867.0
1996	1,223.0	825.0	67.46	198.0	27.0	19.0	16.0	1,025.0
1997	417.0	186.0	44.60	834.0	94.0	19.0	662.0	-417.0
1998	388.0	149.0	38.40	889.0	374.0	17.0	469.0	-501.0
1999	339.0	45.0	13.27	758.0	271.0	20.0	431.0	-419.0
2000	315.0	88.0	27.94	1,399.0	295.0	21.0	1047.0	-1,084.0

Source: 'Yearbook of China's foreign economy and trade', and 'China Statistical yearbook', Various issues from 1984. Cited from Wu, Laping, 2001. 'Price comparison between world and domestic grain markets', prepared for the ACIAR project on China's grain market, The Australian National University, Canberra.

Table A5.4   Comparison of the quota, above-quota, rural market, domestic wholesale, imported and exported prices of rice, 1985-2000 (yuan/tonne)

	Quota	Above-quota	Rural market	Wholesale	FOB	CIF
1985	540	557	640	..	..	..
1986	547	680	763	..	..	..
1987	587	784	896	..	..	..
1988	618	944	1,133	..	..	..
1989	741	1,340	1,582	..	..	..
1990	787	1,257	1,259	..	..	..
1991	786	1,120	1,126	..	..	..
1992	854	997	1,052	..	..	..
1993	949	1,140	1,268	..	..	..
1994	1,375	1,747	2,057	..	2,918	..
1995	1,694	2,588	2,838	..	2,712	..
1996	2,054	2,634	2,818	2,398	3,637	3,278
1997	2,277	2,267	2,244	2,088	2,552	3,631
1998	2,246	2,103	2,132	2,105	2,084	4,145
1999	2032		2,081	1,934	1,923	3,957
2000	1849		1,846	1,466	1,499	3,859

Note: Prices are in nominal terms. They are arithmetic averages from monthly data. The quota, above-quota, and rural market prices of rice are converted from paddy prices at a ratio of 0.65. The wholesale prices of rice are from Heilongjiang and Hunan markets. Source: The quota, above-quota, and rural market prices are from Ministry of Agriculture database. The wholesale, FOB, CIF prices are cited from Wu, Laping, 2001. 'Price comparison between world and domestic grain markets', prepared for the ACIAR project on China's grain market, Australian National University; and calculated from National Bureau of Statistics, various years. *Statistical Yearbook of China*, China Statistics Press, Beijing.

Table A5.5   Comparison of the quota, above-quota, rural market, domestic wholesale, imported and exported prices of wheat, 1985–2000 (yuan/tonne)

	Quota	Above-quota	Rural market	Wholesale	CIF
1985	426	428	462	..	..
1986	436	512	537	..	..
1987	442	545	620	..	..
1988	467	629	763	..	..
1989	505	890	1,066	..	..
1990	508	846	890	..	..
1991	512	772	783	..	..
1992	594	734	776	..	..
1993	659	749	809	..	..
1994	895	1,044	1,141	..	1,134
1995	1,080	1,528	1,688	..	1,460
1996	1,312	1,649	1,741	1,774	1,891
1997	1,461	1,442	1,479	1,545	1,653
1998	1,440	1,317	1,357	1,384	1,514
1999	1,270		1,286	1,343	1,465
2000	1,199		1,136	1,004	1,375

Note: Prices are in nominal terms. They are arithmetic averages from monthly data. The wholesale prices are from Zhengzhou and Hubei markets.
Source: The quota, above-quota, and rural market prices are from Ministry of Agriculture database. The wholesale, FOB, CIF prices are cited from Wu, Laping, 2001. 'Price comparison between world and domestic grain markets', prepared for the ACIAR project on China's grain market, Australian National University; and calculated from National Bureau of Statistics, various years. *Statistical Yearbook of China*, China Statistics Press, Beijing.

Table A5.6   Comparison of the quota, above-quota, rural market, domestic wholesale, imported and exported prices of corn, 1985–2000 (yuan/tonne)

	Quota	Above-quota	Rural market	Wholesale	FOB
1985	312	327	373	..	..
1986	317	402	453	..	..
1987	332	445	503	..	..
1988	343	471	571	..	..
1989	370	643	782	..	..
1990	376	626	690	..	..
1991	375	546	596	..	..
1992	416	548	628	..	..
1993	459	644	731	..	..
1994	688	904	1,009	..	..
1995	855	1,385	1,580	..	..
1996	1,058	1,389	1,487	1,344	1,364
1997	1,235	1,089	1,145	1,179	1,122
1998	1,230	1,052	1,103	1,275	937
1999	984		986	1,071	850
2000	863		828	970	822

Note: Prices are in nominal terms. They are arithmetic averages from monthly data. The wholesale prices are from Heilingjiang and Hubei markets.
Source: The quota, above-quota, and rural market prices are from Ministry of Agriculture database. The wholesale, FOB, CIF prices are cited from Wu, Laping, 2001. 'Price comparison between world and domestic grain markets', prepared for the ACIAR project on China's grain market, Australian National University; and calculated from National Bureau of Statistics, various years. *Statistical Yearbook of China*, China Statistics Press, Beijing.

Table A5.7   Comparison of the quota, above-quota, rural market, domestic wholesale, imported and exported prices of soybeans, 1985-2000 (yuan/tonne)

	Quota	Above-quota	Rural market	Wholesale	FOB	CIF
1985	668	762	877	..	..	..
1986	704	877	1,001	..	..	..
1987	738	933	1,102	..	..	..
1988	748	1,026	1,296	..	..	..
1989	785	1,396	1,785	..	..	..
1990	832	1,335	1,591	..	..	..
1991	883	1,256	1,493	..	..	..
1992	909	1,476	1,806	..	..	..
1993	1,044	1,842	2,206	..	..	..
1994	1,539	2,125	2,451	..	2,310	2,286
1995	1,805	2,422	2,665	..	2,190	2,144
1996	1,954	2,920	3,212	3,018	3,009	2,380
1997	2,293	3,103	3,437	3,134	3,461	2,559
1998	..	..	2,439	2,546	3,373	2,232
1999	..	..	1,741	2,202	2,749	1,729
2000	2,366	..	1,811	2,217	2,817	1,785

Note: Prices are in nominal terms. They are arithmetic averages from monthly data The wholesale prices in this table are from Heilongjiang and Fujian markets.
Source: The quota, above-quota, and rural market prices are from Ministry of Agriculture database. The wholesale, FOB, CIF prices are cited from Wu, Laping, 2001. 'Price comparison between world and domestic grain markets', prepared for the ACIAR project on China's grain market, The Australian National University, Canberra; and calculated from National Bureau of Statistics, various years. *Statistical Yearbook of China*, China Statistics Press, Beijing.

06 The impact of China's WTO accession on regional economies

Tingsong Jiang

China was admitted to the WTO in November 2001, after making commitments far beyond those most member economies agreed to when they joined (Lardy 2002). The accession ended a 15-year long and difficult negotiation process. However, the discussion of the impact of China's accession on the domestic economy and on the world economy has only just begun.

Many studies discuss the impact of China's accession using general equilibrium models, because these models enable panoramic analysis of economy-wide effects.[1] These studies share the view that, overall, China will achieve gains in economic efficiency but that agriculture, the auto industry and the banking system are vulnerable sectors.

Few studies have tried to investigate the impact of WTO accession on regional development. Yang and Huang (1997) and Jiang (2002a) use different types of representative households to approximate the regional impacts of trade liberalisation and WTO accession. Diao, Fan and Zhang (2002) and Diao et al (2002) present a general equilibrium model with partial disaggregation, that is, distinguishing nine different regions but only for the agricultural sectors. Fan and Zheng (2000, 2001) discuss the regional impact of trade liberalisation in their PRCGEM model following the top-down approach. However, their analysis is incomplete because it attributes the regional impact only to the difference in sectoral composition. One reason for this limitation is that constructing a multi-regional CGE model of the Chinese economy requires detailed regional input-output, income, consumption and trade data, all of which are often difficult to obtain.

Although difficulties exist, the regional impacts of WTO accession deserve the same attention as the sectoral impacts. China is a big country in many senses, with huge regional differences in geographic and economic terms. Regional income disparity worsened along with economic reform and growth during the 1990s. Income disparity has become so large that the central government announced its 'West Development Strategy' in 2000. Analysis of sectoral effects is a part of the investigation into the regional effects of WTO accession; therefore, a regional analysis could provide a more comprehensive picture of the impacts.

Impact of WTO accession on China's regional economies

China has made a WTO-plus commitment. It promised not only to reduce significantly tariff and non tariff barriers but also to open up sectors such as telecommunications, banking, insurance, asset management and distribution of foreign investment. It also agreed to abide by all WTO rules. Moreover, China has been forced to accept discriminatory treatment in two important rule-based areas: safeguards and antidumping (Lardy 2002).

It is difficult to accommodate all of China's commitments in one simulation. Rather, this study investigates the impact of the most obvious and simplest commitment: the required tariff cut. Even this is not as easy as it may seem. China's import tariffs are often subject to exemption and reduction under special arrangements that make the effective tariff rates significantly different from the statutory rates. For example, the average statutory tariff rate was 16.4 per cent in 2000, while tariff revenue accounted for only 4.03 per cent of the value of imports (National Bureau of Statistics of China 2001).

The database of the model represents the Chinese economy in 2000, with the average tariff rate being 16.2 per cent, which is close to the statutory tariff rate.[2] The tariff rates used in this study (Table A6.2) are mainly drawn from the GTAP Database 5 with some revisions based on other studies (for example Wang 2000, Ianchovichina and Martin 2001 and Anderson, Huang and Ianchovichina 2002). The WTO tariff rates are only approximately consistent with the actual commitments (Annex 8: Schedule CLII of Protocol on the Accession of the People's Republic of China, 2001).

Closures of the model for the simulations

Endowments of primary factors are treated as exogenous. There are slack labour variables to allow unemployment. But in the simulations described below these slack variables are set to be zero, leaving wages to adjust for full employment.[3]

All tax rates, including tariff rates, and technological shifters are set exogenously. Exogenous tax rates imply that government revenues will change along with changes in production, income and trade after shocks. This form of closure differs from a closure where tax rates are adjusted so as to raise a fixed amount of revenues. This particular form of closure is chosen because we want to identify the impact of WTO accession from other policy changes. The shares of transfer payments from the central government to regional governments do not change, that is, payments to each region change at the same rate.

The propensities to save are fixed, although they vary across households and regions. The difference between national savings and aggregate investment is the net capital inflow, which is equal to the trade deficit. There are two closures in the following simulation.

- No control on the trade balance. In this closure, the nominal exchange rate is fixed, the trade balance is endogenous, and foreign capital flows automatically match the balance.

- No change in the trade balance. In this closure, a floating exchange rate regime is assumed so that the change in the trade balance can be exogenously fixed at zero.[4]

One might think that the first closure is the more natural one, involving only tariff cuts in the simulation. However, it may still require some other policy changes to validate the closure. For example, it requires capital inflows to match the trade deficit at whatever level the model generates, implying that there is no control on foreign investment. This is clearly not the case. But as it is expected that foreign investment will increase after WTO accession (Chen 2002), and zero change seems too extreme, one would expect that the real situation lies somewhere between the two closures, although perhaps closer to the former.

Simulation results of the tariff cut can be found in Tables 6.1 and 6.2 and Tables A6.3 to A6.5. Table 6.1 reports the macroeconomic effects of tariff cuts under different closures, and Table 6.2 reports the impact of tariff cuts on regional output, imports and exports of aggregated commodities or sectors. The disaggregated sectoral results are reported in Tables A6.3 to A6.5.

Table 6.1   Macroeconomic effects of tariff cuts

Indications	No control on trade balance				No change in trade balance			
	Eastern	Central	Western	National	Eastern	Central	Western	National
Real GDP(%)	0.87	-0.06	0.33	0.56	0.85	-0.07	0.33	0.54
GDP deflator(%)	-1.90	-2.03	-2.79	-2.07	3.35	3.04	2.28	3.11
CPI (%)								
Rural households	-2.66	-2.50	-3.11	-2.70	2.42	2.51	1.90	2.35
Urban households	-2.56	-2.27	-2.77	-2.53	2.55	2.82	2.33	2.57
Government	-0.67	-1.30	-1.33	-0.95	4.72	3.98	3.97	4.40
Regional average	-2.61	-2.40	-2.95	-2.62	2.48	2.64	2.10	2.45
Total utility (%)								
Rural households	0.67	0.22	0.17	0.46	0.35	0.01	0.02	0.20
Urban households	1.85	1.16	1.36	1.62	1.30	0.69	1.05	1.13
Government	0.49	-1.94	-2.47	-0.37	1.41	-2.01	-2.54	0.23
Regional average	1.19	0.30	0.48	0.86	0.98	0.02	0.25	0.63
Equivalent variation (billion yuan)								
Rural households	11.86	2.32	0.83	15.01	6.23	0.13	0.1	6.46
Urban households	38.68	7.89	9.79	56.35	27.08	4.66	7.53	39.26
Government	5.57	-6.84	-6.40	-7.67	15.97	-7.07	-6.58	2.32
Regional sum	56.11	3.36	4.22	63.69	49.28	-2.28	1.04	48.04
Savings (nominal, per cent)								
Rural households	-1.47	-2.03	-2.64	-1.76	3.58	2.97	2.28	3.26
Urban households	-0.14	-1.07	-1.04	-0.48	5.29	4.25	4.29	4.91
Government	-3.45	-3.26	-3.78	-3.44	1.63	1.85	1.32	1.64
Regional average	-1.02	-1.66	-1.41	-1.22	4.24	3.47	3.84	4.01
Nominal exchange rate			-					5.45
Change in trade balance (billion yuan)			-42.55					-
Terms of trade (%)			-0.46					-0.47

Source: Author's CERD simulation.

Simulating with no control on the trade balance

For this simulation it can be seen from Table 6.1 that China has a net gain from WTO accession. By cutting the tariff rates as listed in Table A6.2, China's real GDP would increase by 0.56 per cent, utility by 0.86 per cent, and the equivalent variation, a welfare indicator, would reach 63.69 billion yuan. Higher welfare comes from higher real incomes, thus higher real consumption and saving. But the tariff cut has an adverse impact on the trade balance and the terms of trade. China's trade balance declines by 42.55 billion yuan because imports increase more than exports and the terms of trade decreases by 0.46 per cent. This is because the tariff cut does not affect the border price (CIF). However, Chinese exports become cheaper because tariff cuts help to lower production costs.

It can be seen from Tables 6.2 and A6.3 that the impact will not be evenly distributed. The motor vehicle and other transport equipment sector is the biggest loser, with output declining by 16.37 per cent. It is followed by the food and tobacco processing, machinery and chemicals sectors, where output would decline by more than 3 per cent. The impact on the agricultural sectors would not be as severe as some studies have suggested. Crop output declines by 1.2 per cent and total agricultural output falls by less than 0.1 per cent. These results are in line with the results in Ianchovichina and Martin (2001). The smaller decline in agricultural output may be partly due to the smaller extent of the tariff cuts. It may be also partly attributed to the aggregation of agricultural sectors. If the crop sector could be disaggregated to individual crops, it may be that the output of some crops would fall significantly.

Table A6.3 shows that the sector with the highest growth would be apparels, with a more than 14 per cent increase in output; followed by electronics (9.93 per cent), textiles (6.34 per cent) and instruments and cultural and office machinery (3.95 per cent).

The changes in imports and exports are consistent with the changes in output. Imports of other agricultural products, food and tobacco processing, motor vehicles and other transportation equipment are more than doubled after WTO accession and crops imports increase by 88 per cent.[5] Increases in exports of apparel, textiles and electronics are of a smaller magnitude. In general, these changes reflect the comparative advantage and disadvantage of Chinese industries.

Table 6.2   Impact of tariff cuts on regional output, exports and imports

Sector[a]	No control on trade balance				No change in trade balance			
	Eastern	Central	Western	National	Eastern	Central	Western	National
Output								
agri	-	-0.13	-0.08	-0.06	-0.06	-0.17	-0.12	-0.10
mine	-1.73	1.24	2.60	-0.03	-1.88	1.40	2.76	-0.02
fprc	-4.20	-1.71	-5.35	-3.59	-4.33	-1.80	-5.53	-3.71
lind	7.68	0.04	2.79	5.86	7.77	0.20	3.37	6.00
chem	-2.02	-0.96	-0.75	-1.63	-2.04	-0.90	-0.69	-1.63
motr	-17.88	-14.87	-11.54	-16.37	-18.05	-14.58	-11.10	-16.36
mche	-3.86	-0.95	0.29	-2.76	-3.92	-0.75	0.49	-2.73
elen	5.72	-0.85	3.10	4.98	5.75	-0.76	3.38	5.03
cnst	0.46	1.09	1.18	0.72	0.43	1.11	1.20	0.72
svce	0.82	-0.13	0.15	0.52	0.87	-0.22	0.07	0.50
Exports								
agri	17.19	15.07	17.41	16.56	20.21	17.83	19.90	19.44
mine	1.73	8.15	9.24	4.74	1.83	8.78	9.76	5.05
fprc	9.58	9.67	6.92	9.19	10.95	11.09	7.85	10.50
lind	21.70	8.87	19.21	20.03	22.37	9.93	21.78	20.82
chem	4.89	5.97	6.65	5.15	5.16	6.49	7.20	5.48
motr	-10.31	-5.47	0.80	-7.86	-10.37	-4.81	1.59	-7.70
mche	1.20	5.82	7.47	2.60	1.32	6.45	8.01	2.84
elen	14.94	5.68	12.77	14.39	15.23	6.16	13.50	14.72
cnst	2.53	5.26	5.37	3.04	2.58	5.54	5.59	3.11
svce	2.83	4.06	4.90	3.32	2.86	4.29	5.11	3.41
Imports								
agri	47.47	39.20	63.54	47.97	44.21	36.35	60.84	44.82
mine	-4.45	-4.95	-4.38	-4.53	-4.72	-5.14	-4.53	-4.77
fprc	155.48	173.44	158.17	158.30	151.10	169.02	154.93	154.15
lind	22.54	11.02	5.03	19.02	21.96	9.85	4.02	18.31
chem	22.45	19.27	20.82	21.75	22.17	18.85	20.41	21.44
motr	110.89	107.67	87.98	106.08	110.06	106.43	87.27	105.17
mche	12.55	13.41	12.59	12.69	12.30	13.05	12.33	12.42
elen	6.63	3.88	4.39	5.93	6.31	3.40	4.03	5.58
cnst	-2.19	-4.06	-4.03	-3.17	-2.31	-4.34	-4.25	-3.36
Svce	-2.18	-4.93	-5.14	-3.44	-2.31	-5.37	-5.52	-3.70

Note: [a] sector code: agri: agriculture; mine: mining; fprc: food processing; lind: light industry; chem: chemicals; motr: motor vehicle and other transportation equipment; mche: machinery and equipment; elen: electronics and electric equipment; cnst: construction; svce: services.
Source: Author's CERD simulation.

The simulation results also show that the eastern coastal region gains more from the WTO accession than the inland regions. The order of increase in total utility and equivalent variation is: eastern, western and central. This result suggests that regional income disparity would worsen after the accession, although all regions may gain. The eastern region gains most of the benefit from the expanding sectors. For example, apparel output in the eastern region increases by over 17 per cent, while this sector's outputs in the central and western regions increase by only 0.3 per cent and 1.5 per cent, respectively (Table A6.3). Similarly, textile output in the eastern region increases the most, while it falls in the central region.

The results show that the output level in the central region may decline after WTO accession, as indicated by the negative change in real GDP. This result seems surprising because it is generally perceived that the western region is the least developed region and thus should be the most adversely affected. However, the result may be justified in the following way. First, the western region has the cheapest labour, which helps in the development of labour-intensive sectors. Second, the western region has relatively abundant natural resource endowments, which leads to its comparative advantage in resource-intensive products. Finally, the industrial base in the western region may not be as poor as people think. The Chinese government has made huge investments in the so-called 'third line' program which brought about development in some sectors.

The results of this simulation show that WTO accession would worsen rural-urban income inequality in all regions. That rural households experience a smaller increase in utility and welfare is understandable because rural households receive part of their income from agricultural sectors, which decline in all regions following WTO accession.

Simulating with no change in the trade balance

This closure would have a similar impact on regional economies, despite some differences in the macroeconomic dimensions. The fixed exchange rate in the previous closure brings about domestic deflation, while the fixed trade balance with a floating exchange rate leads to depreciation of the RMB by 5.5 per cent—which in turn causes domestic inflation of 2.45 per cent. In both cases, the real exchange rate increases; but in the present form of closure the increase is slightly greater (3.0 per cent versus 2.6 per cent) (Table 6.1).

This form of closure gives a smaller welfare gain than the previous one because it restrains the gains from trade by fixing the trade balance. Moreover, it causes an even wider welfare gap. In the previous closure, the eastern region's share in the total equivalent variation is 88 per cent. In the current closure, all the gains are absorbed by the eastern region while the inland regions have a net loss. The gap between rural and urban households is also likely to widen. Urban households' share in the total household welfare gain increases from 79 per cent to 86 per cent (Table 6.1).

Why does WTO accession worsen regional disparity?

To some, it may be surprising, even unacceptable, that trade liberalisation should lead to worsening regional disparity in China. However, further analysis may reveal that it is a necessary result.

First, it is a natural extension of the historical trend. As shown in Figure 6.1, the income gap between China's coastal and inland regions has been widening since the economic reforms in 1978. Of the three regions defined here, the richest eastern coastal region has experienced the fastest economic growth over the past two decades. The average per capita GDP in the eastern region has increased sevenfold, while per capita GDP in the poorest western region has increased by less than five times. Consequently, the gap in per capita GDP between the eastern and western regions increased from 380.6 yuan in 1978 to 3354.9 yuan in 2000 in real terms. This phenomenon has been well documented by many authors (for example Wu 1999, Sun 2000, Sun and Parikh 2001 and Jiang 2002).

Second, the regional economies in China are segregated, and this segregation is embodied in both factor and commodity markets. The labour factor market segregation takes legal form in the household registration system. The general equilibrium CERD model captures this feature. Because labour is not freely mobile, wage differentials between regions increase after the WTO accession, which leads to worsening regional income disparity. The model assumes perfect mobility of capital, which may not be the case in

142

Figure 6.1   Per capita GDP by region, 1978–2000

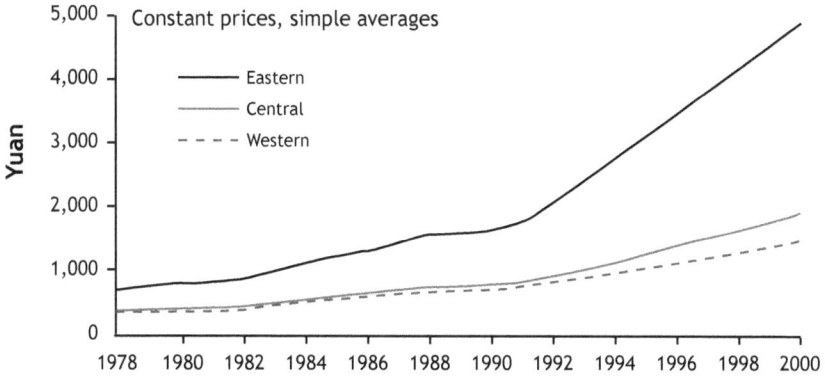

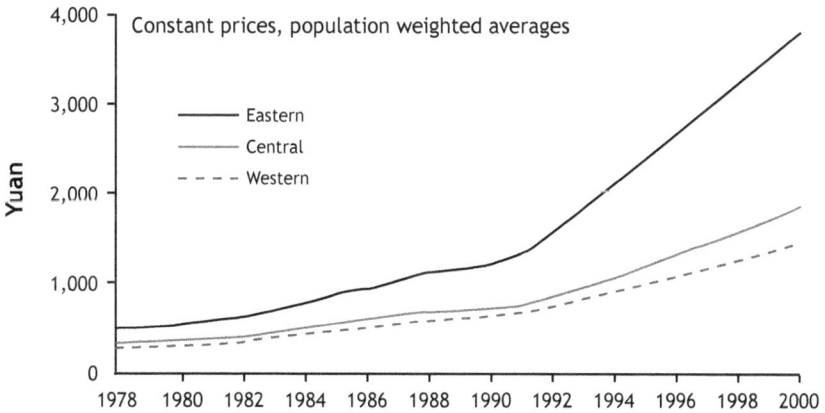

Source: Author's construction based on data from National Bureau of Statistics of China, 2001. *Statistical Yearbook of China*, China Statistics Press, Beijing.

reality. However, it captures some effects on regional disparity. Although the rate of return to capital is the same across regions, the coastal region still gains more than the inland regions as it owns a higher proportion of the capital stock.

It is commonly recognised that regional protectionism prevails in China, which prevents the integration of commodity markets. Because commodities are not to be freely traded between regions within China even after the WTO accession, factor price equalisation cannot be achieved. The CERD model captures this effect through distinguishing regional commodity markets and assuming imperfect substitution between the same commodity sourced from different regions.

Third, the technological levels are different in the three regions, which violates another underlying assumption of the factor price equalisation hypothesis. The differences in technology are embodied in the database of the CERD model. The coastal region has more advanced technology than the inland regions, and, thus, accrues higher welfare gains after the WTO accession.

It is argued that simulating only tariff cuts tends to underestimate the impact of trade liberalisation (Kehoe 2002). Considering the high probability of higher technological progress following the WTO accession, the regional income disparity may be even worse than presented above. This is because the coastal region could well experience faster technological progress than the inland regions. This is evident as the foreign direct investment has been and will be concentrated in the coastal area.

Fourth, because the economy is regionally segregated, the difference in regional economic structure also results in worsening income disparity. As shown above, the coastal region has a higher proportion of most of the expanding sectors following the WTO accession, while the inland regions have a higher proportion of contracting sectors.

Policy simulations

From the above discussion, some policy recommendations may be raised. This section presents the simulation results of three important policies: additional protection to the agricultural sector; domestic market liberalisation; and government transfer payments to inland regions.

Additional protection to agricultural sectors

Many studies of China's WTO accession have projected that agriculture will be one of the hardest hit sectors. The simulations reported above confirm that agriculture will be adversely affected by the WTO accession, leading to a worsening of regional income disparities as the inland regions have a higher proportion of agricultural activities in their economies. Therefore, the Chinese government (and academic circles) have been worrying about this issue and adopted some measures to anticipate the detrimental outcomes.[6,7] These measures provide additional protection to agriculture.

To capture this additional protection to agriculture, the simulations presented in the previous section are revised with the tariff cuts in the agricultural sectors being only half of the level in the previous section, *ceteris paribus*. The simulation results are reported in Tables 6.3 and 6.4. Several points are evident from these results.

First, such policy actions do provide some cushioning effects to agricultural sectors. The increase in agricultural imports is smaller, falling from 47.3 per cent to 15.2 per cent with the closure of no control on the trade balance, and from 44.2 per cent to 13.1 per cent with the closure of no change in the trade balance. Consequently, agricultural output declines by only 0.03 and 0.07 per cent, respectively. Second, because the central region is the major agricultural production area, this protection helps the central region achieve higher welfare (up from 3.36 billion yuan to 4.12 billion yuan) with the closure of no control on the trade balance, or lower welfare loss (from -2.28 billion yuan to -0.50 billion yuan). Third, it also helps to ease the worsening rural-urban inequality. Rural households have a higher growth in utility than in the previous scenario while urban households have a lower growth. Fourth, the negative impact on the trade balance and the terms of trade is now smaller. Finally, however, the total welfare gain is smaller than for the full tariff cuts set by the WTO agreement.

Domestic market reform

It is often argued that China's domestic economy is not well integrated and heavy regional protection exists. One prominent example is the restriction on the movement of people. Also, in many regions the use of

Table 6.3 Macroeconomic effects of WTO tariff cuts with agricultural protection

Indicators	No control on trade balance				No change in trade balance			
	Eastern	Central	Western	National	Eastern	Central	Western	National
Real GDP (%)	0.77	-0.10	0.28	0.48	0.75	-0.10	0.27	0.46
GDP deflator (%)	-1.74	-1.82	-2.53	-1.89	2.50	2.27	1.56	2.29
CPI (%)								
Rural households	-2.24	-2.12	-2.62	-2.28	1.86	1.92	1.42	1.79
Urban households	-2.23	-2.05	-2.51	-2.24	1.89	2.05	1.60	1.88
Government	-0.77	-1.39	-1.44	-1.04	3.57	2.86	2.83	3.27
Regional average	-2.24	-2.09	-2.57	-2.26	1.87	1.98	1.50	1.83
Total utility (%)								
Rural households	0.91	0.44	0.49	0.70	0.64	0.27	0.36	0.48
Urban households	1.58	0.91	1.09	1.35	1.14	0.53	0.84	0.96
Government	0.24	-1.92	-2.42	-0.53	0.99	-1.97	-2.48	-0.04
Regional average	1.10	0.33	0.44	0.81	0.93	0.10	0.25	0.62
Equivalent variation (billion yuan)								
Rural households	16.02	4.68	2.34	23.04	11.34	2.85	1.71	15.89
Urban households	33.05	6.20	7.84	47.08	23.72	3.60	6.03	33.36
Government	2.67	-6.76	-6.28	-10.36	11.19	-6.95	-6.44	-2.19
Regional sum	51.75	4.12	3.89	59.76	46.25	-0.50	1.31	47.06
Savings (nominal, per cent)								
Rural households	-1.03	-1.59	-1.98	-1.30	3.05	2.44	2.00	2.75
Urban households	-0.30	-1.19	-1.19	-0.64	4.07	3.08	3.10	3.70
Government	-3.59	-3.32	-3.84	-3.58	0.50	0.80	0.27	0.51
Regional average	-0.97	-1.46	-1.38	-1.14	3.26	2.67	2.84	3.07
Nominal exchange rate			-					4.39
Change in trade balance (billion yuan)			-34.42					-
Terms of trade (%)			-0.42					-0.43

Source: Author's CERD simulation.

Table 6.4 Impact on output, exports and imports of WTO tariff cuts with agricultural protection

Sector[a]	No control on trade balance				No change in trade balance			
	Eastern	Central	Western	National	Eastern	Central	Western	National
Output								
Agri	-	-0.06	-0.05	-0.03	-0.05	-0.09	-0.07	-0.07
Mine	-1.49	1.24	2.78	0.12	-1.62	1.37	2.91	0.12
Fprc	-5.33	-1.80	-5.41	-4.24	-5.42	-1.87	-5.55	-4.33
Lind	7.35	-0.22	1.34	5.47	7.43	-0.08	1.81	5.59
Chem.	-1.92	-0.95	-0.75	-1.57	-1.94	-0.91	-0.69	-1.56
Motr	-17.62	-14.71	-11.12	-16.10	-17.76	-14.48	-10.76	-16.09
Mche	-3.61	-0.90	0.49	-2.56	-3.66	-0.73	0.65	-2.54
Elen	6.43	-0.79	3.49	5.61	6.44	-0.71	3.73	5.65
Cnst	0.44	1.04	1.16	0.69	0.42	1.06	1.17	0.69
Svce	0.75	-0.13	0.14	0.48	0.80	-0.21	0.08	0.48
Exports								
Agri	11.45	10.88	12.57	11.39	13.88	13.11	14.6	13.72
Mine	2.63	8.55	10.06	5.48	2.69	9.05	10.48	5.73
Fprc	5.25	7.75	5.08	5.86	6.38	8.92	5.86	6.95
Lind	20.99	8.12	14.57	19.14	21.53	8.98	16.6	19.79
Chem.	5.25	6.29	6.89	5.50	5.47	6.71	7.33	5.76
Motr	-9.64	-4.95	1.71	-7.17	-9.70	-4.42	2.36	-7.04
Mche	1.92	6.32	8.21	3.29	2.01	6.83	8.65	3.48
Elen	16.33	6.14	13.8	15.72	16.55	6.53	14.39	15.97
Cnst	2.85	5.50	5.67	3.35	2.88	5.73	5.85	3.41
Svce	3.05	4.29	5.17	3.54	3.08	4.47	5.33	3.62
Imports								
Agri	15.25	11.36	20.17	15.21	13.08	9.42	18.43	13.11
Mine	-4.86	-5.37	-4.76	-4.93	-5.07	-5.52	-4.88	-5.12
Fprc	161.79	177.93	162.41	163.94	158.03	174.18	159.65	160.40
Lind	22.54	11.45	5.62	19.15	22.06	10.49	4.78	18.56
Chem.	22.27	19.05	20.63	21.57	22.05	18.71	20.30	21.32
Motr	109.31	106.26	86.76	104.61	108.68	105.28	86.20	103.90
Mche	12.11	12.86	12.06	12.22	11.92	12.58	11.86	12.01
Elen	6.33	3.31	3.88	5.57	6.07	2.93	3.60	5.28
Cnst	-2.64	-4.44	-4.45	-3.60	-2.72	-4.66	-4.62	-3.73
Svce	-2.54	-5.21	-5.45	-3.77	-2.64	-5.56	-5.76	-3.98

Note: [a] sector code: agri: agriculture; mine: mining; fprc: food processing; lind: light industry; chem: chemicals; motr: motor vehicle and other transportation equipment; mche: machinery and equipment; elen: electronics and electric equipment; cnst: construction; svce: services.
Source: Author's CERD simulation.

land is controlled by governments. It is also frequently reported that local governments erect barriers to bar products from other regions. However, there are no complete and accurate data to enable the calculation of tariff-like protection in Chinese regions. For this reason, a hypothetical scenario is constructed to investigate this issue. Specifically, a set of elasticities are chosen to imitate domestic market reforms.

Elasticities of transformation or substitution usually describe the difference between two goods or factors. However, they may also reflect how easily one good or factor can be transformed or substituted into another. For example, a higher elasticity of transformation between rural and urban labour also describes the higher mobility of labour from rural to urban areas. In this sense, the domestic market reform can be represented by increases in selected elasticities. These elasticities include the elasticity of transformation between agricultural, non agricultural and migrant labour (from 0.1 to 1); the elasticity of transformation between different land uses (from 0.05 to 0.5); the elasticity of substitution between commodities produced locally and in other regions (Table 6.5); and the elasticity of substitution between different types of labour (from 4 to 14).

Table 6.5   Elasticities of substitution between local and other products

Comm	Original	New	Comm	Original	New	Comm	Original	New	Comm	Original	New
crop	4.4	10.4	aprl	8.8	10.4	eltn	5.6	10.4	trad	3.8	10.4
Frst	5.6	10.4	furn	5.6	10.4	inst	5.6	10.4	cate	3.8	10.4
Live	4.6	10.4	papr	3.6	10.4	main	5.6	10.4	past	3.8	10.4
Fish	5.6	10.4	ptpc	3.8	10.4	omnp	5.6	10.4	fina	3.8	10.4
Otha	5.6	10.4	chem	3.8	10.4	scrp	5.6	10.4	rest	3.8	10.4
coal	5.6	10.4	nmmp	5.6	10.4	powr	5.6	10.4	sser	3.8	10.4
Petr	5.6	10.4	mtsp	5.6	10.4	gasp	5.6	10.4	heth	3.8	10.4
mtom	5.6	10.4	mtlp	5.6	10.4	watr	5.6	10.4	educ	3.8	10.4
nmtm	5.6	10.4	mach	5.6	10.4	cons	3.8	10.4	scir	3.8	10.4
Fdtp	4.9	10.4	treq	10.4	10.4	tran	3.8	10.4	teks	3.8	10.4
Txtl	4.4	10.4	eltc	5.6	10.4	ptlc	3.8	10.4	padm	3.8	10.4

Note: See sector classification abbreviations at Table A6.1.
Source: Author's construction.

With these new values, the inter-regional factor and commodity flows become more intensive. For example, in the closure of no control on the trade balance, the price of migrant labour declines by 0.18 per cent when the elasticities are higher while it increases by 0.18 per cent with the former elasticities. Consequently, migrant labour supply increases by 1.47 per cent (previously the increase was only 0.25 per cent). The migrant labour supply increases in all regions, with the western region increasing the most (1.98 per cent versus 0.31 per cent), followed by the central region (1.68 per cent versus 0.26 per cent) and the eastern region (0.83 per cent versus 0.20 per cent). Similar changes can be observed in respect of inter-regional trade of commodities. These results are understandable and do not need more explanation. It is more important to look at the welfare implications. The summary results are reported in Tables 6.6 and 6.7 while the detailed sectoral results are in Tables A6.9–A6.11.

It can be seen from the table that with the new values of parameters, WTO accession leads to higher welfare gains. The total welfare gain increases from 63.69 billion yuan in the closure with no control on the trade balance and 48.04 billion yuan in the closure with no change of the trade balance to 69.04 and 56.79 billion yuan, respectively. Although there is still a worsening of the regional income disparity, its magnitude is smaller when factor and commodities are allowed to move more freely. Every region has a higher welfare gain than before, and the shares of inland regions in the total gain increase from 12 per cent to 14 per cent. The increment in the rural households' utility and welfare after the WTO accession is higher with the revised elasticities than with former ones, while the increment in the urban households' utility and welfare is now smaller—implying a smaller degree of the rural-urban inequality.

Transfer payments

A counterfactual transfer payment program is also simulated using the CERD model. The program simply involves a 10 per cent increase in the central government's transfer payments to the central and western regions. The results are reported in Tables 6.8, 6.9 and A6.12–A6.14.

The simulation results do not give a clear justification for such a program. Although the regional and rural-urban inequality improves slightly with

Table 6.6   Macroeconomic effects of WTO tariff cuts: higher elasticities

Indicators	No control on trade balance				No change in trade balance			
	Eastern	Central	Western	National	Eastern	Central	Western	National
Real GDP (%)	0.93	-0.12	0.29	0.57	0.91	-0.12	0.28	0.56
GDP deflator (%)	-1.88	-1.76	-2.47	-1.94	2.52	2.53	1.82	2.41
CPI (%)								
Rural households	-2.37	-2.14	-2.70	-2.36	1.93	2.13	1.56	1.92
Urban households	-2.34	-2.05	-2.53	-2.31	1.97	2.25	1.77	2.00
Government	-0.84	-1.39	-1.39	-1.07	3.63	3.01	3.03	3.37
Regional average	-2.35	-2.10	-2.62	-2.34	1.95	2.18	1.66	1.96
Total utility (%)								
Rural households	1.02	0.46	0.51	0.77	0.81	0.33	0.43	1.92
Urban households	1.65	0.93	1.16	1.41	1.15	0.50	0.87	2.00
Government	0.62	-1.80	-2.33	-0.24	1.41	-1.83	-2.37	3.37
Regional average	1.24	0.36	0.50	0.91	1.07	0.13	0.31	0.72
Equivalent variation (billion yuan)								
Rural households	17.97	4.93	2.41	25.32	14.22	3.54	2.05	19.81
Urban households	34.42	6.34	8.35	49.11	23.94	3.37	6.23	33.54
Government	7.03	-6.36	-6.05	-5.39	16.02	-6.44	-6.15	3.43
Regional sum	59.42	4.92	4.71	69.04	54.18	0.47	2.13	56.79
Savings (nominal, %)								
Rural households	-1.00	-1.58	-2.02	-1.29	3.32	2.69	2.22	3.01
Urban households	-0.30	-1.17	-1.12	-0.62	4.20	3.25	3.32	3.85
Government	-3.36	-3.20	-3.70	-3.36	0.89	1.09	0.57	0.90
Regional average	-0.94	-1.45	-1.33	-1.11	3.47	2.89	3.06	3.28
Nominal exchange rate				-				4.56
Change in trade balance (billion yuan)				-36.74				-
Terms of trade (%)				-0.47				-0.48

Source: Author's CERD simulation.

Table 6.7   Impact on output, exports and imports of WTO tariff cuts: higher elasticities

Sector[a]	No control on trade balance				No change in trade balance			
	Eastern	Central	Western	National	Eastern	Central	Western	National
Output								
agri	-0.83	-0.47	-0.46	-0.64	-1.00	-0.58	-0.53	-0.77
mine	-1.81	1.69	3.15	0.18	-1.93	1.90	3.32	0.22
fprc	-4.84	-2.05	-5.53	-4.07	-5.05	-2.18	-5.70	-4.25
lind	8.18	-0.71	1.39	5.99	8.29	-0.61	1.72	6.12
chem	-1.73	-1.01	-0.79	-1.46	-1.73	-0.93	-0.72	-1.43
motr	-17.31	-14.64	-11.80	-15.98	-17.37	-14.33	-11.39	-15.91
mche	-3.60	-0.58	0.77	-2.45	-3.60	-0.33	0.99	-2.37
elen	7.06	-1.48	2.08	5.97	7.25	-1.34	2.38	6.17
cnst	0.56	1.22	1.32	0.83	0.55	1.26	1.35	0.84
svce	0.90	0.04	0.28	0.63	0.96	-0.01	0.23	0.65
Exports								
agri	10.23	10.45	12.59	10.54	11.64	11.80	13.87	11.92
mine	2.02	9.19	10.30	5.37	2.18	9.91	10.88	5.75
fprc	6.54	7.53	5.16	6.58	7.21	8.31	5.70	7.26
lind	22.39	7.29	14.79	20.22	23.00	8.05	16.38	20.89
chem	5.62	6.15	6.54	5.75	5.93	6.69	7.04	6.11
motr	-9.20	-4.91	0.67	-7.03	-9.10	-4.19	1.45	-6.74
mche	2.06	6.66	8.32	3.45	2.30	7.38	8.92	3.80
elen	17.28	4.77	11.16	16.33	17.83	5.31	11.90	16.90
cnst	3.08	5.68	5.71	3.55	3.21	6.01	5.96	3.70
svce	3.29	4.45	5.16	3.74	3.40	4.72	5.39	3.89
Imports								
agri	53.53	43.24	68.15	53.58	51.52	41.57	66.50	51.65
mine	-4.90	-4.81	-4.28	-4.80	-5.18	-4.97	-4.44	-5.05
fprc	160.52	177.43	161.94	162.94	157.50	174.43	159.66	160.08
lind	22.77	11.32	5.72	19.31	22.28	10.39	4.93	18.73
chem	22.43	18.90	20.56	21.65	22.18	18.50	20.18	21.36
motr	109.13	106.67	87.49	104.73	108.18	105.39	86.72	103.73
mche	12.21	12.97	12.26	12.33	11.93	12.59	11.97	12.04
elen	6.52	3.53	3.79	5.73	6.21	3.05	3.42	5.38
cnst	-2.64	-4.36	-4.26	-3.52	-2.81	-4.66	-4.50	-3.74
Svce	-2.48	-5.12	-5.22	-3.68	-2.64	-5.52	-5.57	-3.94

Note: [a] sector code: agri: agriculture; mine: mining; fprc: food processing; lind: light industry; chem: chemicals; motr: motor vehicle and other transportation equipment; mche: machinery and equipment; elen: electronics and electric equipment; cnst: construction; svce: services.
Source: Author's CERD simulation.

Table 6.8   Macroeconomic effects of WTO tariff cuts: with transfer
payments

Indicators	No control on trade balance				No change in trade balance			
	Eastern	Central	Western	National	Eastern	Central	Western	National
Real GDP (%)	0.88	-0.07	0.28	0.55	0.85	-0.07	0.27	0.54
GDP deflator (%)	-1.93	-1.98	-2.72	-2.07	3.22	2.99	2.26	3.01
CPI (%)								
Rural households	-2.67	-2.48	-3.10	-2.69	2.32	2.43	1.82	2.26
Urban households	-2.57	-2.25	-2.75	-2.53	2.44	2.75	2.25	2.48
Government	-0.70	-1.27	-1.28	-0.95	4.59	3.92	3.93	4.31
Regional average	-2.62	-2.39	-2.94	-2.61	2.38	2.56	2.02	2.36
Total utility (%)								
Rural households	0.69	0.22	0.17	0.47	0.38	0.02	0.02	0.21
Urban households	1.84	1.23	1.42	1.64	1.30	0.76	1.11	1.16
Government	0.04	-1.38	-1.89	-0.49	0.95	-1.45	-1.97	0.10
Regional average	1.11	0.40	0.58	0.85	0.91	0.11	0.36	0.62
Equivalent variation (billion yuan)								
Rural households	12.15	2.35	0.82	15.33	6.62	0.21	0.10	6.93
Urban households	38.50	8.33	10.24	57.07	27.11	5.16	8.01	40.28
Government	0.47	-4.86	-4.91	-9.31	10.73	-5.10	-5.10	0.52
Regional sum	51.12	5.82	6.15	63.09	44.45	0.26	3.01	47.72
Savings (nominal, %)								
Rural households	-1.47	-2.01	-2.62	-1.03	3.49	2.89	2.21	4.13
Urban households	-0.16	-0.97	-0.95	-1.60	5.17	4.24	4.29	3.43
Government	-3.43	-2.68	-3.16	-1.33	1.56	2.37	1.87	3.82
Regional average	-1.75	-0.46	-3.39	-1.20	3.18	4.83	1.60	3.93
Nominal exchange rate			-					5.35
Change in trade balance (billion yuan)			-41.79					-
Terms of trade (%)			-0.46					-0.47

Source: Author's CERD simulation.

Table 6.9   Impact on output, exports and imports of WTO tariff cuts: with transfer payment

Sector[a]	No control on trade balance				No change in trade balance			
	Eastern	Central	Western	National	Eastern	Central	Western	National
Output								
agri	-	-0.13	-0.08	-0.06	-0.05	-0.17	-0.12	-0.10
mine	-1.57	1.15	2.30	-0.03	-1.72	1.30	2.45	-0.02
fprc	-4.19	-1.72	-5.37	-3.59	-4.32	-1.80	-5.54	-3.71
lind	7.78	-0.07	2.58	5.90	7.87	0.09	3.15	6.03
chem	-1.95	-1.01	-0.90	-1.62	-1.98	-0.96	-0.84	-1.61
motr	-17.73	-14.99	-12.13	-16.38	-17.89	-14.71	-11.69	-16.37
mche	-3.75	-1.05	0.01	-2.74	-3.81	-0.85	0.20	-2.71
elen	5.84	-0.97	2.69	5.04	5.87	-0.88	2.97	5.10
cnst	0.47	1.07	1.14	0.72	0.45	1.09	1.16	0.71
svce	0.73	-0.05	0.19	0.49	0.78	-0.14	0.11	0.49
Exports								
agri	17.12	15.04	17.45	16.52	20.08	17.74	19.89	19.34
mine	1.94	7.93	8.69	4.71	2.03	8.54	9.21	5.02
fprc	9.60	9.59	6.84	9.17	10.94	10.99	7.76	10.46
lind	21.89	8.61	18.68	20.13	22.54	9.65	21.19	20.90
chem	5.01	5.81	6.31	5.20	5.27	6.32	6.85	5.52
motr	-10.08	-5.71	-0.05	-7.87	-10.15	-5.06	0.73	-7.71
mche	1.34	5.61	6.98	2.61	1.46	6.23	7.51	2.85
elen	15.15	5.44	12.09	14.52	15.44	5.91	12.81	14.84
cnst	2.59	5.16	5.21	3.06	2.64	5.44	5.43	3.13
svce	2.88	3.95	4.70	3.31	2.91	4.18	4.91	3.40
Imports								
agri	47.57	39.20	63.47	48.03	44.36	36.41	60.81	44.94
mine	-4.40	-4.96	-4.49	-4.51	-4.66	-5.15	-4.63	-4.74
fprc	155.46	173.65	158.35	158.35	151.16	169.31	155.16	154.28
lind	22.54	11.11	5.13	19.04	21.96	9.95	4.14	18.34
chem	22.45	19.33	20.84	21.77	22.19	18.91	20.45	21.46
motr	110.72	107.98	88.18	106.10	109.91	106.75	87.48	105.20
mche	12.58	13.47	12.62	12.72	12.34	13.12	12.37	12.46
elen	6.63	3.92	4.34	5.93	6.31	3.45	3.98	5.58
cnst	-2.24	-3.99	-3.94	-3.15	-2.35	-4.26	-4.15	-3.33
Svce	-2.37	-4.75	-4.97	-3.47	-2.50	-5.18	-5.35	-3.73

Note: [a] sector code: agri: agriculture; mine: mining; fprc: food processing; lind: light industry; chem: chemicals; motr: motor vehicle and other transportation equipment; mche: machinery and equipment; elen: electronics and electric equipment; cnst: construction; svce: services.
Source: Author's CERD simulation.

additional transfer payment to the inland regions, the total welfare gains decline from 63.69 and 48.04 billion yuan, respectively, to 63.09 and 47.72 billion yuan. The reduction in total welfare comes from the much sharper decline in government utility in the eastern region. It could be argued that government spending in the eastern region has higher returns, therefore a switch from eastern to inland regions causes efficiency losses. However, one cannot claim too much just based on this simulation, because the use of government spending is not specifically modelled and the transfer payment decision is not governed by an optimisation process.

Conclusions

Regional income disparity in China has been worsening since economic reforms began. Using a general equilibrium model of the Chinese economy with regional details (CERD), this chapter finds that this trend will be reinforced rather than eased by the WTO accession. The eastern coastal region will have much higher gains than the inland regions. The two inland regions will experience similar gains with the western region being only marginally better off than the central region. It is also found that the rural-urban inequality will worsen in all regions.

The simulation results are robust to whether the trade balance is fixed or not as the form of model closure. However, keeping the trade balance unchanged leads to smaller overall welfare gains and a wider regional income gap than when the trade balance is made endogenous.

Lowering the tariff cuts in agriculture reduces the total welfare gains, although it modifies the trend of worsening inequality between rural and urban households and between regions. Similarly, increasing transfer payments to the inland regions could marginally improve the regional and rural-urban inequality at the cost of a smaller overall welfare gain. However, allowing freer movement of factors and commodities across regions could improve the regional and rural-urban inequality and achieve higher total welfare gains.

Although most of the results derived by the analysis are consistent with other studies and people's perceptions, one should be cautious in accepting these results. First, this paper discusses only tariff cuts, rather than the whole framework of China's WTO commitments. The analysis could be extended to cover other issues, such as non tariff barriers, tariff rate quotas (TRQs) and domestic support. For example, the baseline tariff rates of the service sector are set at zero. Clearly, in reality this is not the case. The simulation using the closure of no control on trade balance shows that imports of crops increase by 84 to 88 per cent, which implies that the tariff rate quota for some crops may be binding. So the introduction of the TRQ in the model is necessary to obtain more realistic results.

Second, the CERD model is a national model, which suppresses international linkages and may omit some important information. For example, it predicts that China's apparel sector will increase following WTO accession. However, this result is very much dependent on whether other countries initiate the special textile safeguards. Therefore, it would be appropriate to link CERD with a global model to reflect these international relationships.

Third, the database and parameters need to be refined. For example, regional protection measures should be introduced. Also, the agricultural sectors could be disaggregated further as the present aggregation may hide significant impacts on some crops.

Notes

1 For example, see Li et al (1998), Wang (2000), Ianchovichina and Martin (2001), Lloyd and Zhang (2001), Anderson, Huang and Ianchovichina (2002), Diao, Fan and Zhang (2002), Diao et al (2002), Francois and Spinanger (2002), and Yu and Frandsen (2002).

2 Although this is higher than the effective rate, it may represent the actual protection level if non- tariff barriers are taken into consideration. Moreover, the effective rate tends to underestimate the actual protection level as it is weighted by import volumes.

3 It should be noted that the treatment here implies no change in the employment (or unemployment) level embodied in the database.

4 It could be set at any level, but zero change is an obvious target.

5 It should be noted that China's crop imports account for only 1.3 per cent of total imports in the baseline, and that even after the 88 per cent increase crop imports are still small in absolute terms.

6 The problem of so-called *san long* (agriculture, farmers and rural development) has been a popular topic.

7 For example, the newly implemented reporting requirement for genetically modified food is interpreted as an important tool to protect China's traditional soybean growing areas in the northeast region against competition from the US. However, it should be pointed out that these practices have been learnt from other countries. China often complains that its exports face even stricter technical barriers.

References

Anderson, K., Jikun Huang, and Ianchovichina, E, 2002. *'Impact of China's WTO accession on rural-urban income inequality'*, Paper presented at the Australian Agricultural and Resource Economics Society Pre-Conference Workshop on WTO: Issues for Developing Countries, Canberra, 12 February 2002.

Chen, Chunlai, 2001. 'Foreign direct investment: prospects and policies', in OECD (ed.), *China in the World Economy: domestic policy challenges*, OECD, Paris:323-57.

Diao, Xinshen, Shenggen Fan and Xiaobo Zhang, 2002. *How China's WTO accession affects rural economy in the less-developed regions: a multi-region, general equilibrium analysis*, TMD Discussion Paper 87, International Food Policy Research Institute, Washington, DC.

Diao, Xinshen, Sherman Robinson, Agapi Somwaru, and Francis Tuan, 2002. 'Regional and national perspectives of China's integration into the WTO: a computable general equilibrium inquiry', Paper presented at the 5th conference on global economic analysis, Taipei, 5-7 June.

Fan, Mingtai and Yuxin Zheng. 2000. 'The impact of China's trade liberalisation for WTO accession: a computable general equilibrium analysis', Paper presented at the Third Annual Conference on Global Economic Analysis, 27-30 June, Monash University, Melbourne. Available at http://www.monash.edu.au/policy/conf/42fan.pdf.

——, 2001. 'China's tariff reduction and WTO accession: a computable general equilibrium analysis', in P. Lloyd and Xiao-guang Zhang (eds), *Models of the Chinese Economy*, Edward Elgar, Cheltenham:211-35.

Francois, J.F., and Spinanger, D., 2002. 'Greater China's accession to the WTO: implication for international trade and for Hong Kong', Paper presented at the 5th conference on global economic analysis, Taipei, 5-7 June.

Ianchovichina, E., and Martin, W., 2001. Trade liberalisation in China's accession to WTO', World Bank, Washington, DC (unpublished).

Jiang, Tingsong. 2002. 'WTO accession and regional incomes', in Garnaut, R. and Ligang Song (eds), *China 2002: WTO Entry and World Recession*, Asia Pacific Press, Canberra:45-62.

Kehoe, T.J., 2002. 'An evaluation of the performance of applied general equilibrium models of the impact of NAFTA', Keynote speech at the fifth conference on global economic analysis, Taipei, 5-7 June.

Lardy, N.R., 2002. *Integrating China into the Global Economy*, Brookings Institution, Washington, DC.

Li, Shantong, Zhi Wang, Fan Zhai, and Lin Xu, 1998. *The global and domestic impact of China joining the World Trade Organisation*, Research Report, Washington Centre for China Studies and Development Research Center, the State Council, People's Republic of China.

Lloyd, Peter, and Xiao-guang Zhang, eds., 2001. *Models of the Chinese Economy*, Edward Elgar, Cheltenham.

National Bureau of Statistics of China, 2001. *Statistical Yearbook of China*, China Statistics Press, Beijing.

Sun, Haishun. 2000. 'Economic growth and regional disparity in China', *Regional Development Studies*, 6:43-66.

Sun, Haishun and Ashor Parikh, 2001. 'Exports, inward foreign direct investment (FDI) and regional economic growth in China', *Regional Studies*, 35(3):187-96.

Wang, Zhi, 2000. The economic impact of China's WTO accession on the world economy, Economic Research Service, US Department of Agriculture, Washington, DC (unpublished).

World Trade Organization, 2001. *China's WTO Accession Protocol*, World Trade Organization, Geneva. Available online at http://www.wto.org/english/thewto_e/acc_e/completeacc_e.htm

Wu, Yanrui, 1999. *Income disparity and convergence in China's regional economies*, Discussion Paper 99-15, Department of Economics, University of Western Australia. Available at http://www.econs.ecel.uwa.edu.au/economics/dpapers/DP1999/9.15.pdf, last accessed on 2 January 2002.

Yang, Yongzheng and Yiping Huang, 1997. *The impact of trade liberalisation on income distribution in China,* Economics Division Working Papers, China Economy 97/1, Research School of Pacific and Asian Studies, The Australian National University, Canberra.

Yu, Wusheng, and Frandsen, Søren E., 2002. 'China's WTO commitments in agriculture: does the impact depend on OECD agricultural policies', Paper presented at the 5th conference on global economic analysis, Taipei, 5-7 June.

Appendix

Table A6.1   Sector classification in the CERD model

Sectors in the CERD model	Code	Sectors in the CERD model	Code
Agriculture			
01 Crops	crop	04 Fishery	fish
02 Forestry	frst	05 Other agricultural products	otha
03 Livestock and livestock products	live		
Industry and construction			
06 Coal mining and processing	coal	20 Machinery and equipment	mach
07 Crude petroleum and natural gas products	petr	21 Transport equipment	trep
		22 Electric equipment and machinery	eltc
08 Metal ore mining	mtom	23 Electronic and telecommunication equipment	eltn
09 Non metal mineral mining	nmtm		
10 Manufacture of food products and tobacco processing	fdtp	24 Instruments, meters, cultural and office machinery	inst
11 Textile goods	txtl	25 Maintenance and repair of machine and equipment	main
12 Wearing apparel, leather, furs, down and related products	aprl		
		26 Scrap and waste	scrp
13 Sawmills and furniture	furn	27 Electricity, steam and hot water production supply	powr
14 Paper and products, printing and record medium reproduction	papr		
		28 Gas production and supply	gasp
15 Petroleum processing and coking	ptpc	29 Water production and supply	watr
16 Chemicals	chem	30 Construction	cons
17 Non metal mineral products	nmmp	31	
18 Metals smelting and pressing	mtsp		
19 Metal products	mtlp		
Services			
32 Transport and warehousing	tran	40 Health services, sports and social welfare	heth
33 Post and telecommunication	ptlc		
34 Wholesale and retail trade	trad	41 Education, culture and arts, radio, film and television	educ
35 Eating and drinking places	cate		
36 Passenger transport	past	42 Scientific research	scir
37 Finance and issurance	fina	43 General technical services	teks
38 Real estate	rest	44 Public administration and other sectors	padm
39 Social services	sser		

Table A6.2   Baseline and WTO commitment tariff rates

Sector	Baseline rate	WTO rate	Change (per cent)	Sector	Baseline rate	WTO rate	Change (per cent)
crop	27.70	16.88	-39.06	eltn	13.54	10.00	-26.15
frst	2.65	2.00	-24.41	inst	13.54	10.00	-26.15
live	17.63	15.00	-14.93	main	-	-	-
fish	16.59	15.00	-9.59	omnp	26.25	15.00	-42.86
otha	59.01	17.00	-71.19	scrp	-	-	-
coal	4.97	1.26	-74.65	powr	-0.01	-	-
petr	-	-	-46.24	gasp	-	-	-
mtom	-	-	-	watr	-	-	-
nmtm	0.44	-	-	cons	-	-	-
fdtp	52.90	20.00	-62.20	tran	-	-	-
txtl	33.49	20.00	-40.28	ptlc	-	-	-
aprl	25.55	25.00	-2.15	trad	-	-	-
furn	13.81	10.00	-27.60	cate	-	-	-
papr	13.19	10.00	-24.17	past	-	-	-
ptpc	9.09	7.00	-23.01	fina	-	-	-
chem	15.41	5.00	-67.55	rest	-	-	-
nmmp	22.12	20.00	-9.60	sser	-	-	-
mtsp	9.85	7.00	-28.91	heth	-	-	-
mtlp	15.37	7.00	-54.46	educ	-	-	-
mach	15.62	11.00	-29.57	scir	-	-	-
treq	23.26	10.00	-57.01	teks	-	-	-
eltc	13.54	10.00	-26.15	padm	-	-	-

Table A6.3   Percentage change in output — full WTO tariff cuts

Sector	No control on trade balance				No change in trade balance			
	Eastern	Central	Western	National	Eastern	Central	Western	National
crop	-1.89	-0.57	-0.81	-1.22	-1.90	-0.58	-0.78	-1.22
frst	1.84	1.34	1.52	1.60	2.20	1.54	1.66	1.86
live	2.96	0.35	1.01	1.63	2.88	0.27	0.88	1.54
fish	-0.09	0.29	0.28	-	-0.34	0.11	-0.07	-0.23
otha	1.18	-0.04	-0.09	0.61	1.07	-0.25	-0.12	0.49
coal	-1.94	0.54	0.51	-0.44	-2.21	0.61	0.53	-0.52
petr	-1.05	7.74	4.36	0.89	-1.10	8.47	4.59	0.96
mtom	-3.99	1.08	3.30	-0.92	-4.11	1.39	3.62	-0.82
nmtm	-0.75	1.46	1.52	0.39	-0.90	1.57	1.54	0.36
fdtp	-4.20	-1.71	-5.35	-3.59	-4.33	-1.80	-5.53	-3.71
txtl	8.11	-1.20	7.46	6.34	8.40	-0.66	9.23	6.79
aprl	17.14	0.27	1.49	14.42	17.29	0.18	1.32	14.53
furn	0.51	1.36	1.16	0.81	0.35	1.35	1.05	0.69
papr	1.92	-0.11	-0.80	1.32	1.96	-0.11	-0.75	1.35
ptpc	-1.30	0.66	2.34	-0.41	-1.41	0.80	2.48	-0.43
chem	-3.05	-3.89	-3.07	-3.21	-3.01	-3.85	-2.96	-3.16
nmmp	-0.16	1.40	1.27	0.52	-0.29	1.45	1.24	0.46
mtsp	-4.89	-0.51	1.29	-2.73	-5.02	-0.24	1.51	-2.71
mtlp	-2.82	-1.95	-2.63	-2.66	-2.83	-1.86	-2.57	-2.65
mach	-3.98	-1.10	-0.78	-3.15	-4.00	-0.91	-0.54	-3.10
treq	-17.88	-14.87	-11.54	-16.37	-18.05	-14.58	-11.10	-16.36
eltc	0.52	0.23	0.73	0.50	0.37	0.31	0.84	0.39
eltn	11.04	-2.99	5.62	9.93	11.22	-2.92	6.04	10.13
inst	5.24	-2.38	-0.93	3.95	5.45	-2.15	-0.63	4.17
main	-0.46	-0.13	0.86	-0.16	-0.51	-0.10	0.87	-0.18
omnp	-0.87	0.94	-0.06	-0.30	-1.11	1.08	-0.02	-0.41
scrp	-3.22	-1.08	4.86	-1.88	-3.34	-1.01	5.12	-1.91
powr	-0.45	0.10	0.40	-0.19	-0.62	0.13	0.46	-0.28
gasp	1.16	1.25	0.89	1.15	0.48	0.38	0.75	0.49
watr	0.54	0.03	0.24	0.38	0.36	-0.16	0.09	0.20
cons	0.46	1.09	1.18	0.72	0.43	1.11	1.20	0.72
tran	0.06	0.41	0.99	0.25	-0.05	0.48	1.05	0.20
ptlc	1.19	0.30	0.52	0.93	0.98	0.08	0.32	0.72
trad	0.59	0.08	1.14	0.56	0.40	0.07	1.14	0.44
cate	1.71	0.63	0.67	1.32	1.46	0.34	0.35	1.05
past	0.76	0.47	0.97	0.71	0.66	0.33	0.91	0.60
fina	0.17	0.06	0.49	0.20	-0.09	0.08	0.58	0.04
rest	1.17	0.45	0.69	0.98	0.79	-0.03	0.37	0.59
sser	1.50	0.30	0.31	1.16	1.59	0.20	0.21	1.19
heth	2.05	-0.80	-0.66	0.84	2.47	-1.02	-1.02	0.96
educ	1.49	-0.79	-0.72	0.61	1.95	-1.11	-1.06	0.75
scir	0.55	-1.81	-1.89	-0.26	1.18	-1.86	-1.92	0.14
teks	0.68	-1.44	-1.67	-0.14	1.63	-1.51	-1.73	0.44
padm	1.87	-1.97	-2.53	0.16	3.33	-2.04	-2.60	0.98

Table A6.4 Percentage change in exports–full WTO tariff cuts

Sector	No control on trade balance				No change in trade balance			
	Eastern	Central	Western	National	Eastern	Central	Western	National
crop	17.01	15.18	17.08	16.52	20.01	17.91	19.52	19.39
frst	13.40	12.93	15.23	13.39	15.89	15.46	17.73	15.89
live	20.60	15.60	18.67	18.20	23.79	18.38	21.22	21.11
fish	17.36	15.38	17.51	16.60	20.84	18.62	20.30	19.99
otha	16.50	13.18	17.15	15.71	19.40	15.40	19.79	18.39
coal	0.85	6.67	7.22	5.84	0.67	7.19	7.56	6.24
petr	1.81	13.01	9.47	4.15	1.95	14.14	10.01	4.44
mtom	-0.54	7.74	9.24	4.41	-0.54	8.47	9.86	4.79
nmtm	1.90	8.22	8.72	5.60	1.86	8.79	9.10	5.91
fdtp	9.58	9.67	6.92	9.19	10.95	11.09	7.85	10.50
txtl	20.88	9.54	23.13	19.45	21.75	11.20	26.51	20.66
aprl	30.28	9.15	13.92	28.59	31.07	9.92	14.74	29.37
furn	5.84	8.34	9.47	6.60	5.97	8.88	9.88	6.84
papr	8.40	7.05	8.29	8.26	8.72	7.62	8.93	8.62
ptpc	1.54	5.90	8.33	3.39	1.60	6.41	8.81	3.59
chem	5.39	4.10	5.84	5.31	5.71	4.64	6.45	5.66
nmmp	3.43	8.00	8.06	5.44	3.45	8.50	8.38	5.66
mtsp	-1.16	5.89	7.91	3.24	-1.16	6.56	8.45	3.57
mtlp	2.13	4.97	4.46	2.34	2.28	5.47	4.88	2.52
mach	1.12	6.15	7.30	2.31	1.25	6.74	7.88	2.53
treq	-10.31	-5.47	0.80	-7.86	-10.37	-4.81	1.59	-7.70
eltc	6.15	7.41	8.15	6.31	6.19	7.92	8.61	6.40
eltn	20.14	3.82	14.32	19.16	20.54	4.23	15.11	19.60
inst	11.68	4.44	6.28	11.18	12.07	5.05	6.88	11.58

main	-	-	-	-	-	-	-	-
omnp	5.80	8.42	8.52	6.50	5.78	9.15	9.19	6.68
scrp	-	-	-	-	-	-	-	-
powr	2.34	5.57	5.51	4.04	2.35	6.00	5.87	4.26
gasp	3.17	5.25	4.86	3.41	2.58	4.63	4.92	2.91
watr	-	-	-	-	-	-	-	-
cons	2.53	5.26	5.37	3.04	2.58	5.54	5.59	3.11
tran	1.56	4.21	4.85	2.56	1.51	4.51	5.10	2.63
ptlc	2.78	3.36	4.12	3.08	2.68	3.36	4.12	3.02
trad	2.23	3.69	5.05	3.04	2.15	3.94	5.28	3.08
cate	6.27	5.77	6.98	6.13	6.49	6.08	7.13	6.37
past	2.72	4.30	4.79	3.57	2.70	4.39	4.94	3.62
fina	1.90	3.34	3.86	1.95	1.75	3.62	4.18	1.81
rest	-	-	-	-	-	-	-	-
sser	3.65	4.15	4.62	3.71	3.84	4.33	4.75	3.90
heth	5.92	4.27	4.16	5.88	6.42	4.29	3.98	6.37
educ	2.64	2.59	3.03	2.64	3.12	2.49	2.84	3.10
scir	-	-	-	-	-	-	-	-
teks	-	-	-	-	-	-	-	-
padm	3.57	1.50	0.87	1.83	5.08	1.64	0.95	2.41

Table A6.5   Percentage change in imports — full WTO tariff cuts

Sector	No control on trade balance				No change in trade balance			
	Eastern	Central	Western	National	Eastern	Central	Western	National
crop	86.02	95.08	91.48	87.77	81.98	91.39	88.52	83.91
frst	-8.21	-8.22	-9.74	-8.27	-9.98	-10.30	-11.72	-10.12
live	-3.11	-3.11	-5.17	-3.35	-5.65	-5.31	-7.16	-5.79
fish	-10.90	-7.02	-9.95	-10.53	-14.25	-10.16	-12.83	-13.82
otha	299.33	306.77	285.04	301.77	286.96	296.73	275.51	290.42
coal	14.83	12.28	12.32	14.34	14.38	11.71	11.94	13.87
petr	-4.11	-3.58	-4.70	-4.07	-4.38	-3.70	-4.94	-4.32
mtom	-7.04	-7.01	-3.76	-6.15	-7.27	-7.21	-3.80	-6.33
nmtm	-3.71	-7.18	-6.67	-4.75	-3.99	-7.60	-7.03	-5.07
fdtp	155.48	173.44	158.17	158.30	151.10	169.02	154.93	154.15
txtl	42.07	34.88	27.90	40.36	41.50	33.63	26.62	39.68
aprl	-19.38	-15.66	-18.98	-18.51	-20.60	-17.36	-20.58	-19.90
furn	11.77	10.26	9.61	11.35	11.18	9.41	8.86	10.71
papr	5.18	3.65	0.97	4.28	4.95	3.13	0.49	3.97
ptpc	3.84	2.79	2.77	3.37	3.62	2.59	2.55	3.15
chem	25.91	27.27	26.95	26.19	25.64	26.75	26.49	25.87
nmmp	5.17	1.68	2.44	4.52	4.88	1.13	2.01	4.19
mtsp	5.34	5.04	5.46	5.31	5.05	4.72	5.21	5.02
mtlp	34.92	34.93	33.34	34.74	34.57	34.41	32.87	34.34
mach	12.25	12.64	10.70	12.10	12.04	12.29	10.48	11.87
treq	110.89	107.67	87.98	106.08	110.06	106.43	87.27	105.17
eltc	8.81	7.10	8.02	8.45	8.36	6.52	7.65	7.98
eltn	6.66	2.70	3.51	5.71	6.37	2.21	3.11	5.37
inst	2.97	4.50	3.26	3.29	2.67	4.18	3.05	3.00

main	-	-8.53	-9.61	-7.65	-3.47	-8.86	-9.90	-8.48
omnp	47.91	49.51	47.19	47.95	47.24	48.15	46.07	47.19
scrp	-3.17	-3.48	-	-2.01	-3.39	-3.66	-3.09	-3.36
powr	-3.66	-6.17	-5.35	-4.51	-4.02	-6.56	-5.60	-4.86
gasp	-2.49	-5.92	-6.41	-4.70	-3.24	-7.14	-6.78	-5.42
watr	-2.84	-6.34	-5.83	-4.50	-3.17	-6.89	-6.26	-4.92
cons	-2.19	-4.06	-4.03	-3.17	-2.31	-4.34	-4.25	-3.36
tran	-1.85	-5.08	-4.41	-3.61	-2.02	-5.33	-4.60	-3.82
ptlc	-1.13	-4.05	-4.46	-2.68	-1.48	-4.55	-4.89	-3.09
trad	-	-5.34	-5.15	-4.64	-1.92	-5.69	-5.48	-5.15
cate	-4.42	-6.38	-7.08	-4.90	-5.23	-7.40	-7.89	-5.72
past	-1.78	-5.47	-4.87	-3.08	-1.98	-5.94	-5.22	-3.36
fina	-1.71	-3.98	-3.66	-3.52	-2.09	-4.24	-3.82	-3.77
rest	-0.87	-1.99	-2.18	-1.39	-1.38	-2.67	-2.70	-1.96
sser	-2.14	-4.87	-5.34	-3.37	-2.22	-5.32	-5.71	-3.60
heth	-2.80	-7.11	-6.67	-5.25	-2.48	-7.62	-7.25	-5.46
educ	-0.05	-5.04	-5.40	-2.93	0.35	-5.64	-5.94	-3.09
scir	-2.06	-7.03	-6.64	-5.68	-1.49	-7.40	-6.90	-5.78
teks	-0.57	-6.11	-	-2.28	0.37	-6.50	-3.47	-2.83
padm	-0.25	-6.20	-6.82	-3.23	1.08	-6.58	-7.14	-2.75

Table A6.6 Percentage change in output — WTO tariff cuts with protection to agriculture

Sector	No control on trade balance				No change in trade balance			
	Eastern	Central	Western	National	Eastern	Central	Western	National
crop	-0.99	-0.33	-0.47	-0.66	-1.00	-0.34	-0.45	-0.67
frst	1.04	0.94	1.01	1.00	1.35	1.11	1.13	1.22
live	1.93	0.27	0.62	1.07	1.88	0.21	0.51	1.01
fish	-0.78	-0.14	-0.02	-0.62	-0.97	-0.28	-0.30	-0.80
otha	0.46	0.14	-0.24	0.23	0.38	-0.03	-0.26	0.14
coal	-1.85	0.52	0.52	-0.42	-2.08	0.58	0.54	-0.48
petr	-0.63	7.88	4.61	1.25	-0.69	8.47	4.80	1.30
mtom	-3.61	1.12	3.70	-0.66	-3.72	1.38	3.96	-0.57
nmtm	-0.63	1.45	1.51	0.44	-0.76	1.54	1.52	0.41
fdtp	-5.33	-1.80	-5.41	-4.24	-5.42	-1.87	-5.55	-4.33
txtl	7.46	-1.87	3.79	5.48	7.71	-1.42	5.18	5.85
aprl	16.59	0.03	0.60	13.90	16.72	-0.04	0.48	14.00
furn	0.54	1.35	1.11	0.82	0.40	1.34	1.03	0.72
papr	1.94	-0.17	-1.00	1.31	1.97	-0.17	-0.96	1.33
ptpc	-1.20	0.65	2.50	-0.32	-1.29	0.77	2.61	-0.34
chem	-2.94	-3.86	-3.12	-3.13	-2.91	-3.82	-3.03	-3.09
nmmp	-0.08	1.37	1.29	0.56	-0.19	1.42	1.27	0.51
mtsp	-4.55	-0.45	1.53	-2.48	-4.66	-0.23	1.72	-2.46
mtlp	-2.59	-1.91	-2.58	-2.47	-2.60	-1.83	-2.52	-2.47
mach	-3.77	-1.04	-0.58	-2.97	-3.79	-0.88	-0.38	-2.93
treq	-17.62	-14.71	-11.12	-16.10	-17.76	-14.48	-10.76	-16.09
eltc	0.77	0.27	0.85	0.72	0.65	0.34	0.94	0.63
eltn	12.15	-2.94	6.21	10.95	12.28	-2.88	6.56	11.10
inst	6.24	-2.17	-0.45	4.82	6.40	-1.99	-0.21	4.98

main	-0.47	-0.14	0.87	-0.17	-0.51	-0.11	0.88	-0.19
omnp	-0.74	0.86	-0.36	-0.26	-0.94	0.97	-0.31	-0.35
scrp	-2.95	-1.07	4.80	-1.71	-3.05	-1.01	5.02	-1.74
powr	-0.43	0.07	0.41	-0.18	-0.57	0.10	0.46	-0.26
gasp	1.00	1.04	0.89	0.99	0.45	0.33	0.78	0.46
watr	0.49	0.02	0.17	0.34	0.34	-0.14	0.06	0.20
cons	0.44	1.04	1.16	0.69	0.42	1.06	1.17	0.69
tran	0.07	0.40	0.96	0.25	-0.02	0.45	1.01	0.21
ptlc	1.13	0.27	0.51	0.88	0.96	0.09	0.35	0.71
trad	0.58	0.07	1.12	0.55	0.43	0.07	1.12	0.46
cate	1.50	0.54	0.59	1.15	1.30	0.30	0.33	0.93
past	0.78	0.49	0.99	0.73	0.69	0.38	0.95	0.64
fina	0.14	0.02	0.43	0.17	-0.07	0.03	0.50	0.03
rest	1.11	0.45	0.73	0.95	0.81	0.06	0.47	0.63
sser	1.48	0.31	0.30	1.14	1.55	0.23	0.23	1.17
heth	1.80	-0.82	-0.66	0.70	2.15	-1.00	-0.95	0.80
educ	1.30	-0.76	-0.70	0.50	1.67	-1.01	-0.98	0.62
scir	0.55	-1.81	-1.82	-0.25	1.06	-1.85	-1.84	0.07
teks	0.48	-1.41	-1.58	-0.25	1.25	-1.47	-1.63	0.22
padm	1.57	-1.91	-2.46	0.01	2.76	-1.98	-2.52	0.68

Table A6.7 Percentage change in export – WTO tariff cuts with protection to agriculture

Sector	No control on trade balance				No change in trade balance			
	Eastern	Central	Western	National	Eastern	Central	Western	National
crop	11.29	11.02	12.40	11.33	13.71	13.23	14.40	13.65
frst	8.96	9.07	10.83	9.13	10.99	11.13	12.87	11.17
live	13.35	11.20	13.29	12.44	15.89	13.44	15.37	14.79
fish	12.71	10.98	12.50	12.05	15.49	13.59	14.75	14.76
otha	10.62	9.99	12.59	10.73	12.95	11.81	14.74	12.90
coal	1.44	7.02	7.75	6.24	1.28	7.43	8.02	6.55
petr	2.79	13.69	10.33	5.09	2.88	14.60	10.77	5.31
mtom	0.32	8.21	10.17	5.15	0.31	8.79	10.68	5.46
nmtm	2.51	8.58	9.14	6.06	2.47	9.04	9.46	6.31
fdtp	5.25	7.75	5.08	5.86	6.38	8.92	5.86	6.95
txtl	19.61	7.85	16.87	17.68	20.33	9.21	19.52	18.67
aprl	29.35	8.52	11.87	27.67	30.00	9.15	12.57	28.32
furn	6.06	8.53	9.53	6.81	6.16	8.98	9.87	7.01
papr	8.67	7.06	7.81	8.48	8.93	7.53	8.34	8.76
ptpc	2.14	6.31	9.09	3.99	2.18	6.71	9.48	4.14
chem	5.70	4.33	5.85	5.59	5.96	4.76	6.34	5.88
nmmp	4.03	8.39	8.56	5.95	4.03	8.79	8.82	6.12
mtsp	-0.34	6.39	8.68	3.96	-0.35	6.93	9.12	4.22
mtlp	2.83	5.45	4.99	3.02	2.94	5.85	5.32	3.16
mach	1.82	6.66	8.02	2.98	1.91	7.13	8.49	3.15
treq	-9.64	-4.95	1.71	-7.17	-9.70	-4.42	2.36	-7.04
eltc	6.89	7.88	8.76	7.03	6.91	8.29	9.14	7.08
eltn	21.83	4.25	15.43	20.76	22.13	4.57	16.08	21.10
inst	13.20	5.02	7.23	12.64	13.49	5.50	7.72	12.94

main	-	-	-	-	-	-	-	-
omnp	6.29	8.57	8.06	6.88	6.26	9.16	8.62	7.01
scrp	-	-	-	-	-	-	-	-
powr	2.91	6.03	6.12	4.58	2.91	6.37	6.40	4.74
gasp	3.32	5.30	5.22	3.59	2.84	4.80	5.27	3.18
watr	-	-	-	-	-	-	-	-
cons	2.85	5.50	5.67	3.35	2.88	5.73	5.85	3.41
tran	1.90	4.50	5.18	2.89	1.85	4.74	5.38	2.94
ptlc	3.09	3.65	4.46	3.39	3.01	3.65	4.45	3.33
trad	2.48	3.96	5.28	3.29	2.41	4.16	5.47	3.32
cate	5.50	5.29	6.50	5.45	5.69	5.55	6.64	5.66
past	3.03	4.62	5.11	3.88	3.01	4.69	5.23	3.92
fina	2.22	3.59	4.12	2.27	2.09	3.81	4.37	2.15
rest	-	-	-	-	-	-	-	-
sser	3.90	4.37	4.86	3.96	4.05	4.51	4.96	4.11
heth	5.93	4.50	4.45	5.90	6.34	4.51	4.30	6.30
educ	2.79	2.93	3.42	2.80	3.17	2.84	3.26	3.16
scir	-	-	-	-	-	-	-	-
teks	-	-	-	-	-	-	-	-
padm	3.57	1.86	1.33	2.12	4.80	1.97	1.38	2.59

Table A6.8 Percentage change in import — WTO tariff cuts with protection to agriculture

Sector	No control on trade balance				No change in trade balance			
	Eastern	Central	Western	National	Eastern	Central	Western	National
crop	30.33	32.80	31.06	30.71	27.91	30.69	29.32	28.41
frst	-6.59	-6.60	-8.01	-6.64	-8.11	-8.39	-9.71	-8.23
live	-3.59	-4.80	-6.38	-4.06	-5.72	-6.61	-8.04	-6.09
fish	-11.12	-6.70	-9.07	-10.61	-13.90	-9.34	-11.51	-13.34
otha	83.18	83.46	75.49	83.03	78.42	79.67	71.81	78.68
coal	14.24	11.80	11.68	13.76	13.88	11.35	11.38	13.40
petr	-4.56	-4.02	-5.25	-4.53	-4.77	-4.11	-5.44	-4.72
mtom	-7.25	-7.41	-4.01	-6.41	-7.43	-7.56	-4.04	-6.55
nmtm	-4.20	-7.57	-7.13	-5.22	-4.41	-7.90	-7.42	-5.46
fdtp	161.79	177.93	162.41	163.94	158.03	174.18	159.65	160.40
txtl	42.22	35.78	29.19	40.67	41.76	34.74	28.11	40.10
aprl	-18.99	-15.10	-17.96	-17.97	-19.99	-16.50	-19.29	-19.12
furn	11.49	9.94	9.42	11.08	11.01	9.25	8.81	10.56
papr	4.91	3.54	1.04	4.08	4.73	3.13	0.65	3.83
ptpc	3.47	2.45	2.34	3.00	3.30	2.30	2.17	2.84
chem	25.78	27.11	26.85	26.07	25.56	26.69	26.48	25.80
nmmp	4.60	1.19	1.92	3.96	4.37	0.76	1.58	3.70
mtsp	5.03	4.50	4.88	4.95	4.81	4.26	4.69	4.73
mtlp	34.25	34.25	32.62	34.06	33.98	33.84	32.25	33.75
mach	11.78	12.11	10.22	11.62	11.61	11.83	10.04	11.44
treq	109.31	106.26	86.76	104.61	108.68	105.28	86.20	103.90
eltc	8.25	6.54	7.49	7.89	7.89	6.08	7.20	7.52
eltn	6.48	2.09	2.98	5.42	6.25	1.70	2.66	5.15
inst	2.43	4.11	2.89	2.81	2.20	3.85	2.73	2.58

main	-	-9.00	-10.10	-8.05	-3.74	-9.25	-10.31	-8.86
omnp	47.27	49.06	47.25	47.37	46.73	47.96	46.33	46.77
scrp	-3.67	-4.18	-	-2.37	-3.82	-4.30	-3.22	-3.74
powr	-4.25	-6.71	-5.97	-5.08	-4.52	-7.01	-6.16	-5.36
gasp	-3.19	-6.56	-6.97	-5.34	-3.78	-7.53	-7.27	-5.91
watr	-3.47	-6.85	-6.52	-5.10	-3.72	-7.29	-6.85	-5.43
cons	-2.64	-4.44	-4.45	-3.60	-2.72	-4.66	-4.62	-3.73
tran	-2.29	-5.47	-4.90	-4.05	-2.43	-5.67	-5.05	-4.21
ptlc	-1.70	-4.50	-4.91	-3.19	-1.97	-4.90	-5.25	-3.51
trad	-	-5.73	-5.52	-4.98	-2.22	-6.00	-5.78	-5.46
cate	-3.91	-5.98	-6.70	-4.41	-4.58	-6.83	-7.37	-5.10
past	-2.19	-5.83	-5.25	-3.48	-2.35	-6.20	-5.53	-3.70
fina	-2.18	-4.37	-4.10	-3.94	-2.48	-4.58	-4.22	-4.14
rest	-1.31	-2.46	-2.61	-1.84	-1.72	-3.00	-3.02	-2.29
sser	-2.59	-5.13	-5.64	-3.75	-2.65	-5.49	-5.94	-3.92
heth	-3.35	-7.37	-6.99	-5.65	-3.08	-7.78	-7.46	-5.81
educ	-0.67	-5.36	-5.80	-3.40	-0.33	-5.84	-6.24	-3.51
scir	-2.49	-7.27	-6.93	-5.99	-2.02	-7.56	-7.14	-6.06
teks	-0.97	-6.33	-	-2.52	-0.18	-6.64	-3.61	-3.15
padm	-0.91	-6.52	-7.20	-3.72	0.17	-6.81	-7.45	-3.32

Table A6.9   Percentage change in output — WTO tariff cuts with higher elasticities

Sector	No control on trade balance				No change in trade balance			
	Eastern	Central	Western	National	Eastern	Central	Western	National
crop	-2.87	-0.99	-1.22	-1.90	-3.01	-1.07	-1.26	-2.00
frst	0.91	0.94	1.08	0.95	1.07	1.02	1.12	1.06
live	2.20	0.07	0.69	1.14	2.02	-0.08	0.53	0.97
fish	-0.62	0.02	0.09	-0.46	-0.92	-0.21	-0.23	-0.74
otha	0.57	0.09	-0.28	0.27	0.39	-0.09	-0.32	0.12
coal	-2.18	0.84	0.70	-0.38	-2.44	0.95	0.72	-0.42
petr	-1.23	10.76	5.19	1.18	-1.27	11.77	5.46	1.29
mtom	-3.74	1.64	3.96	-0.49	-3.79	2.01	4.29	-0.33
nmtm	-0.77	1.73	1.90	0.53	-0.89	1.88	1.93	0.53
fdtp	-4.84	-2.05	-5.53	-4.07	-5.05	-2.18	-5.70	-4.25
txtl	8.68	-2.78	4.16	6.25	8.94	-2.45	5.24	6.58
aprl	17.72	-0.96	-0.66	14.67	17.87	-1.12	-0.95	14.77
furn	0.75	1.39	1.32	0.98	0.65	1.40	1.24	0.91
papr	2.31	-0.32	-0.93	1.55	2.39	-0.31	-0.89	1.62
ptpc	-1.58	1.61	3.28	-0.28	-1.67	1.86	3.46	-0.28
chem	-2.62	-4.34	-3.54	-3.03	-2.55	-4.29	-3.47	-2.97
nmmp	0.06	1.52	1.44	0.70	-0.02	1.60	1.44	0.68
mtsp	-4.71	0.04	1.86	-2.40	-4.78	0.38	2.11	-2.31
mtlp	-2.41	-1.81	-2.55	-2.32	-2.36	-1.69	-2.47	-2.26
mach	-3.69	-0.83	-0.28	-2.84	-3.66	-0.60	-	-2.75
treq	-17.31	-14.64	-11.80	-15.98	-17.37	-14.33	-11.39	-15.91
eltc	0.97	0.35	0.74	0.88	0.90	0.49	0.87	0.85
eltn	13.23	-5.98	3.61	11.52	13.64	-5.90	4.03	11.91
inst	6.85	-2.29	-0.89	5.27	7.26	-2.02	-0.54	5.66

main	-0.52	0.06	1.16	-0.11	-0.56	0.13	1.18	-0.11
omnp	-0.29	0.66	-0.44	-0.03	-0.44	0.80	-0.42	-0.09
scrp	-3.28	-1.33	7.12	-1.76	-3.33	-1.24	7.39	-1.75
powr	-0.36	0.28	0.54	-0.08	-0.49	0.35	0.60	-0.13
gasp	1.10	1.28	1.17	1.14	0.53	0.55	1.07	0.58
watr	0.60	0.21	0.26	0.47	0.46	0.09	0.14	0.34
cons	0.56	1.22	1.32	0.83	0.55	1.26	1.35	0.84
tran	0.05	0.83	1.38	0.39	-0.04	0.94	1.45	0.36
ptlc	1.28	0.38	0.65	1.02	1.12	0.21	0.50	0.86
trad	0.68	0.22	1.27	0.67	0.54	0.25	1.28	0.58
cate	1.73	0.57	0.65	1.32	1.52	0.32	0.39	1.09
past	0.83	0.68	1.14	0.84	0.76	0.60	1.11	0.78
fina	0.25	0.16	0.53	0.28	0.04	0.23	0.62	0.15
rest	1.28	0.48	0.79	1.08	0.98	0.10	0.53	0.77
sser	1.72	0.39	0.38	1.34	1.83	0.34	0.31	1.40
heth	2.06	-0.70	-0.62	0.88	2.42	-0.86	-0.92	0.99
educ	1.48	-0.52	-0.47	0.70	1.87	-0.77	-0.76	0.83
scir	0.84	-1.63	-1.78	-0.02	1.41	-1.65	-1.78	0.35
teks	0.66	-1.29	-1.61	-0.12	1.48	-1.34	-1.65	0.39
padm	1.99	-1.80	-2.39	0.30	3.24	-1.83	-2.43	1.01

Table A6.10 Percentage change in export — WTO tariff cuts with higher elasticities

Sector	No control on trade balance				No change in trade balance			
	Eastern	Central	Western	National	Eastern	Central	Western	National
crop	9.61	10.42	12.20	10.10	11.00	11.76	13.48	11.46
frst	8.49	9.11	10.90	8.87	9.74	10.42	12.21	10.14
live	13.84	10.98	13.81	12.64	15.30	12.31	15.07	14.01
fish	11.63	10.59	12.55	11.24	13.36	12.23	13.99	12.93
otha	10.56	9.73	12.72	10.64	11.94	10.87	14.11	11.96
coal	1.38	7.31	7.70	6.45	1.31	7.88	8.08	6.91
petr	2.00	16.53	10.62	4.77	2.19	18.01	11.22	5.13
mtom	0.28	8.68	10.19	5.30	0.41	9.50	10.85	5.78
nmtm	2.57	8.83	9.34	6.24	2.66	9.45	9.75	6.63
fdtp	6.54	7.53	5.16	6.58	7.21	8.31	5.70	7.26
txtl	21.32	6.73	17.43	18.88	22.05	7.81	19.51	19.78
aprl	31.01	7.22	10.27	29.08	31.72	7.70	10.64	29.77
furn	6.48	8.48	9.53	7.09	6.68	9.01	9.90	7.39
papr	9.26	6.79	7.67	8.96	9.64	7.31	8.18	9.35
ptpc	1.68	7.29	9.60	3.93	1.79	7.93	10.14	4.19
chem	6.15	3.71	5.17	5.89	6.50	4.22	5.67	6.26
nmmp	4.31	8.48	8.45	6.13	4.45	9.03	8.82	6.45
mtsp	-0.47	6.87	8.82	4.07	-0.34	7.65	9.42	4.50
mtlp	3.05	5.51	4.80	3.23	3.33	6.07	5.25	3.53
mach	1.98	6.85	8.15	3.14	2.22	7.52	8.80	3.47
treq	-9.20	-4.91	0.67	-7.03	-9.10	-4.19	1.45	-6.74
eltc	7.16	7.88	8.41	7.26	7.34	8.48	8.92	7.48
eltn	23.15	0.72	12.17	21.55	23.87	1.14	12.96	22.26
inst	13.99	4.77	6.50	13.32	14.65	5.45	7.19	13.99

173

main	-	-	-	-	-	-	-	-
omnp	6.95	8.23	7.77	7.27	7.06	8.94	8.28	7.54
scrp	-	-	-	-	-	-	-	-
powr	3.01	6.10	5.91	4.62	3.11	6.59	6.30	4.91
gasp	3.64	5.54	5.38	3.89	3.22	5.09	5.52	3.55
watr	-	-	-	-	-	-	-	-
cons	3.08	5.68	5.71	3.55	3.21	6.01	5.96	3.70
tran	1.99	4.91	5.44	3.08	2.03	5.27	5.72	3.22
ptlc	3.18	3.64	4.42	3.44	3.17	3.72	4.48	3.46
trad	2.70	4.10	5.26	3.45	2.70	4.40	5.50	3.56
cate	5.92	5.31	6.52	5.74	6.05	5.49	6.60	5.88
past	3.15	4.77	5.08	3.99	3.22	4.93	5.27	4.11
fina	2.35	3.63	4.00	2.39	2.29	3.96	4.31	2.34
rest	-	-	-	-	-	-	-	-
sser	4.25	4.40	4.76	4.28	4.51	4.62	4.92	4.53
heth	6.33	4.62	4.35	6.28	6.83	4.71	4.24	6.77
educ	3.16	3.14	3.48	3.16	3.66	3.13	3.38	3.64
scir	-	-	-	-	-	-	-	-
teks	-	-	-	-	-	-	-	-
padm	4.21	1.97	1.22	2.28	5.60	2.17	1.37	2.86

Table A6.11   Percentage change in import — WTO tariff cuts with higher elasticities

Sector	No control on trade balance				No change in trade balance			
	Eastern	Central	Western	National	Eastern	Central	Western	National
crop	93.64	100.38	96.67	94.80	91.11	98.21	94.86	92.41
frst	-4.76	-5.24	-6.51	-4.94	-5.77	-6.45	-7.69	-6.01
live	1.25	-0.19	-2.18	0.68	-0.30	-1.52	-3.40	-0.80
fish	-6.61	-3.11	-5.87	-6.29	-8.77	-5.08	-7.73	-8.41
otha	320.37	320.59	300.12	319.76	312.69	314.54	294.48	312.79
coal	14.32	11.97	12.04	13.87	13.84	11.43	11.62	13.39
petr	-4.61	-3.24	-4.39	-4.37	-4.90	-3.32	-4.63	-4.62
mtom	-7.31	-7.02	-3.73	-6.30	-7.54	-7.20	-3.79	-6.47
nmtm	-4.21	-7.46	-6.88	-5.17	-4.52	-7.87	-7.24	-5.51
fdtp	160.52	177.43	161.94	162.94	157.50	174.43	159.66	160.08
txtl	42.82	35.27	29.18	41.11	42.40	34.35	28.25	40.61
aprl	-19.57	-14.99	-18.08	-18.33	-20.61	-16.36	-19.34	-19.48
furn	11.42	10.12	9.71	11.07	10.87	9.36	9.06	10.49
papr	5.01	3.55	1.30	4.20	4.78	3.10	0.90	3.92
ptpc	3.58	2.60	2.59	3.14	3.35	2.40	2.35	2.92
chem	25.96	26.81	26.67	26.15	25.71	26.33	26.25	25.84
nmmp	4.57	1.36	2.26	3.99	4.23	0.82	1.83	3.63
mtsp	5.20	4.63	5.09	5.12	4.92	4.30	4.80	4.83
mtlp	34.39	34.37	32.90	34.21	34.00	33.82	32.40	33.78
mach	11.82	12.20	10.40	11.69	11.57	11.83	10.14	11.42
treq	109.13	106.67	87.49	104.73	108.18	105.39	86.72	103.73
eltc	8.31	6.68	7.62	7.97	7.85	6.11	7.22	7.50
eltn	6.74	2.35	2.74	5.63	6.49	1.86	2.33	5.32
inst	2.41	4.26	3.08	2.85	2.08	3.93	2.85	2.54

main	-	-8.86	-9.81	-7.86	-3.85	-9.21	-10.12	-8.77
omnp	47.31	49.28	47.56	47.45	46.65	48.06	46.64	46.73
scrp	-3.42	-3.71	-	-2.15	-3.63	-3.86	-3.07	-3.49
powr	-4.13	-6.50	-5.56	-4.91	-4.51	-6.89	-5.84	-5.28
gasp	-3.42	-6.57	-6.87	-5.39	-4.18	-7.69	-7.34	-6.13
watr	-3.50	-6.78	-6.26	-5.05	-3.88	-7.32	-6.71	-5.50
cons	-2.64	-4.36	-4.26	-3.52	-2.81	-4.66	-4.50	-3.74
tran	-2.30	-5.37	-4.67	-3.96	-2.52	-5.64	-4.90	-4.19
ptlc	-1.47	-4.29	-4.63	-2.95	-1.81	-4.75	-5.04	-3.34
trad	-	-5.73	-5.29	-4.85	-2.29	-6.09	-5.61	-5.40
cate	-3.92	-5.93	-6.61	-4.41	-4.54	-6.70	-7.24	-5.03
past	-2.14	-5.72	-4.97	-3.37	-2.37	-6.15	-5.30	-3.66
fina	-2.03	-4.33	-3.89	-3.83	-2.41	-4.60	-4.07	-4.09
rest	-1.09	-2.21	-2.25	-1.59	-1.56	-2.82	-2.71	-2.11
sser	-2.53	-5.04	-5.41	-3.66	-2.67	-5.44	-5.75	-3.90
heth	-3.28	-7.29	-6.75	-5.51	-3.09	-7.75	-7.26	-5.73
educ	-0.69	-5.26	-5.54	-3.32	-0.46	-5.78	-6.04	-3.51
scir	-2.40	-7.13	-6.76	-5.84	-1.99	-7.44	-7.01	-5.96
teks	-1.05	-6.27	-	-2.54	-0.35	-6.61	-3.57	-3.20
padm	-0.78	-6.41	-6.92	-3.60	0.25	-6.76	-7.23	-3.26

Table A6.12   Percentage change in output — WTO tariff cuts with transfer payments

Sector	No control on trade balance				No change in trade balance			
	Eastern	Central	Western	National	Eastern	Central	Western	National
crop	-1.89	-0.58	-0.81	-1.22	-1.89	-0.58	-0.79	-1.22
frst	1.84	1.33	1.51	1.59	2.19	1.53	1.65	1.85
live	2.96	0.35	1.03	1.64	2.89	0.27	0.89	1.55
fish	-0.10	0.31	0.32	–	-0.35	0.13	-0.03	-0.23
otha	1.18	-0.05	-0.11	0.60	1.08	-0.25	-0.13	0.49
coal	-1.78	0.48	0.37	-0.43	-2.05	0.54	0.39	-0.51
petr	-0.87	7.39	3.95	0.89	-0.92	8.10	4.18	0.96
mtom	-3.78	0.95	2.82	-0.95	-3.91	1.26	3.13	-0.85
nmtm	-0.64	1.38	1.38	0.39	-0.80	1.49	1.40	0.36
fdtp	-4.19	-1.72	-5.37	-3.59	-4.32	-1.80	-5.54	-3.71
txtl	8.27	-1.35	7.05	6.41	8.56	-0.83	8.77	6.84
aprl	17.27	0.12	1.27	14.50	17.42	0.03	1.10	14.61
furn	0.56	1.30	1.06	0.82	0.40	1.29	0.95	0.70
papr	1.91	-0.12	-0.77	1.31	1.95	-0.12	-0.72	1.34
ptpc	-1.22	0.57	2.08	-0.41	-1.32	0.71	2.22	-0.43
chem	-2.98	-3.93	-3.21	-3.18	-2.94	-3.89	-3.11	-3.13
nmmp	-0.10	1.33	1.17	0.52	-0.23	1.39	1.14	0.46
mtsp	-4.70	-0.63	0.96	-2.71	-4.83	-0.36	1.18	-2.68
mtlp	-2.75	-2.01	-2.77	-2.63	-2.77	-1.93	-2.71	-2.62
mach	-3.91	-1.19	-1.07	-3.14	-3.92	-1.00	-0.83	-3.09
treq	-17.73	-14.99	-12.13	-16.38	-17.89	-14.71	-11.69	-16.37
eltc	0.59	0.10	0.47	0.53	0.44	0.18	0.58	0.42
eltn	11.22	-3.12	5.07	10.04	11.40	-3.05	5.49	10.23
inst	5.37	-2.46	-1.27	4.02	5.58	-2.24	-0.97	4.23

main	-0.45	-0.14	0.80	-0.17	-0.50	-0.11	0.81	-0.19
omnp	-0.79	0.76	-0.24	-0.31	-1.03	0.89	-0.19	-0.42
scrp	-3.07	-1.13	4.50	-1.83	-3.19	-1.06	4.76	-1.87
powr	-0.40	0.04	0.23	-0.20	-0.57	0.07	0.29	-0.29
gasp	1.16	1.31	0.74	1.14	0.49	0.45	0.61	0.50
watr	0.52	0.03	0.18	0.36	0.34	-0.16	0.04	0.19
cons	0.47	1.07	1.14	0.72	0.45	1.09	1.16	0.71
tran	0.14	0.33	0.83	0.26	0.03	0.39	0.88	0.21
ptlc	1.13	0.34	0.58	0.90	0.92	0.12	0.39	0.69
trad	0.66	0.02	1.02	0.58	0.48	0.02	1.02	0.46
cate	1.65	0.64	0.72	1.29	1.40	0.36	0.41	1.02
past	0.75	0.51	0.94	0.70	0.64	0.37	0.88	0.60
fina	0.21	-0.02	0.30	0.19	-0.05	-	0.39	0.03
rest	1.16	0.46	0.65	0.97	0.78	-0.01	0.33	0.59
sser	1.40	0.37	0.36	1.10	1.48	0.28	0.27	1.13
heth	1.67	-0.41	-0.32	0.78	2.08	-0.63	-0.67	0.91
educ	1.11	-0.45	-0.40	0.51	1.56	-0.76	-0.74	0.65
scir	0.28	-1.46	-1.69	-0.34	0.90	-1.52	-1.73	0.05
teks	0.26	-1.03	-1.31	-0.27	1.19	-1.10	-1.37	0.30
padm	1.17	-1.43	-1.97	-0.02	2.60	-1.51	-2.05	0.78

Table A6.13   Percentage change in export — WTO tariff cuts with transfer payments

Sector	No control on trade balance				No change in trade balance			
	Eastern	Central	Western	National	Eastern	Central	Western	National
crop	16.94	15.15	17.10	16.47	19.88	17.83	19.51	19.29
frst	13.33	12.92	15.28	13.34	15.77	15.40	17.74	15.80
live	20.52	15.57	18.72	18.16	23.65	18.29	21.23	21.02
fish	17.30	15.34	17.54	16.55	20.72	18.52	20.28	19.87
otha	16.43	13.12	17.22	15.66	19.28	15.31	19.82	18.29
coal	1.10	6.50	6.88	5.72	0.92	7.00	7.22	6.11
petr	2.02	12.55	8.90	4.16	2.16	13.66	9.43	4.44
mtom	-0.27	7.51	8.58	4.31	-0.27	8.22	9.20	4.68
nmtm	2.07	8.03	8.38	5.55	2.04	8.59	8.76	5.86
fdtp	9.60	9.59	6.84	9.17	10.94	10.99	7.76	10.46
txtl	21.10	9.26	22.48	19.53	21.96	10.89	25.79	20.71
aprl	30.49	8.86	13.50	28.76	31.27	9.61	14.31	29.53
furn	5.95	8.17	9.20	6.63	6.08	8.70	9.60	6.86
papr	8.45	6.94	8.17	8.29	8.76	7.50	8.80	8.64
ptpc	1.66	5.73	7.87	3.37	1.72	6.22	8.34	3.57
chem	5.50	3.96	5.53	5.37	5.82	4.49	6.13	5.73
nmmp	3.56	7.82	7.78	5.42	3.58	8.31	8.10	5.64
mtsp	-0.91	5.67	7.40	3.15	-0.91	6.33	7.93	3.47
mtlp	2.23	4.79	4.13	2.41	2.38	5.27	4.54	2.59
mach	1.25	5.94	6.79	2.33	1.37	6.53	7.37	2.56
treq	-10.08	-5.71	-0.05	-7.87	-10.15	-5.06	0.73	-7.71
eltc	6.27	7.16	7.69	6.39	6.31	7.66	8.15	6.47
eltn	20.39	3.59	13.59	19.32	20.79	3.99	14.36	19.74
inst	11.87	4.24	5.72	11.32	12.25	4.84	6.31	11.71

main	-	-	-	-	-	-	-	-
omnp	5.95	8.08	8.18	6.52	5.93	8.80	8.84	6.69
scrp	-	-	-	-	-	-	-	-
powr	2.44	5.42	5.19	3.98	2.44	5.84	5.55	4.20
gasp	3.21	5.23	4.60	3.41	2.63	4.63	4.66	2.92
watr	-	-	-	-	-	-	-	-
cons	2.59	5.16	5.21	3.06	2.64	5.44	5.43	3.13
tran	1.70	4.06	4.57	2.58	1.64	4.35	4.81	2.65
ptlc	2.74	3.34	4.09	3.05	2.65	3.35	4.09	2.99
trad	2.37	3.55	4.82	3.05	2.28	3.80	5.05	3.09
cate	6.25	5.72	6.94	6.09	6.46	6.02	7.09	6.33
past	2.75	4.28	4.65	3.56	2.74	4.37	4.81	3.60
fina	1.98	3.19	3.57	2.02	1.83	3.46	3.87	1.88
rest	-	-	-	-	-	-	-	-
sser	3.61	4.15	4.57	3.67	3.79	4.33	4.70	3.85
heth	5.60	4.62	4.40	5.57	6.10	4.64	4.22	6.06
educ	2.36	2.87	3.23	2.38	2.83	2.77	3.04	2.83
scir	-	-	-	-	-	-	-	-
teks	-	-	-	-	-	-	-	-
padm	2.93	1.96	1.30	1.88	4.42	2.09	1.37	2.45

Table A6.14 Percentage change in import — WTO tariff cuts with transfer payments

Sector	Eastern	No control on trade balance			Eastern	No change in trade balance		
		Central	Western	National		Central	Western	National
crop	86.14	95.10	91.40	87.85	82.17	91.47	88.49	84.05
frst	-8.14	-8.23	-9.82	-8.22	-9.88	-10.28	-11.76	-10.04
live	-3.03	-3.09	-5.22	-3.30	-5.53	-5.25	-7.17	-5.69
fish	-10.86	-6.95	-9.91	-10.49	-14.16	-10.04	-12.74	-13.73
otha	299.69	306.89	284.41	302.00	287.53	297.02	275.06	290.84
coal	14.88	12.36	12.42	14.40	14.44	11.80	12.04	13.94
petr	-4.06	-3.62	-4.71	-4.04	-4.33	-3.74	-4.95	-4.28
mtom	-6.95	-7.00	-3.98	-6.16	-7.17	-7.19	-4.01	-6.33
nmtm	-3.71	-7.12	-6.59	-4.73	-3.98	-7.52	-6.95	-5.04
fdtp	155.46	173.65	158.35	158.35	151.16	169.31	155.16	154.28
txtl	42.16	34.88	27.91	40.44	41.61	33.65	26.66	39.77
aprl	-19.51	-15.50	-18.89	-18.54	-20.70	-17.18	-20.46	-19.90
furn	11.71	10.40	9.71	11.33	11.13	9.56	8.97	10.70
papr	5.09	3.75	1.14	4.25	4.86	3.24	0.67	3.95
ptpc	3.86	2.79	2.79	3.38	3.64	2.60	2.58	3.17
chem	25.92	27.34	26.97	26.21	25.65	26.83	26.52	25.89
nmmp	5.14	1.80	2.56	4.52	4.85	1.26	2.14	4.19
mtsp	5.44	5.06	5.45	5.39	5.16	4.75	5.20	5.11
mtlp	34.92	35.02	33.44	34.76	34.57	34.50	32.98	34.37
mach	12.24	12.71	10.73	12.12	12.04	12.36	10.52	11.89
treq	110.72	107.98	88.18	106.10	109.91	106.75	87.48	105.20
eltc	8.78	7.18	8.05	8.44	8.33	6.61	7.69	7.98
eltn	6.67	2.72	3.42	5.71	6.38	2.24	3.02	5.38
inst	2.91	4.58	3.27	3.27	2.61	4.26	3.07	2.99

main	-	-8.44	-9.45	-7.54	-3.51	-8.76	-9.74	-8.37
omnp	47.80	49.78	47.30	47.87	47.14	48.44	46.20	47.14
scrp	-3.04	-3.50	-	-1.98	-3.26	-3.67	-3.18	-3.37
powr	-3.66	-6.12	-5.36	-4.49	-4.01	-6.50	-5.61	-4.84
gasp	-2.58	-5.76	-6.35	-4.68	-3.32	-6.95	-6.71	-5.38
watr	-2.92	-6.24	-5.71	-4.49	-3.25	-6.79	-6.14	-4.91
cons	-2.24	-3.99	-3.94	-3.15	-2.35	-4.26	-4.15	-3.33
tran	-1.87	-5.05	-4.40	-3.61	-2.04	-5.30	-4.59	-3.81
ptlc	-1.23	-3.93	-4.26	-2.66	-1.57	-4.43	-4.68	-3.05
trad	-	-5.26	-5.09	-4.58	-1.94	-5.61	-5.41	-5.09
cate	-4.54	-6.27	-6.93	-4.96	-5.33	-7.28	-7.72	-5.77
past	-1.90	-5.31	-4.73	-3.10	-2.10	-5.77	-5.07	-3.37
fina	-1.73	-3.98	-3.74	-3.55	-2.11	-4.24	-3.89	-3.79
rest	-0.92	-1.92	-2.14	-1.39	-1.42	-2.59	-2.64	-1.95
sser	-2.34	-4.71	-5.15	-3.41	-2.41	-5.15	-5.52	-3.63
heth	-3.27	-6.65	-6.19	-5.15	-2.95	-7.16	-6.77	-5.36
educ	-0.54	-4.61	-4.91	-2.90	-0.14	-5.20	-5.46	-3.04
scir	-2.43	-6.63	-6.30	-5.49	-1.87	-7.00	-6.57	-5.59
teks	-1.19	-5.61	-	-2.38	-0.26	-6.01	-3.19	-2.87
padm	-1.03	-5.59	-6.09	-3.31	0.27	-5.96	-6.41	-2.84

07 WTO accession and food security in China

Tingsong Jiang

After 15 years of long and difficult negotiations, China was admitted to WTO in November 2001. China made the most exceptional accession commitment in the history of the WTO and its predecessor, the General Agreement on Tariff and Trade (GATT). For example, China agreed to lower its average statutory tariff on industrial products to 8.9 per cent, while the rates for Argentina, Brazil, India and Indonesia were set as high as 30.9, 27.0, 32.4 and 30.9 per cent, respectively (Lardy 2002). For agricultural goods, China has agreed to have no agricultural export subsidies, and to limit its domestic support to farmers to 8.5 per cent of the production value, while other developing countries were allowed to have domestic support at a level of 10 per cent of production value.

Because of the expectation of possible compromise in the agricultural commitments and the comparative disadvantage of land intensive products, the adverse impact on agriculture of WTO accession caused great concern for Chinese policymakers and academics leading up to the accession. These concerns were all based on the prediction that the agricultural sector would be one of the most hard hit sectors and have been focused on two issues: food security and farmer incomes.

According to the World Food Summit, food security exists when all people, at all times, have physical and economic access to sufficient, safe and nutritious food to meet their dietary needs and food preferences for an active and healthy life. It means achieving four major components

- ensuring that sufficient food is available
- maintaining relatively stable food supplies
- allowing access to food for those in need of it
- ensuring the biological utilisation of food.

Like other Asian countries, however, food security in China almost exclusively has meant food self-sufficiency, or grain self-sufficiency. A 95 per cent grain self-sufficiency level has been set as the target. The followings are some of the arguments made for China to maintain a certain rate of grain self-sufficiency and responses to these arguments.

First, China is a very large country in population terms and cannot rely on the world market for its food supplies. Brown (1995) painted a terrifying picture of China's food supply and demand to the effect that a food shortage in China would severely deplete world food supplies and hurt other developing countries. Although Brown's arguments were dismissed by the Chinese government and scholars, they increased concerns about China's food supply capacity.

Second, food is a special good and the independence of food supply has political and economic importance. Some people even fear the possibility of a food embargo as China has many ideological, political and strategic differences with major western powers. However, Lu (1997) and Yang (2000) point out that a food embargo against China is unlikely. Globally, food was often excluded from lists of embargoed commodities in the past because of humanitarian considerations. Moreover, a food embargo is difficult to apply and often means even higher economic costs for the countries that initiate the embargo.

Third, as many Chinese farmers earn their income exclusively from producing food, they may be badly hurt by cheaper food prices after WTO accession (Zhou 2001). This kind of impact has been seen previously in China and other developing countries. For example, in order to solve their food surplus problem in the early 1930s, a number of countries, including the United States, dumped a huge amount of food into the Chinese market and significantly depressed food prices. As a result, many Chinese farmers went bankrupt and China's grain sector was seriously injured (Xu 1996).

Chapter 6 (this volume) discusses the income effects, especially the regional income disparity impacts, of WTO accession. It was concluded that although China's WTO accession raises the real income and welfare of rural and urban households in all regions, regional income disparity and rural-urban inequality will worsen as the richer eastern region receives most of the welfare gains of accession. This chapter focuses on the food security issue.

China's achievements in food security and food self sufficiency policies

China's achievements in food security

China has made great progress in providing food security for its people since the economic reforms began in 1978. According to FAO statistics, the average per capita daily food consumption in China was only 2017 calories in 1979, well below the world average of 2500 and below the average of 2200 for other developing countries at that time. By 2001, it had increased by about 47 per cent to 2963 calories—above the world average of 2800. Nutritional intake and food quality have also improved. China had made the greatest achievement in reducing undernourishment in both absolute and relative terms. During the period from 1990-92 to 1998-2000, China reduced the number of undernourished people by 70 million, or by 7 percentage points (FAO 2002).

Grain production and household incomes have increased as well. Total and per capita grain production increased at annual rates of 1.9 and 0.7 per cent, respectively, from 1978 to 2002. Per capita incomes of rural and urban households increased by 7.4 and 6.3 per cent per annum, respectively, over the same period (Table 7.1). The food self-sufficiency target has also been well maintained. In 2001, both total grain and total food self-sufficiency rates were above 95 per cent (Table 7.2).

These achievements were brought about by the agricultural economic reforms, especially the rural household responsibility system and market reform for production factors and agricultural commodities (Lin 1997). Between 1979 and 1984, when the household responsibility system began, the growth in agricultural total factor productivity (TFP) jumped to 5.1

per cent per annum, compared to the -0.25 per cent per annum prevailing between 1952 and 1978. Agricultural TFP growth was kept at 3.9 per cent per annum from the late 1980s through to the early 1990s (Fan 1997). Market reform has also been remarkable. In 1978, the prices of 93 per cent of agricultural commodities were fixed by government and the prices of another 2 per cent were 'guided' by the government, leaving only 6 per cent to the market. In 1999, the composition was virtually reversed. The prices of 83 per cent of agricultural commodities were determined by the market (Table 7.3).

Although the growth in agricultural productivity has been remarkable, food consumption measured in terms of daily per capita calorie consumption has grown at an even higher rate (see Table 7.1). This difference implies that China's achievements in food security are mainly due to increases in household incomes.

Food security policies in China

China's food security policies have aimed at increasing grain self-sufficiency and food availability to households, especially urban households. These policies have included self-sufficiency policies, grain marketing policies, and grain reserve policies (Lohmar 2002).[1]

Grain self-sufficiency has been vigorously targeted since 1995, partly because of the 'wakeup call' by Brown (1995). Grain self-sufficiency is not only required at the national level, but also at local level. Under the Governor Grain Bag Policy (GGBP) established in 1995, provincial governors are required to be responsible for grain self-sufficiency in their jurisdictions. In order to fulfil their targets, governors have instructed their subordinates to increase sown areas, yields and grain production. As a result, almost every locality has its own grain production quota to fill.

The side effects of attempting to achieve grain self-sufficiency at local level are obvious. Increases in grain production have been achieved at the cost of efficiency losses as the practice ignores regional comparative advantages. This policy has been relaxed following WTO accession. At the national level, provinces are classified into three categories: major

Table 7.1   Grain production, food consumption and household income in China, 1978 and 2000

	Total grain production (million ton)	Per capita grain production (kg)	Per capita daily calories*	Per capita income of rural households (yuan)	Per capita disposable income of urban households (yuan)
1978	304.8	316.6	2,017	133.6	343.4
2000	462.2	365.1	2,963	646.0	1,317.6
annual growth rate (per cent)	1.9	0.7	1.8	7.4	6.3

Note: * per capita daily calories are for 1979 and 2001.
Source: National Bureau of Statistics, 2002. *Statistical Yearbook of China 2001*, Beijing: China Statistics Press; Food and Agriculture Organisation, 2002. *The State of Food Insecurity in the World 2002*, Rome:

Table 7.2   Food balance and self-sufficiency in China, 2001

	Production	Imports	Stock change	Exports	Total domestic utilisation	Self-sufficiency rate
Rice	119,596	694	4,353	2,109	122,535	97.6
Wheat	93,876	2,286	7,530	1,052	102,640	91.5
Corn	114,254	5,392	5,830	6,138	119,338	95.7
Pulses	5,127	263	8	739	4,659	110.0
Other grains	194,709	11,660	-45	799	205,524	94.7
Oilseeds	52,844	18,370	-346	1,202	6,9667	75.9
Vegetable and fruits	425,088	3,861	-39	7,618	421,292	100.9
Other crops	88,854	282	0	4	89,133	99.7
Forest food	2,444	377	-3	791	2,027	120.6
Livestock	107,239	4,793	5	2,366	109,671	97.8
Fishery	48,982	6,047	15	3,174	43,969	111.4
Food processing	56,821	6,021	-563	1,407	60,873	93.3
Total grains	527,562	20,295	17,676	10,837	554,696	95.1
Total food	1,309,834	60,046	16,745	27,399	1,351,328	96.9

Note: Production, imports, exports, stock change and utilisation are in million metric tons, and the self-sufficiency rates are in per centages.
Source: Food and Agriculture Organisation 2003. *FAO STAT*, Rome: Food and Agriculture Organisation, available online at http://www.fao.org.

Table 7.3 Agricultural commodity price reform in China, 1978-99

Year	Market determined	State guided	State fixed
1978	6	2	93
1985	40	23	37
1987	54	17	29
1991	58	20	22
1995	79	4	17
1999	83	7	9

Source: Lardy, Nicholas R., 2002. *Integrating China into the Global Economy*, Washington D.C.: Brookings Institution, Washington, DC:25.

grain production provinces, provinces with balanced grain production and consumption, and net grain importing provinces. Major grain production provinces have comparative advantage in grain production and are thus encouraged to increase grain production. Other provinces are allowed to import grains from other regions and overseas to meet the local demand.

Associated with the grain self-sufficiency and availability policies are grain marketing and pricing policies. The Chinese government has intervened in grain production and markets using procurement quotas and contracts and different types of prices through the network of state owned grain bureaus and stations. Before the rural reforms began in 1979, the goals of these policies were to produce ample and cheap food for urban residents and to export farm products, to earn hard currency for importing advanced technology and equipment, and to develop industries in urban areas.

Since the economic reforms began, grain marketing and pricing policies have been significantly changed.[2] Policy has been focused on improvement of farmers' incomes and long term food security and self-sufficiency (Ke 1999). The government attempts to counter price volatility through government procurement prices and sales prices: in a year of bumper crops the procurement price sets a floor price so that farmers are not hurt by falling market prices; while in a year of poor harvests, the government sells reserves to cap the soaring market price.

However, such price stabilisation practices send the wrong signal to farmers and make the situation even worse, as evidenced by the 'grain surplus crisis' in recent years. In a bid to increase grain production and farmers' incomes, the government significantly raised the procurement prices in 1994 and 1996—by 40 per cent each time; as a result, grain production kept increasing. In 1998, the total grain output reached 504.5 million tons, well above the upper range of the target set for the year 2000. Although the open market price started falling from the end of 1997, as the higher government procurement price is 'guaranteed', output failed to adjust. As a result, on the one hand some farmers faced difficulties in selling their grains, resulting in political difficulties. On the other hand, the agency implementing the price protection policy, the state owned grain sector, accumulated huge losses. An estimate by Zhang and Li (1998) indicates that the amount of non-performing loans owed by the state grain sector to the Chinese Agricultural Development Bank reached 21.4 billion yuan.

These huge losses by the state grain sector were completely unacceptable, and grain marketing and distribution system reforms initiated in May 1998 were designed to reduce the losses. Despite the continuing assertion that the government buys the grain from farmers at protected prices, the new policies required state grain agencies to sell grain at a price that covers the full costs incurred and preferably with a small profit margin. The state grain agencies were also asked to improve their operational efficiency through redundancies and other measures.

The original idea was gradually to liberalise the grain markets so that government procurement prices would move close to market prices. However, due to the fundamental flaw in the design of the policy, that is, that the state grain sector was asked to carry out two conflicting duties—to act as a government agency implementing the protection price policy and to achieve financial soundness as a commercial entity—after one year the declining trend in grain prices increased and the government subsidy continued to soar. The government realised the problem and slightly changed the policy in mid May 1999. The remedies included reducing the scope of the price support coverage and the support price levels and allowing grain processing and feed industries to purchase grain directly from farmers.

Another direct approach to achieving food security has been the state grain reserve policy. The national government controls a large amount of reserve grain stocks. However the volume of stocks held is a state secret. The state stocks are managed by the State Administration for Grain Reserves (SAGR) and held by the grain bureaus. When there is a grain shortage, reserves are sold to put downwards pressure on prices. However, the management of state held stocks is seen as being too bureaucratically constrained to effectively reduce price volatility (Lohmar 2002). Moreover, maintaining the stockpile is expensive. Nyberg and Rozelle (1999) estimate that the costs of carrying over one ton of wheat, rice and corn were about US$42, US$6 and US$39, respectively, being equivalent to over 20 per cent of the price of each commodity on the world market.

The impact on agricultural production and food security of WTO accession

China's WTO accession commitments in agriculture

China's commitments in agriculture are more far reaching than those in manufactured goods. They include the reduction in tariff rates, adoption of a tariff rate quota system, limits on domestic support and on export subsidies for agricultural products, and the elimination of some technical trade barriers.

China agreed to reduce the average statutory tariff rate for agricultural products from 22 per cent to 15 per cent by January 2004. It also agreed to adopt the tariff rate quota system for key agricultural products such as wheat, rice and corn. Within the specified import quota, which is scheduled to increase over time, the tariff rate for these three grain products is only 1 per cent (Table 7.4). The state owned agricultural trading firms no longer enjoy the monopoly power they previously had, as private firms have been allowed to enter the foreign trade market and to import any unfulfilled quota left by the state owned firms.

However, these commitments do not necessarily lead to significant adverse impacts on agricultural production as China already had relatively low protection on agriculture. For example, the nominal rates of protection for rice, vegetables and fruits and meats are negative (Table 7.5). This situation was partly due to the history of taxing agriculture to subsidise industrial development and partly due to the effort of reducing protection along with the WTO accession negotiations.

Overall impacts of WTO accession

Simulations of the impacts of WTO accession were carried out using the closure of no control on the trade balance; that is, the exchange rate is fixed and the trade balance is endogenous. The macroeconomic impacts of China's WTO accession are summarised in Table 7.6. Several points can be drawn from the simulation results.

First, China has an overall gain from WTO accession. Real GDP, consumption and the utility level increase, by 0.4, 1.0 and 0.8 per cent,

Table 7.4   Tariff rate quotas and imports of major agricultural products, million metric tons

Commodity	Initial quota	Ultimate quota	In quota tariff rate	Out quota tariff rate	1998	1999	2000
Wheat	7.884	9.636	1	65	1.55	0.45	0.88
Corn	5.175	7.200	1	65	0.25	0.07	
Rice	3.325	5.320	1	65	0.26	0.17	0.24
Soybean oil	2.118	3.587	9	63.3→9	0.83	2.08	1.79
Palm oil	2.100	3.168	9	63.3→9			
Rapeseed oil	0.739	1.243	9	63.3→9			
Sugar	1.680	1.945	20→15	71.6→50		0.42	0.64
Wool	0.253	0.287	1	38		0.199	
0.301							
Cotton	0.781	0.894	1	61.6→40	0.21	0.05	0.05

Source: Protocol on the Accession of the People's Republic of China, Schedule CLII; National Bureau of Statistics, 2002. *Statistical Yearbook of China 2001*, China Statistics Press, Beijing.

Table 7.5   Nominal rates of protection (tariff or tariff equivalent) for agricultural products in China, 1995-01

	1995	1997	2001
Rice	-5	-5	-3
Wheat	25	17	12
Coarse grains	20	28	20
Vegetables & fruits	-10	-8	-4
Oilseeds	30	28	32
Sugar	44	42	40
Cotton	20	17	17
Meats	-20	-19	-15
Milk	30	30	30

Source: Anderson, Kym, Jikun Huang, and Ianchovichina, E., 2002b, Long-run impacts of China's WTO accession on farm-nonfarm income inequality and rural poverty, Memo, Centre for International Economic Studies, University of Adelaide, Table 6.

respectively. The welfare improvement, measured by the equivalent variation (EV), amounts to 58.2 billion yuan.

Second, the tariff reductions lead to lower prices of imports, which, in turn, result in lower overall prices. The consumer price index (CPI) and GDP deflator decline by 2.2 and 1.8 per cent, respectively. Because of the cheaper domestic prices, the terms of trade decline by 0.4 per cent.

Third, both imports and exports increase after the tariff cut by 19.0 and 11.1 per cent, respectively. As imports increase more than the exports, the trade balance declines by about 34 billion yuan.

Fourth, the impacts of WTO accession are not evenly distributed. Crops, food processing, motor vehicle and parts, and the machinery sectors are

Table 7.6   Macroeconomic impact of WTO accession

	Eastern	Central	Western	National
Real GDP (per cent)	0.703	-0.114	0.275	0.436
Real consumption (per cent)	1.694	0.19	0.218	1.021
Utility (per cent)	1.093	0.317	0.402	0.790
EV (billion yuan)	51.397	3.627	3.179	58.203
CPI (per cent)	-2.118	-2.019	-2.477	-2.157
GDP deflator (per cent)	-1.673	-1.783	-2.500	-1.831
Changes in trade balance (billion yuan)				-33.955
Terms of trade (per cent)				-0.416
Imports (per cent)				19.042
Exports (per cent)				11.096
Sectoral output				
Crops	-0.674	-0.294	-0.399	-0.490
Other agriculture	0.637	0.257	0.466	0.487
Mining	-1.484	1.270	2.943	0.160
Food processing	-5.679	-1.751	-5.332	-4.404
Light industry	7.188	-0.217	1.129	5.341
Chemicals	-1.922	-0.931	-0.680	-1.556
Motor vehicles and parts	-17.612	-14.658	-10.814	-16.035
Machinery	-3.590	-0.865	0.641	-2.516
Electronic and electrics	6.549	-0.736	3.736	5.737
Construction	0.426	1.033	1.162	0.687
Service	0.774	-0.159	0.132	0.481

Source: Author's CERD simulation

adversely affected while the other sectors benefit. The motor vehicle and parts sector is the hardest hit while light industry benefits most.

In terms of regional distribution, almost all the benefits go to the eastern coastal region. This is because the eastern region has the highest proportion of booming sectors. This pattern of impact implies that the regional income disparity worsens after WTO accession. Chapter 6 (this volume) discusses this issue in more detail.

Impact of WTO accession on agriculture and food security

The impacts of WTO accession on production, exports, imports and self-sufficiency of agricultural products are reported in Tables 7.7-7.10. Although agricultural imports and exports change significantly, the imported volume is still within the quota, and the contraction of output is limited. For most grains, the reduction in output is less than 1 per cent. Consequently, the impact on grain and food self-sufficiency is also limited and the target rate of 95 per cent self-sufficiency is likely to be achieved.

Grains production in the eastern region declines the most, while that in the central region declines the least. This result is in line with the comparative advantage of regional agricultural production. Compared to the eastern region, the central region has a relatively abundant labour force and land resources. Compared to the western region, the central region has more favourable weather conditions, and arable land resources are also relatively abundant.

However, food security may still be an issue even if the self-sufficiency target is achieved. This is because the benefits and costs of accession are not evenly distributed across regions and households. Even if the overall benefit is larger than the cost, for those engaged mainly in the contracting sectors, food insecurity caused by the reduction in income may be a problem. This is especially important for rural households in inland regions. For example, the real wage in agricultural sectors in the central region declines after the WTO accession (Table 7.11).[3] Table 7.11 also shows that rural household income increases to a smaller extent than urban household income; that is, the rural-urban income inequality worsens.

Table 7.7   Impact of WTO accession on agricultural production (per cent)

	Eastern	Central	Western	National
Rice	-0.318	-0.169	-0.227	-0.242
Wheat	-1.072	-0.367	-0.523	-0.690
Corn	-1.800	-0.563	-0.813	-1.072
Pulse	-1.125	-0.379	-0.544	-0.588
Other grains	-1.136	-0.385	-0.549	-0.720
Cotton	-1.226	-0.327	-0.499	-0.624
Oilseeds	-1.981	-0.537	-0.801	-1.313
Vegetable & fruit	-0.019	0.006	-0.024	-0.015
Other crops	-0.478	-0.127	-0.205	-0.356
Forestry	0.696	0.768	0.854	0.753
Livestock	1.523	0.298	0.555	0.892
Fishery	-0.989	-0.228	-0.100	-0.798
Other agriculture	0.223	0.222	-0.243	0.126
Processed food	-5.679	-1.751	-5.332	-4.404

Source: Author's CERD simulation

Table 7.8   Impact of WTO accession on agricultural exports (per cent)

	Eastern	Central	Western	National
Rice	9.253	9.629	10.787	9.511
Wheat	9.898	10.037	11.357	10.118
Corn	10.075	9.994	11.475	10.201
Pulse	9.913	10.028	11.367	10.143
Other grains	9.888	10.008	11.339	10.134
Cotton	9.955	10.018	11.344	10.193
Oilseeds	10.209	10.051	11.546	10.187
Vegetable & fruit	9.574	10.027	11.090	9.740
Other crops	9.741	10.048	11.210	9.915
Forestry	8.018	8.247	9.740	8.219
Livestock	11.295	10.169	11.981	10.937
Fishery	11.477	9.894	11.253	10.868
Other agriculture	9.013	9.240	11.394	9.405
Processed food	3.931	7.377	4.739	4.938

Source: Author's CERD simulation

Table 7.9   Impact of WTO accession on agricultural imports (per cent)

	Eastern	Central	Western	National
Rice	-7.908	-7.743	-8.635	-7.969
Wheat	31.737	33.813	32.069	32.055
Corn	66.536	71.405	68.605	67.630
Pulse	32.460	34.700	32.904	33.142
Other grains	32.521	34.760	32.970	32.880
Cotton	53.627	56.787	54.684	54.474
Oilseeds	83.812	90.133	86.939	84.771
Vegetable & fruit	-7.466	-7.576	-8.326	-7.543
Other crops	15.097	15.871	14.705	15.099
Forestry	-4.753	-4.531	-5.692	-4.730
Livestock	2.255	0.647	-0.857	1.702
Fishery	-7.215	-2.421	-4.781	-6.632
Other agriculture	325.428	323.875	307.226	324.185
Processed food	163.876	179.087	163.574	165.720

Source: Author's CERD simulation

Table 7.10   Impact of WTO accession on food self-sufficiency

	Base rate (per cent)	Change (per cent)	Post accession rate (per cent)
Rice	97.6	0.282	97.9
Wheat	91.5	-0.162	91.3
Corn	95.7	-0.553	95.2
Pulse	110.0	-0.122	109.9
Other grains	94.7	-0.197	94.6
Oilseeds	75.9	-0.983	75.1
Vegetable & fruit	100.9	0.349	101.3
Other crops	99.7	-0.002	99.7
Livestock	97.8	0.158	97.9
Fishery	111.4	0.105	111.5
Processed food	93.3	-4.333	89.3
Total grain	95.1	-0.063	95.0
Total food	96.9	-1.548	95.4

Source: Author's CERD simulation

Policy implications and conclusion

The simulation results reported above show that agricultural sectors will be adversely affected by the WTO accession: agricultural output falls, grain and total food self-sufficiency rates decline, and rural-urban income inequality worsens. Although the magnitudes of the impacts are smaller than widely anticipated, the results may still give concern to policymakers. This section discusses policy options to tackle these adverse impacts.

A direct response to the impact may be to provide support to agricultural sectors and farmers. Support can be provided in many forms, with some conforming the WTO rules and others not. Chapter 6 (this volume) considers a scenario where the level of tariff cut on agricultural commodities is halved because of various forms of new barriers such as labelling requirements, reporting procedures, and so on. The simulation of this scenario using CERD indicates that although such a policy may provide some cushioning of agricultural sectors and help to ease the worsening trend of rural-urban income inequality, the overall welfare gain from WTO accession is smaller— the equivalent variation is 6 per cent less than that of the full tariff cut of the WTO accession commitments.

Here, support in the form of a production subsidy is considered. It is assumed that the government subsidises farmers at a level that would keep

Table 7.11   Changes in household income and welfare after WTO accession (per cent)

Indicators	Eastern	Central	Western
Real wages			
Agricultural labour	0.042	-0.103	0.148
Non agricultural labour	2.528	0.675	1.163
Real income			
Rural household	1.196	0.539	0.681
Urban household	1.772	0.742	1.174
Utility			
Rural household	0.972	0.520	0.615
Urban household	1.502	0.834	1.003

Source: Author's CERD simulation

the grain self-sufficiency rate constant after the WTO accession. Simulation of CERD shows that it would cost 7.2 billion yuan to make up for the tiny reduction of 0.06 per cent in the grain self-sufficiency rate (Table 7.12). If the target were to restore the full decline in the total food self-sufficiency rate (1.5 per cent), the cost would be as high as 180 billion yuan.

Increasing transfer payments to inland regions was also simulated in Chapter 6. The rationale behind this policy option is that because the economy as a whole benefits from the WTO accession, there is the possibility of transferring income from one group to another to make all sectors better off. However, the simulation gives a similar result to the scenario of halving the cut in agricultural tariffs. The improvement in regional and rural-urban household income disparity is achieved at the cost of lower overall welfare gains—1 per cent less welfare gain than the full tariff cut scenario. Increases in transfers to the inland regions are made possible through penalising the coastal regions that have a higher rate of return.

Another policy option is to increase agricultural research and development to improve agricultural productivity in China, as suggested by many authors, for example, Fan, Fang and Zhang (2001) and Hazell and Haddad (2001). Simulations of CERD were conducted to find out how much productivity improvement is required to keep the grain and food self-sufficiency rates constant after WTO accession. It was found that the required productivity improvement is 0.32 and 2.85 per cent, respectively, for constant grain and total food self-sufficiency rates. To give an idea about what these figures mean, agricultural TFP in the past may be used as a benchmark. According to Fan (1997), agricultural TFP was 5.1 and 3.91 per cent per annum during the periods 1978-84 and 1984-96. A 2.85 per

Table 7.12   Cost of restoring grain self-sufficiency to the pre-accession level (million yuan)

	Eastern	Central	Western	National
EV (million yuan)	2,117.987	2,368.282	2,686.207	7,172.476

Source: Author's CERD simulation.

cent productivity improvement implies that China should almost double the level of agricultural research and development.

Agricultural productivity may be improved through other channels. One such approach could be reforming the current land tenure system. It is often argued that the average farm size in China is too small to take full advantage of available economies of scale. Without well defined land ownership, farmers are reluctant to invest in land, and the lack of a land market impedes needed adjustments in China's rural economy (Lohmar and Somwaru 2002). However, secure land tenure is also important so that low-income farmers can have food security. Zhou (1998) finds that China's 'dual land system'[4] is superior in terms of avoiding generating new landlessness and inefficient landholding. Further, Ho (2001) finds that many Chinese farmers prefer the current system, especially in poor villages, as it guarantees households access to land. FAO (2002, p.26) also shows that the poverty index is negatively related to land holding, and productivity is negatively related to farm size, that is, smaller farms have higher productivity. These findings may prompt second thoughts about the popular idea of increasing farm size, especially in poor regions.

Finally, further liberalisation is also an option. It was suggested in Chapter 6 that domestic market reform characterised by freer movement of primary factors and commodities may enhance overall welfare and improve regional income disparity.

In addition, China should promote trade liberalisation globally. China has made significant commitments in trade liberalisation of agricultural products and its protection of agriculture is among the lowest in the world. This puts China in a good position in the new round of trade liberalisation negotiations. GTAP simulations show that the removal of agricultural protection in OECD countries improves the competitiveness of Chinese agricultural products in world markets and increases the grain self-sufficiency rate (Yu and Frandsen 2002).

Notes

1 For a review of broader agricultural policies in China, see Colby, Diao and Tuan (2001).
2 For an introduction to the evolution of these policies, see Jiang and Duncan (2001).
3 This seems a puzzle at first glance as the central region has the smallest reduction in almost all agricultural sectors among the three regions. However, as the magnitude of impact varies across sectors and the sectoral composition differs across regions, it turns out that the central region becomes the hardest hit region in the agricultural sector as a whole.
4 Land is divided into self-sufficiency land and responsibility land. Self-sufficiency land, or grain rations land, was equally contracted to households on a per capita basis for planting mainly grains for self-consumption. The responsibility land was contracted on the conditions of fulfilling state output quotas and paying agricultural taxes and fees.

References

Anderson, K., Jikun Huang, and Ianchovichina, E., 2002. 'Long-run impacts of China's WTO accession on farm-nonfarm income inequality and rural poverty', Memo, Centre for International Economic Studies, University of Adelaide.

Brown, L., 1995. *Who Will Feed China? Wake up Call for a Small Planet*, Norton and Company, Washington, DC.

Colby, H., Xingshen Diao and Tuan, F. 2001. *China's WTO accession: conflicts with domestic agricultural policies and institutions*, TMD Discussion Paper No. 68, International Food Policy Research Institute, Washington, DC.

Fan, Shenggeng, 1997. 'Production and productivity growth in Chinese agriculture: new measurement and evidence', *Food Policy*, 22(3):213–28.

Fan, Shenggeng, Cheng Fang and X. Zhang, 2001. *How agricultural research affects urban poverty in developing countries: the case of China*, Environment and Production Technology Discussion Paper 83, International Food Policy Research Institute, Washington, DC.

Food and Agriculture Organisation, 2002. *The State of Food Insecurity in the World 2002*, Food and Agriculture Organisation, Rome.

Food and Agriculture Organisation 2003. *FAO STAT*, Rome: Food and Agriculture Organisation, available at http://www.fao.org.

Hazell, P. and Haddad, L., 2001. *Agricultural research and poverty reduction*, Food, Agriculture and the Environment Discussion Paper 34, International Food Policy Research Institute, Washington, DC.

Ho, P., 2001. 'Who owns China's land? Property rights and deliberate institutional ambiguity', *China Quarterly*, 166:394–421.

Jiang, Tingsong and Duncan, R. 2001. WTO accession and China food policy: a literature survey, Memo, National Centre for Development Studies, The Australian National University, Canberra.

Ke, Bingsheng, 1999. *China's agriculture and policy facing WTO entry*, Paper presented at International Agricultural Trade Research Consortium, China's Agricultural Trade and Policy: Issues, Analysis, and Global Issues, San Francisco, 25–26 June.

Lardy, Nicholas R., 2002. *Integrating China into the Global Economy*, Brookings Institution, Washington, DC.

Lin, Justin Yifu, 1997. 'Institutional reform and dynamics of agricultural growth in China', *Food Policy*, 22(3):201–12.

Lohmar, Bryan, 2002. 'Market reforms and policy initiatives: rapid growth and food security in China', *Food Security Assessment*, GFA-13, Economic Research Service, United States Department of Agriculture, Washington, DC:22–30.

Lohmar, Bryan and Somwaru, A., 2002. 'Does China's land-tenure system discourage structural adjustment?', *China's Food and Agriculture: Issues for the 21st Century*, AIB-775, Economic Research Service, US Department of Agriculture, Washington, DC:39–40.

Lu, Feng, 1997. *Food trade policy adjustment in China and an evaluation of the risk of a food embargo*, Working Paper C1997007, China Centre for Economic Research, Peking University, Beijing.

Nyberg, Albert J. and Rozelle, S., 1999. *Accelerating China's Rural Transformation*, World Bank, Washington, DC.

National Bureau of Statistics, 2002. *Statistical Yearbook of China 2001*, China Statistics Press, Beijing.

Yang, Yongzheng, 2000. Are food embargoes a real threat to China?, Memo, National Centre for Development Studies, The Australian National University, Canberra.

Yu, Wusheng, and Frandsen, S.E., 2002. China's WTO commitments in agriculture: does the impact depend on OECD agricultural policies, Paper presented at the 5th conference on global economic analysis, Taipei, 5-7 June.

Zhang, Ci and Yong Li, 1998. '214 *yi, haoda ge liangkuang kulong* (21.4 billion: what a big hole for grain funds)', *Cai Jing* (Finance), October:12-17.

Zhou, Jian-Ming, 1998. *Is nominal public but de facto private land ownership appropriate? Comparative study among Cambodia, Laos, Vietnam; Japan, Taiwan Province of China, South Korea; China, Myanmar; and North Korea*, EUI Working Paper ECO No. 98/12, Economics Department, European University Institute, Florence.

Zhou, Zhangyue, 2001. 'Joining WTO and China's food security', *Association for Chinese Economic Studies (Australia) Newsletter*, 3.

08 Revisiting the economic costs of food self-sufficiency in China

Ron Duncan, Lucy Rees and Rod Tyers

The comparatively rapid economic growth experienced in the economies of East Asia has been associated with declines in food self-sufficiency and increases in agricultural protection. This has been most noteworthy in Japan, Korea and Taiwan, where the relative decline in economic importance of the agricultural sectors has not been accompanied by a similar decline in the political influence of farmers (Anderson et al. 1986; Krueger 1992). Rapid expansion in these countries' manufacturing and services sectors meant that the relative cost of protecting agriculture, as distinct from the expanding sectors, declined. Moreover, as agriculture shed workers to the modern sectors and farms consolidated, the number of farmers declined, reducing the cost of coordinating their political activity (Downs 1957; Olson 1965). The net effect has been the growth of protection to levels so extreme that the cost burden on their economies is large in spite of their small agricultural sectors.

China's very rapid economic expansion of the past two decades now threatens to yield a political economy similar in this respect to those driving the rises in protection in Japan, Korea and Taiwan. Today, along with widening of per capita income disparities between urban and rural areas, the rhetoric of self-sufficiency is the most prominent weapon of China's protectionists. While WTO accession lessens the risk that they will succeed, the agricultural ministry has been assigned a prominent role in trade policy formation and negotiation, with the power to press for further agricultural protection on self-sufficiency as well as distributional grounds (Anderson et al. 2002; Tong 2003).

Anderson et al. (2002) also put the view that closing the rural-urban per capita income gap will not be achieved by the use of protection of the agricultural sector in that the protection will raise land rents and land values without reducing the numbers of the rural poor. Growth in rural per capita incomes will come, as it has in the industrialised countries, from productivity increases in the agricultural sector, the development of off-farm income opportunities for rural households and the migration of workers to urban areas (Chang and Tyers 2003). In this paper we examine the costs that would be borne by the Chinese economy if it attempted to maintain or increase its levels of self-sufficiency in agricultural commodities, rather than continuing to open its agricultural markets. We also address the Anderson et al. view about urban-rural income distribution.

In an early analysis, Yang and Tyers (1989) used a global agricultural sector model to examine the implications of rapid income growth on the composition of food consumption and the implications of this for food self-sufficiency. They found that the anticipated redistribution of consumption toward livestock products would raise import demand for feed grains and that this would make the maintenance of self-sufficiency through protection very costly. Because their analysis was restricted to the agricultural sector, however, they could not examine the redistributive and overall contractionary effects of the protection needed to maintain self-sufficiency. In this paper we do this using a more general global model, the scope of which is the entire economy. Our model is adapted originally from GTAP,[1] which allows for a multi-region, multi-product general equilibrium analysis. Following Yang and Tyers (2000), to this GTAP base is added independent representations of governments' fiscal regimes, with both direct and indirect taxation.[2]

We begin by using this model to project the world economy to 2010,[3] noting the trends in the self-sufficiency rates for agricultural products in China. We then ask two questions. First, if China's food self-sufficiency rates are to be held constant to 2010, will increases in protection be required? Second, what increases in protection would be required to achieve self-sufficiency by 2010 and what would be the contractionary and distributional effects of this protection? Consistent with Yang and Tyers (1989) our projections to 2010 show substantial declines in Chinese food self-sufficiency, particularly for livestock products and feed grains, so that substantial increases in protection are needed to maintain the 2001 levels. To achieve self-sufficiency in all agricultural products by 2010 considerable further protection would be required. Moreover, this protection would be both contractionary and redistributive, and it would retard growth in other sectors. The strength of the results notwithstanding, a sensitivity analysis shows they rest quite heavily on the precision with which some parameters are measured, particularly the income elasticity of demand for livestock products.

Modelling the Chinese and world economies, 2001-2010

The model is a modified version of that introduced by Hertel (1997), which is global in scope. It offers the following useful properties

- a capital goods sector in each region to service investment
- explicit savings in each region, combined with open regional capital accounts that permit savings in one region to finance investment in others
- multiple trading regions, goods and primary factors
- product differentiation by country of origin
- empirically based differences in tastes and technology across regions
- non-homothetic preferences
- explicit transportation costs and indirect taxes on trade, production and consumption.

All individual goods and services entering final and intermediate demand are constant elasticity of substitution (CES) blends of home products and imports. Government spending is also such a composite, though its mix of goods, as between different product groups and between home products and imports, differs from that embodied in private consumption. In turn, imports are CES composites of the products of all regions, the contents of which depend on regional trading prices. Savings are pooled globally and investment is then allocated between regions from the global pool. Within regions, investment places demands on the domestic capital goods sector, which is also a CES composite of home-produced goods, services and imports in the manner of private consumption and government spending.

Our modifications (Yang and Tyers 2000) make regional governments financially independent by incorporating direct tax regimes. The private saving and consumption decision is represented by a reduced form exponential consumption equation with wealth effects included via the dependence of consumption (and hence savings) on the interest rate. Each region then contributes its total domestic (private plus government) saving, $S_D = S_P + S_G$, to the global pool from which each region's corresponding investment is derived.[4] These relations imply the balance of payments identity at the regional level, which sets the current account surplus equal to the capital account deficit: $X - M = S_P + S_G - I$.[5]

From the global savings pool, investment is allocated across regions and it places demands on capital goods sectors in each region. A high level of global 'capital' mobility is assumed.[6] The allocation to region j (net investment in that region) depends positively on the long-run change in the average rate of return on installed capital, r_j^e, which, in turn, rises when the marginal product of physical capital is increased.[7] This allocation falls when the opportunity cost of financing capital expenditure, the region's real interest rate, r_j, rises. This depends, in turn, on a global capital market clearing interest rate, r^w, calculated such that global savings equals global investment : $\Sigma_j S_j^D = \Sigma_j I_j (r_j^e, r_j)$. Here I_j is nominal gross investment in region j.[8] The region's home interest rate is then $r_j = r^w(1 + \pi_j)$ where π_j is a region-specific interest premium, thought to be driven by risk factors not incorporated in this analysis. The investment demand equation for region j then takes the form

$$I_j = \delta_j K_j + I_j^N = \delta_j K_j + \beta_j K_j \left(\frac{r_j^e}{r_j}\right)^{\varepsilon_j} = K_j \left[\delta_j + \beta_j \left(\frac{r_j^e}{r_j}\right)^{\varepsilon_j}\right] \tag{1}$$

where K_j is the (exogenous) base year installed capital stock, δ_j is the regional depreciation rate, β_j is a positive constant and ε_j is a positive elasticity. Critically, investment in any region responds positively to changes that raise a region's marginal product of physical capital and hence the region's average return on installed capital.[9] Other things equal, then, in the long-run applications in this paper improvements in trans-sectoral efficiency, such as might stem from a trade reform, raise capital returns permanently and hence they raise r_j^e.

The long-run closure adopted for the model in this paper differs from the short-run closure used by Yang and Tyers (2000) and by Rees and Tyers (2002) in that there are no nominal rigidities (no rigidity of nominal wages) and hence full employment is retained and money is neutral;[10] larger production and consumption elasticities are used to reflect the additional time for adjustment;[11] physical capital is not sector specific, it redistributes across sectors to equalise rates of return; China's capital controls are ignored; and changes in government revenue associated with tariff increases are assumed to not be offset via direct (income) tax changes, with the result that the fiscal deficit changes, so that the ratios of government revenue and expenditure to GDP are endogenous while the average direct tax rate is exogenous.[12]

Data and parameters

The regions, primary factors and sectors identified in our analysis are listed in Table A8.1. Considering regions first, we draw on the now well-known GTAP Version 5 global database for 1997, which divides the world into 66 countries and regions. Although this database separates mainland China from Taiwan province, it amalgamates Hong Kong with the mainland.[13] Our further aggregation of mainland China with Taiwan province overlooks effects that are internal to these regions but such effects are not our focus. Instead, we seek to illustrate the strong interaction between self-sufficiency rates, agricultural protection and overall economic performance. These interactions are important for all the economies of East Asia.

Turning to primary factors, skill is separated from raw labour on occupational grounds, with the 'professional' categories of the International Labour Organization (ILO) classification included as skilled.[14] The structure of factor demand has skill and physical capital as complements. This enables the model to represent the links between skill availability, capital returns and investment that are important in China, which has large skilled and unskilled labour forces that are increasingly mobile between sectors.[15] Finally, the sectoral breakdown we have chosen aggregates the 57 sectors in the database to our more manageable 14, offering the most detail in agricultural and marine products.

The most sensitive parameters in determining the trends in self-sufficiency through time prove to be the elasticities of demand. Our model employs the original GTAP CDE (constant difference of elasticities of substitution) system. Its non-homotheticity is a particular asset for our purpose in that it permits a range of income elasticities to exist, either side of unity. While this system is more general than the homothetic ones often used in such models, it is still restrictive in the width of the range compared to still more general systems. The CDE system is employed here because of its parametric economy.[16] The income elasticities thus embodied in our demand parameters are listed in Table A8.2. Because of the restrictiveness of the CDE system the lower bound for the income elasticity of cereal grains cannot be set below 0.1, despite evidence suggesting that it is now negative (Ito et al. 1991; Peterson et al. 1991). As a result, in our analysis the span between the income elasticities of livestock products and processed foods, which are superior goods, and those of cereal grains is likely to be smaller than the truth. One consequence of this is that our results probably underestimate the relative growth in demand for livestock products and processed foods and hence they underestimate the associated derived demand for cereal feeds and other agricultural inputs. This offers a second downward bias in our estimates of the cost of achieving and maintaining agricultural self-sufficiency. Having said this, we did recalibrate the CDE parameters so as to minimise this downward bias.[17] The span between the income elasticities of livestock products and cereal grains was increased from 0.8, which was based on 1997 numbers in the GTAP Version 5 database, to 1.5. The recalibration involved calculating the new values for the expansion parameter and the substitution parameter in the CDE minimum expenditure function to imply income elasticities close to our target values.

Because the simulations are decade-long projections, it was necessary to use long-run elasticities of substitution in product and service demand. These are listed in Table A8.3. They are larger than the standard GTAP demand elasticities, which are designed for simulations over the medium run.

Constructing the 2010 world economy

As indicated earlier, our numerical analysis originates with the GTAP Version 5 global database for 1997. Rees and Tyers (2002) use a short run version of the above model to examine key changes in the Chinese economy between 1997 and 2001, including substantial trade reforms. We commence with their simulated image of the 2001 world economy and proceed to use it as a base from which to construct a reference projection of the 2010 economy. This latter step is, however, a substantial task in itself. Not only does it require assumptions about the exogenous growth rates of primary factor supplies like labour, skill and physical capital, it also rests importantly on assumptions about the pace of technical change.

The pace of technical change is incorporated by constructing a set of region-wide total factor productivity growth rates that are consistent with forecast changes in populations and labour supplies on the one hand and a set of non-controversial regional GDP growth rates on the other. We do this by making GDP growth rates exogenous in the first simulation and a corresponding set of region-wide total factor productivity growth rates endogenous. In the subsequent counterfactual simulations, GDP is made endogenous in each region but the corresponding total factor productivity growth emerging from the reference simulation is held constant. This ensures that, when subsequent simulations incorporate rising agricultural protection to achieve self-sufficiency, total income growth in each region adjusts. This approach to estimating the effects of new agricultural protection is conservative in that, by making total factor productivity coefficients independent of protection rates we expect to underestimate their contractionary effects.[18] The exogenous population, labour force and capital accumulation rates are listed in Table A8.4, along with the implied rates of total factor productivity growth.

Trends in Chinese agricultural self-sufficiency

The dependence of China's domestic markets on trade is most clearly evident from the ratio of exports to domestic value added in each sector, or the corresponding ratio of competing imports to value added. Estimates of these for 2001 are listed in Table A8.5. They show that the most export-

209

oriented food-related sector, the 'beverages' group, also faces the greatest level of import penetration. This sector is characterised by differentiated products and intra-industry trade, which, in China's case, appears roughly to balance out. Livestock products, 'processed food' and the 'other crops' group, which includes the key inputs to the livestock products group, grains and soybeans, are less trade-oriented. Importantly, however, they do not enjoy the balance of exports and competing imports that occurs with beverages. They are import competing and therefore sectors in which China is less than self-sufficient.

To track self-sufficiency, we offer a cruder but more widely used measure: domestic output relative to domestic 'disappearance'. For a particular product group we compare the value of output at producers' prices, Y, to this value of output supplemented by net imports $(M - X)$, where the latter are also valued at domestic prices. Our self-sufficiency ratio is then

$$SSR = \frac{Y}{Y + M - X} \tag{2}$$

The values taken by this ratio in the original database of 1997, and the constructed ones for 2001 and 2010, are listed in Table A8.6. Measured in this way, departures from self-sufficiency are largest amongst agricultural products for the beverages group, 'other crops', and the processed food category. Moreover, the shortfall relative to full self-sufficiency is projected to expand through to 2010. Though smaller in magnitude, expansions in this shortfall also occur for marine products and livestock. If self-sufficiency is a policy objective these results suggest increased agricultural protection will be required in order to prevent deterioration in the ratios through time.

Achieving agricultural self-sufficiency through protection

If the Chinese government were to adopt self-sufficiency as an objective, to be achieved through border protection, it might choose to implement a policy regime that would prevent any negative trend in self-sufficiency ratios for product groups that are already import competing. Alternatively, it might seek a regime that would return the economy to self-sufficiency in all agricultural products. We examine these two possible policy scenarios by constructing alternative simulations of the 2010 global and Chinese economies. In the first, for import-competing agricultural products, the self-sufficiency ratio is held constant at the 2001 level through the implementation of a source-generic tariff in each sector that is additional to existing protection. In the second, these sectors are made to return to full self-sufficiency by 2010 through the more zealous application of such tariffs. The results from these simulations are presented in Table A8.7. To show their power the additional tariffs required are listed in the form of proportional changes to nominal protection coefficients (ratios of domestic post tariff to border prices).

Because the declines in self-sufficiency projected to 2010 are significant, the tariffs necessary to retain 2001 self-sufficiency rates are substantial, particularly for the beverages, 'other crops' and livestock product groups. These taxes on imports are, in fact, taxes on all China's trade which not only reduce food imports. They also reduce China's exports, causing exporting industries to contract. Overall, the increased protection induces a one per cent contraction in GDP along with some restructuring across industrial sectors, favouring the newly protected agricultural industries mostly at the expense of manufacturing.

When the additional tariffs are raised to levels that yield full self-sufficiency in the previously import competing agricultural sectors, further substantial changes occur. The tariffs required by 2010 are very large, particularly on imports in the livestock products, processed food and 'other crops' groups. These distort incentives in the economy substantially, shifting resources into agriculture and contracting both the manufacturing and service sectors, the latter being the primary growth sectors in the economy. Throughout the economy this decline in allocative efficiency reduces returns to installed capital and therefore investment. The level of 2010 GDP is lower by nearly 2 per cent.

The role of the increased tariffs in reducing both imports and exports is clear from Table A8.8. The tariffs that would achieve agricultural self-sufficiency in 2010 also reduce exports from China's growth powerhouse, its light manufacturing industries, by half.[19] While this change is necessary to retain a balance of payments, the mechanism through which it operates is the response of firms and households to the domestic price and associated real exchange rate changes caused by the tariff increases. Domestic resources are reallocated to the agricultural sector, raising costs in manufacturing and reducing the international competitiveness of China's manufacturing industries. The resulting misallocation of labour is particularly striking (Table A8.9). The higher tariffs cause employment in agricultural and food processing activities to be substantially greater, at the expense primarily of light manufacturing.

The implications of the tariffs for domestic income distribution are indicated by the effects on real unit factor rewards summarised in Table A8.10. Higher agricultural tariffs raise land rents by a considerable margin but reduce real wages and capital returns. In China, rural and urban wages are linked by an, albeit imperfect, labour market (Chang and Tyers 2003). The most labour-intensive sector (light manufacturing) is hurt by the tariffs. In the high tariff scenario, therefore, light manufacturing grows less, so that by 2010, fewer workers are employed in it. Real wages grow less in both agriculture and the modern sector. This is true for both production and skilled workers, and it is also true for the owners of physical capital. Again, the capital losses occur because the industries that are hurt by the tariffs are more capital intensive than agriculture. Indeed, the decline in unit capital rewards is serious for China, since this redirects domestic savings abroad and retards future investment and overall growth. In the end, land holders are the only winners from the tariffs.

We might well ask, then, what is gained by the self-sufficiency. Would food be more readily available in China? No. China's 2010 prices of imported foods would be higher with the increased tariffs by up to 60 per cent and even home-produced food products would be more expensive by at least 10 per cent. The key consequence of political significance would be a reduction in interdependence with the global economy — reduced reliance on global markets. But this cuts two ways. Reduced reliance on food imports means curtailing the principal source of China's overall economic growth since the 1980s — access to foreign markets for its labour-intensive goods. Curtailed exports reduce its capital returns, thereby cutting incentives for investment and, ultimately, the growth rate of its overall economy.

Sensitivity analysis

The simulations presented offer just one representation of the Chinese economy and our point estimates of the effects of higher tariffs are subject to substantial error. As it turns out, our estimates of these effects depend most sensitively on just a few parameters, the most critical being the income elasticity of demand for livestock products. This parameter is most important for two reasons. First, livestock products are superior goods the demand for which can be expected to grow disproportionately with income and hence this sector is particularly likely to lose self-sufficiency. Second, livestock feeds are also imported. The faster the demand for livestock products grows the greater is the derived demand for feed grains and the self-sufficiency rate in the import competing feed grain sector can also be expected to decline. The magnitudes of the tariffs required to stem this decline therefore depend critically on the income elasticity of demand for livestock.

By repeating the simulations in Table A8.7, assuming a smaller value for this elasticity, we have calculated the elasticities of sensitivity of key consequences to the income elasticity of demand for livestock products. The results are displayed in Table A8.11. Each of the elasticities shown in the table indicates the per cent by which the nominated variable changes when the income elasticity of demand for livestock products is raised by 1 per cent. Thus, were the income elasticity of demand for livestock products larger by 10 per cent (1.79 instead of 1.63), without new tariffs 2010 livestock imports would be larger by approximately 18 per cent and feed grain imports would be correspondingly larger by 12 per cent. To achieve agricultural self-sufficiency in 2010 this would mean that the nominal protection coefficient on livestock products would have to be higher by 7 per cent (the home price level would have to rise by 7 per cent relative to the import price). This would result in smaller light manufacturing output, by 7 per cent and smaller light manufacturing exports by 14 per cent. The income elasticity of demand for livestock products is clearly a key parameter — one that should be carefully monitored.

Conclusion

A global comparative static model is used to project the world economy to 2010, noting the trends in the self-sufficiency rates for agricultural products in China. If there is no change in China's trade policy regime, agricultural self-sufficiency rates are shown to decline and this decline is shown to be significant in all agricultural sectors except fisheries. Large changes in protection would be needed to hold the line at 2001 self-sufficiency levels. To achieve full self-sufficiency in all agricultural products by 2010 substantial further new protection would be required. This protection would be both contractionary and redistributive, harming worker households and retarding growth in the modern sector of China's economy. Moreover, it would raise domestic prices of food in China, restricting the availability of food products. This, plus slower growth, seems a high price to pay for a modest reduction in China's interdependence with international markets.

Because livestock products, as a group, are very income elastic, these results prove to be particularly sensitive to their income elasticity of demand. This elasticity has received less attention from the consumption literature than others, particularly that of rice, yet small changes in it lead to substantially different projections of agricultural self-sufficiency rates, the tariffs required to achieve self-sufficiency and the export performance of the modern sector of the economy. Further analysis of the consequences of any move to achieve and maintain agricultural self-sufficiency in China needs to be informed by new and high quality estimates of this and related income elasticities.

Finally, we note two downward biases in our estimation of the economic costs of achieving self-sufficiency. First, the links between productivity growth and trade reform are ignored, so that when new protection is applied with the objective of achieving agricultural self-sufficiency, no associated sacrifice of total factor productivity is imposed. Second, even though we have recalibrated the standard GTAP CDE demand system, the range of income elasticities between the superior food items in 'processed food' and 'livestock products' on the one hand and cereals on the other is still likely to be smaller than in reality. As the sensitivity analysis suggests, the economic costs of a tariff regime to restore self-sufficiency are substantially higher if the income elasticity of demand is raised above our estimate of 1.63. These two effects lead to an underestimation of the growth in demand for processed food and livestock products as well as a corresponding underestimation of growth in the consumption of their principal intermediate inputs, namely cereals, soybeans, fruits and vegetables. The result is an overestimation of future self-sufficiency ratios and, thus, an underestimation of the cost of raising these to unity via protection.

Notes

1 A detailed description of the original model is provided by Hertel (1997).
2 The nominal side of the model is not used in this analysis, which focuses on long-run changes. No nominal rigidities are introduced and so money neutrality prevails.
3 Our projection employs a similar approach to that used by Ianchovichina and Martin (2002).
4 Private saving is derived as the difference between disposable income (Y-T) and consumption expenditure, where real consumption is determined in a Keynesian reduced form equation that takes the form

$C = \gamma \; r^{\delta} \; [Y - T]^{\mu}$, where r is the real interest rate.

5 Note that there is no allowance for interregional capital ownership in the starting equilibrium. At the outset, therefore, there are no factor service flows and the current account is the same as the balance of trade.
6 By which it is meant that households can direct their savings to any region in the world without impediment. Installed physical capital, however, remains immobile even between sectors.
7 rje is the expected rental rate on physical capital, adjusted for depreciation and divided by the price of capital goods to yield a unitless net rate of return.
8 Before adding to the global pool, savings in each region is deflated using the regional capital goods price index and then converted into US$ at the initial exchange rate. The global investment allocation process then is made in real volume terms.
9 This investment relation is similar to Tobin's Q in the sense that the numerator depends on expected future returns and the denominator indicates the current cost of capital replacement.
10 The money and asset markets represented by Yang and Tyers (2000) play no role here, as money is neutral and we report only real quantities or relative prices.
11 The long-run elasticity set used is the same as that employed by Tyers and Yang (2000).
12 When tariff rates are raised to achieve food self-sufficiency, this implies that government revenue increases faster than government spending and there is a small fiscal

contractionary effect which tends to lower interest rates and encourage investment in China.

13 Detailed descriptions of the GTAP database's content and sources as they relate to China are available in Gehlhar (2002), which describes the integration of the data for Hong Kong with that of the mainland and discusses the entrepot nature of some of Hong Kong's trade.

14 See Liu et al. (1998) for the method adopted.

15 For further discussion of the role and representation of skill-capital complementarity, see Tyers and Yang (2000).

16 For a discussion of the CDE system and its more complicated alternatives, see Huff et al. (1997).

17 The method for recalibrating the CDE parameters was provided to us by Dr Yongzheng Yang of the IMF, to whom thanks are due.

18 For analyses of the links between productivity and protection, see Chand et al. (1998), Chand (1999) and Stoeckel et al. (1999).

19 Note that China's comparative advantage in light manufacturing declines through time, as does the level of employment in this sector. This is because the growth rates of China's population and production labour forces are slower than those of its populous Asian neighbours. In the reference simulation its production labour to skill ratio declines substantially by 2010.

References

Anderson, K. and Hayami, Y., 1986. *Political Economy of Agricultural Protection: the experience of East Asia*, Allen and Unwin, Sydney, reprinted in Chinese 1996.

Anderson, K., Huang, J. and Ianchovichina, E., 2002. 'Long-run impact of China's WTO accession on farm-non-farm income inequality and rural poverty', Paper presented at the World Bank's Conference on China's WTO Accession, Policy Reform and Poverty Alleviation, Beijing, 28-29 June.

Chand, S., 1999. 'Trade liberalisation and productivity growth: time series evidence from Australian manufacturing', *The Economic Record*, 75(228): 28-36.

Chand, S., McCalman, S.P., and Gretton, P., 1998. 'Trade liberalisation and manufacturing industry productivity growth', in *Microeconomic Reform and Productivity Growth*, Workshop Proceedings, Productivity Commission and The Australian National University, AusInfo, Canberra:239-81.

Chang, J. and Tyers, R., 2003. 'Trade reform, macroeconomic policy and sectoral labour movement in China', in Garnaut, R. and Song, L. (eds) *China 2003: New Engine of World Growth*, Asia Pacific Press, Canberra: 231-75.

Downs, A., 1957. *An Economic Theory of Democracy*, Harper and Row, New York.

Gehlhar, M.J., 1994. Economic growth and trade in the Pacific Rim: an analysis of trade patterns, Doctoral dissertation, Purdue University, West Lafayette (unpublished).

——, 2002. 'Hong Kong's re-exports', in Dimaranan, B.V., and McDougall, R.A. (eds), *Global Trade, Assistance and Production: the GTAP 5 Data Base*, Centre for Global Trade Analysis, Purdue University, West Lafayette. Available online at www.gtap.agecon.purdue.edu/databases/v5/v5_doco.asp.

Graham, B. and Tyers, R., 2002. *Global population forecast errors, economic performance and food demand: preliminary simulations*, Working Papers in Economics and Econometrics No. 418. Also presented at the 46th Annual Conference of the Australian Agricultural and Resource Economics Society, Adelaide.

Hertel, T.W. (ed.), 1997. *Global Trade Analysis Using the GTAP Model*, Cambridge University Press, New York.

Huff, K.M., Hanslow, K., Hertel, T.W., and Tsigas, M.E., 1997. 'GTAP behavioural parameters', in T.W. Hertel (ed.), *Global Trade Analysis Using the GTAP Model*, Cambridge University Press, New York.

Ianchovichina, E. and Martin, W., 2001. 'Trade liberalisation in China's accession to the WTO', *Journal of Economic Integration* 16(4):421-45.

——, 2002. 'Economic impacts of China's accession to the WTO', World Bank, Washington, DC.

Ito, S., Peterson, E.W.F. and Grant, W.R., 1991. 'Rice in Asia: is it becoming an inferior good?', *American Journal of Agricultural Economics*, 73(2):528-32.

Krueger, A.O., 1992. *The Political Economy of Agricultural Pricing Policy: Volume 5, A Synthesis of the Political Economy in Developing Countries*, Johns Hopkins University Press, Baltimore.

Lin, H.C., Chung, L. and Liou, R.W., 2002. 'Taiwan', in Dimaranan, B.V., and McDougall, R.A. (eds), *Global Trade, Assistance and Production: the*

GTAP 5 Data Base, Centre for Global Trade Analysis, Purdue University, West Lafayette. Available online at www.gtap.agecon.purdue.edu/databases/v5/v5_doco.asp.

Liu, H.C., Van Leeuwen, J.N., Vo, T.T., Tyers, R., and Hertel, T.W., 1998. 'Disaggregating labour payments by skill level in GTAP', Technical Paper No.11, Centre for Global Trade Analysis, Purdue University, West Lafayette. Available online at http://www.agecon.purdue.edu/gtap/techpapr/tp-11.htm.

Olson, M., 1965. *The Logic of Collective Action: Public Goods and the Theory of Groups*, Cambridge Massachusetts: Harvard University Press.

Peterson, E.W.F., Jin, L., and Ito, S., 1991. 'An econometric analysis of rice consumption in the People's Republic of China', *Agricultural Economics* 6(1):67-78.

Rees, L. and Tyers, R., 2002. 'Trade reform in the short run: China's WTO accession', *Journal of Asian Economics*, 15(1):1-31. Re-published with permission as Chapter 9 (this volume).

Stoeckel, A., Tang, K.K., and McKibbin, W., 1999. *The gains from trade liberalisation with endogenous productivity and risk premium effects*, Technical Paper, Centre for International Economics, Canberra, presented at the seminar on Reasons versus Emotion: Requirements for a Successful WTO Round, Seattle, 2 December.

Tong, J., 2003. 'WTO commitment: further marketisation and trade liberalisation', in Garnaut, R. and Song, L. (eds) *China 2003: New Engine for World Growth*, Asia Pacific Press, Canberra:141-51.

Tyers, R. and Yang, Y., 2000. 'Capital-skill complementarity and wage outcomes following technical change in a global model', *Oxford Review of Economic Policy*, 16(3):23-41.

Wang, Z., Zhai, F., and Xu, D., 2002. 'China', in Dimaranan, B.V. and McDougall, R.A. (eds), *Global Trade, Assistance and Production: the GTAP 5 Data Base*, Centre for Global Trade Analysis, Purdue University, West Lafayette, www.gtap.agecon.purdue.edu/databases/v5/v5_doco.asp.

Yang, Y. and Tyers, R., 1989. 'The economic cost of food self-sufficiency in China', *World Development*, 17(2):237-53.

——, 2000. *China's post-crisis policy dilemma: multi-sectoral comparative static macroeconomics*, Working Papers in Economics and Econometrics No. 384, The Australian National University, Canberra.

Appendix

Table A8.1 Model structure

Regions	Primary factors
1. China, including Hong Kong and Taiwan	1. Agricultural land
2. Vietnam	2. Natural resources
3. Other ASEAN	3. Skill
4. Japan	4. Labour
5. Korea	5. Physical capital
6. Australia	
7. United States	
8. European Union[a]	
9. Rest of World	

Sectors [b]

1. Paddy rice
2. Beverages (product 8 OCR, 'crops nec')
3. Other crops (wheat, other cereal grains, vegetables, fruits, nuts, oil seeds, sugar cane and sugar beet, plant based fibres and forestry)
4. Livestock products (cattle, sheep, goats, horses, wool, silk-worm cocoons, raw milk, other animal products)
5. Fish (marine products)
6. Energy (coal, oil, gas)
7. Minerals
8. Processed food (meat of cattle, sheep, goats and horses, other meat products, vegetable oils and fats, dairy products, processed rice, processed sugar, processed beverages and tobacco products)
9. Light manufacturing (textiles, wearing apparel, leather products and wood products)
10. Other manufacturing (paper products and publishing, petroleum and coal products, chemicals, rubber and plastic products, other mineral products, ferrous metals, other metals, metal products, motor vehicles and parts, other transport equipment, electronic equipment, other machinery and equipment, other manufactures)
11. Transport (sea transport, air transport and other transport)
12. Infrastructure services (electricity, gas manufacturing and distribution, and water)
13. Construction and dwellings
14. Other services (retail and wholesale trade, communications, insurance, other financial services, other business services, recreation, other private services, public administration, defence, health and education)

---

[a]The European Union of 15.
[b]These are aggregates of the 57 sector GTAP Version 5 database.

Table A8.2   Income elasticities of final demand implied by the model's demand parameters[a]

Merchandise sector	
Rice	0.19
Beverages	0.66
Other crops	0.80
Livestock	1.63
Fish	0.71
Processed food	0.76
Minerals	1.11
Energy	1.24
Light manufacturing	0.85
Heavy manufacturing	1.04
Transport	1.10
Infrastructure services	1.00
Construction	1.05
Other services	1.11

Note: [a] The raw demand parameters are for the CDE (constant difference of elasticities of substitution) system. These income elasticities of demand are implied by those parameters. Note that a major source of demand for all these product categories is as intermediate inputs. These elasticities only characterise the link between the disposable income of households and the final consumption of these product groups.
Source: The original 1997 numbers are aggregated from the 57 commodity categories in the GTAP Version 5 global database, published in 2000. Modifications to the 1997 numbers are detailed in the text.

Table A8.3   Elasticities of substitution in product and service demand[a]

Merchandise sector	Between home goods and generic imports	Between imports according to source
Rice	4.9	9.2
Beverages	4.9	9.2
Other crops	4.9	9.2
Livestock	4.9	9.2
Fish	4.9	9.2
Processed food	4.9	9.2
Minerals	5.6	11.2
Energy	5.6	11.2
Light manufacturing	5.4	11.8
Heavy manufacturing	5.7	11.9
Transport	3.8	7.6
Infrastructure services	3.9	7.6
Construction	3.8	7.6
Other services	4.0	7.7

Note: [a] These long-run elasticities of substitution in product and service demand are larger than the standard GTAP values, reflecting the long-run nature of the simulations.
Source: Values are based on the calibration experiments discussed by Gehlhar, M.J., 1994. Economic growth and trade in the Pacific Rim: an analysis of trade patterns, Unpublished doctoral dissertation, Purdue University and aggregated using the modified GTAP Version 5 database.

Table A8.4   Reference rates of population, labour supply, capital accumulation, productivity and GDP growth, 2001-2010 [a] (per cent/yr)

Region	Population	Production labour	Skilled labour	Physical capital	Total factor productivity [b]	GDP [c]
China	0.84	1.2	2.8	8.8	3.2	8.0
Japan	0.06	-0.19	-0.71	3.0	1.0	2.0
Korea	1.7	2.0	8.9	3.8	0.1	4.0
Vietnam	1.4	2.8	3.1	7.8	2.0	7.0
Other Asia	1.9	2.1	7.2	3.9	3.0	6.5
European Union	0.0	-0.1	0.1	2.7	2.8	4.0
USA	1.1	1.1	1.1	4.1	1.2	3.5
Australia	1.0	1.1	1.0	4.5	1.4	4.0
Rest of world	1.5	1.8	4.8	4.2	.23	3.5

Notes: [a] The rates of growth of population, labour supply and capital accumulation are derived from the sources given below and common to all simulations. In the reference simulation the GDP growth rates are made exogenous targets and the model calculates the sector-generic total factor productivity growth rates consistent with these targets. In the subsequent counterfactual simulations the total factor productivity growth rates are fixed as shown in this table and GDP levels are then endogenous.
[b] Derived in the reference simulation for consistency of factor accumulation and projected GDP growth rates.
[c] These values apply to the reference simulation only. In the subsequent counterfactual simulations GDP is endogenous and departures from these values are reported subsequently.
Source: Factor accumulation rates are drawn from Ianchovichina, E. and Martin, W., 2002. 'Economic impacts of China's accession to the WTO', World Bank, Washington DC, and reference GDP projections from Graham, B. and Tyers, R. 2002. *Global population forecast errors, economic performance and food demand: preliminary simulations,* Working Papers in Economics and Econometrics No. 418. Also presented at the 46th Annual Conference of the Australian Agricultural and Resource Economics Society, Adelaide.

Table A8.5 Trade to value added ratios by industry in 2001[a]

	Exports to value added ratio	Competing imports to value added ratio
Rice	0.01	0.00
Beverages	0.67	0.67
Other crops	0.03	0.11
Livestock	0.04	0.07
Food processing	0.18	0.55
Fish	0.06	0.07
Minerals	0.05	0.21
Energy	0.20	0.58
Light manufacturing	1.58	0.68
Heavy manufacturing	0.94	1.10
Transport	0.24	0.19
Infrastructure services	0.02	0.02
Construction	0.01	0.02
Other services	0.11	0.07

Note: [a] These are quotients of the value of exports or imports at world prices and domestic value added in each industry. They are from the 2001 global database (simulated, based on the trade reforms of 1997-2001 as per Rees, L. and Tyers, R., 2002. *Trade reform in the short run: China's WTO accession*, Working Papers in Economics and Econometrics No. 423, Australian National University. Also presented at the workshop on Agricultural Market Reform in China, Beijing University, 23 September).
Source: The GTAP Version 5 Database, as modified by simulations described in the text.

Table A8.6   Implied Chinese self-sufficiency rates, past and projected, 1997 2001 and 2010 (per cent)[a]

Merchandise sector [b]	1997	2001	2010
Rice	100	100	100
Beverages	98	100	91
Other crops (incl. feed grains)	95	94	89
Livestock	99	99	95
Processed food	92	88	83
Fish	99	99	99
Minerals	95	95	94
Energy	80	80	80
Light manufacturing	123	123	123
Heavy manufacturing	95	95	95

Notes: [a] Self-sufficiency rates are calculated from values of domestic output, Y, imports, M, and exports, X, evaluated at domestic producer prices, from the formula: $SSR=Y/(Y+M-X)$.
[b] The services sectors are represented in the model, as indicated in Table A8.1. Since trade in these is relatively costly, self-sufficiency rates are near unity. They are, in any case, not the focus of this analysis and so they are omitted from this table.
Sources: The original 1997 numbers are aggregated from the 57 commodity categories in the GTAP Version 5 global database, published in 2000. Those for 2001 are based on short run projections from 1997, as conducted by Rees and Tyers (Chapter 9, this volume). For 2010, rates are from the reference simulation discussed in the text.

Table A8.7 Effects of protection to raise agricultural self-sufficiency in 2010

	Reference change, 2001-2010, per cent	Departure from reference 2010, per cent	
		Protection to hold self-sufficiency rates at 2001 levels	Protection to achieve full self-sufficiency
Rise in agricultural nominal protection coefficient			
Rice	-	0.0	0.0
Beverages	-	35.4	50.6
Other crops (incl. feed grains)	-	19.2	72.7
Livestock	-	39.2	78.7
Processed food	-	11.1	67.3
Fish	-	16.1	31.9
Real effective exchange rate	-0.04	2.6	4.3
Terms of trade	-0.38	-0.2	0.3
Return on installed capital	21.2	-0.4	-0.4
Investment	202.6	1.4	1.4
Real gross sectoral output			
Rice	41.4	2.0	8.5
Beverages	61.6	7.1	7.3
Other crops	56.5	3.7	8.9
Livestock	95.2	0.4	0.3
Processed food	52.5	3.8	15.6
Fish	53.0	1.2	2.8
Minerals	108.1	-1.5	-2.2
Energy	78.5	-1.1	-1.7
Light manufacturing	74.7	-22.7	-25.0
Heavy manufacturing	115.5	-3.4	-4.7
Transport	97.7	-1.7	-2.3
Infrastructure services	103.8	-1.6	-1.7
Construction	185.0	1.5	1.4
Other services	92.2	3.9	3.3
GDP	99.9	-1.0	-1.7

Source: Model simulations described in the text.

Table A8.8 Effects of protection to raise agricultural self-sufficiency in 2010 on imports and exports

	Reference change, 2001-2010, per cent	Departure from reference 2010, per cent	
		Protection holds self-sufficiency rates at 2001 levels	Protection to achieves full self-sufficiency
Imports			
Rice	70	0	19
Beverages	101	-75	-87
Other crops (incl. feed grains)	159	-49	-90
Livestock	260	-80	-97
Processed food	99	-28	-87
Fish	65	-43	-67
Minerals	117	-3	-3
Energy	83	-5	-6
Light manufacturing	71	-12	-10
Heavy manufacturing	124	-1	1
Transport	105	3	4
Infrastructure services	80	-1	-1
Construction	208	2	3
Other services	130	-4	-2
Exports			
Rice	-24	-30	-55
Beverages	4	-54	-71
Other crops	-24	-37	4
Livestock	-35	-91	-100
Processed food	25	-32	-59
Fish	36	-41	-58
Minerals	39	-3	-4
Energy	53	3	3
Light manufacturing	53	-51	-53
Heavy manufacturing	79	-14	-17
Transport	66	-14	-17
Infrastructure services	113	-5	-7
Construction	128	-4	-7
Other services	-99	-99	-99

Source: Model simulations described in the text.

Table A8.9   Effects on the employment of production workers of additional protection to raise agricultural self-sufficiency in 2010

	Reference change, 2001–2010, per cent	Departure from reference 2010, per cent	
		Protection to hold self-sufficiency rates at 2001 levels	Protection to achieve full self-sufficiency
Rice	0.4	2.9	11.2
Beverages	6.9	11.0	14.4
Other crops	10.0	4.5	10.7
Livestock	36.8	2.0	3.8
Processed food	3.5	6.8	23.9
Fish	5.7	2.3	4.8
Minerals	32.3	-1.4	-2.3
Energy	6.9	-1.3	-2.2
Light manufacturing	-17.7	-43.0	-45.3
Heavy manufacturing	12.6	-6.1	-8.8
Transport	-14.3	-2.8	-4.7
Infrastructure services	-37.0	-5.4	-6.7
Construction	82.5	4.0	3.5
Other services	-8.9	11.2	9.6

Source: Model simulations described in the text.

Table A8.10   Effects on factor income distribution of protection to raise agricultural self-sufficiency in 2010

	Reference change, 2001–2010, per cent	Departure from reference 2010, per cent	
		Protection to hold self-sufficiency rates at 2001 levels	Protection to achieve full self-sufficiency
Unit factor rewards CPI deflated			
Land	136	7.2	17.1
Unskilled labour (those employed)	39	-1.7	-3.2
Skilled labour	43	-0.8	-2.6
Physical capital	15	-1.8	-3.5
Natural resources	164	-1.8	-2.4

Source: Model simulations described in the text.

227

Table A8.11   Elasticities of sensitivity to the income elasticity of demand for livestock products

	2010 projection without protection increase	2010 projection with protection to achieve self-sufficiency
Imports		Nominal protection coefficient
Beverages	0.76	0.75
Other crops (including feedgrains)	1.18	0.51
Livestock products	1.83	0.66
Processed food	0.79	0.42
Exports		
Light manufactures	-0.38	
Heavy manufactures	-0.36	
Real gross output		
Rice	-0.03	
Beverages	0.18	
Other crops (including feedgrains)	0.06	
Livestock products	0.37	
Processed food	0.02	
Light manufacturing	-0.65	
Heavy manufacturing	-0.10	
GDP	-0.02	
Imports		
Beverages	-1.56	
Other crops (including feedgrains)	-0.71	
Livestock products	-1.71	
Processed food	-0.72	
Exports		
Light manufactures	-1.41	
Heavy manufactures	-0.72	
Unit factor rewards CPI deflated		
Land	0.43	
Unskilled Labour (those employed)	0.00	
Skilled Labour	-0.01	
Physical capital	-0.02	
Natural Resources	0.04	

Source: To construct these elasticities a small deviation in the income elasticity of demand for livestock products is introduced and the standard simulations repeated. Elasticities are deduced from the comparison of simulation results.

09 Trade reform in the short run
China's WTO accession

Rod Tyers and Lucy Rees

China's accession to the WTO was an important event in global economic history and it is fitting that there has been so much quantitative analysis of its implications for trade and growth (Gilbert and Wahl 2001). The most recent quantitative assessments have offered comparatively sophisticated representations of some peculiar trade policies and Chinese labour market conditions.[1] All these studies have, however, focussed on medium to long-run impacts of accession reforms. They have not addressed the issue of 'how do we get there from here' and, in particular, the dependence of the transition on macroeconomic policies. This paper follows on from that by Yang and Tyers (2000) in that it emphasises the short run and the role of the macroeconomic environment, although it departs from that paper in its representation of accession trade policy reforms. Like Ianchovichina and Martin (2002) we make allowance for idiosyncratic trade policies, such as the duty drawbacks on imports used in the manufacture of exported goods. As befits a short-run analysis, we also allow for labour market rigidity and departures from full employment.[2]

Our point of departure is the recognition that the removal of import barriers can be contractionary in the short run in economies where the target of monetary policy is the exchange rate. This is because trade

---

First published in the *Journal of Asian Economics*, 15(1):1-31. Reproduced with permission.

liberalisation, taken alone, reduces the home prices of foreign goods. Households and firms therefore substitute away from home-produced goods, reducing their prices relative to foreign goods abroad, thus causing a real depreciation. If the nominal exchange rate is the target of monetary policy and the home economy is small by comparison with its trading partners, a fall in the home price level (a deflation) is required. This must be brought about by a monetary contraction in defence of the exchange rate. To the extent that wages adjust more sluggishly than product prices, the deflation causes the real wage to increase relatively quickly and hence employment growth to slow. Were the real depreciation the only consequence of the trade liberalisation shock, its effects would therefore be contractionary.

Trade reforms can, however, have positive effects in the short run. These include allocative efficiency gains that emerge, even in the very short run, and that raise aggregate productivity. Trade reforms also tend to raise the expected future net return on installed capital, stimulating investment. If the capital account is sufficiently open, an increase in foreign-financed investment might occur, which contributes substantially to short-run expansion. In essence, then, the issue we address is the robustness of the much anticipated gains from Chinese trade reform in the short run and its dependence on macroeconomic policy settings.

To do this we use a comparative-static, global, macroeconomic model, within which the microeconomic (supply) side is adapted from GTAP,[3] a multi-region, comparative-static model in real variables with price-taking households and all industries comprising identical competitive firms. Following Yang and Tyers (2000), to this microeconomic base is added independent representations of governments' fiscal regimes, with both direct and indirect taxation, as well as separate assets in each region (currency and bonds) and monetary policies with a range of targets. With this model it is possible to conduct trade liberalisation and other experiments under different assumptions about macroeconomic policy regimes.

The model

The microeconomic side of the model (Hertel 1997) offers the following useful properties

- a capital goods sector in each region to service investment
- explicit savings in each region, combined with open regional capital accounts that permit savings in one region to finance investment in others
- multiple trading regions, goods and primary factors
- product differentiation by country of origin
- empirically-based differences in tastes and technology across regions
- non-homothetic preferences and
- explicit transportation costs and indirect taxes on trade, production and consumption.

All goods and services entering final and intermediate demand are constant elasticity of substitution (CES) blends of home products and imports. In turn, imports are CES composites of the products of all regions, the contents of which depend on regional trading prices. Savings are pooled globally and investment is allocated between regions from the global pool. Within regions, investment places demands on the domestic capital goods sector, which is also a CES composite of home-produced goods, services and imports in the manner of government spending.

In constructing the macroeconomic version of the model we have first chosen the regions, primary factors and sectors identified in Table A9.1. Skill is separated from raw labour on occupational grounds, with occupations in the 'professional' categories of the International Labour Organisation (ILO) classification included as skilled.[4] Next, the standard model code is modified to make regional governments financially independent, thus enabling explicit treatment of fiscal policy. Direct taxes are incorporated at the observed average income tax rates for each region. Marginal tax rates are therefore assumed to be constant (say at τ). Regional households receive regional factor income, Y_F, and from this they pay direct tax τY_F. The disposable income that remains is divided between private consumption and saving. Government saving, or the government surplus, $S_G = T - G$, is simply revenue from direct taxes, τY_F, and from the many indirect taxes incorporated in the microeconomic part of the model, T_I,[5] less government spending, G, which could be exogenous or fixed as a proportion of GDP. Thus, $S_G = T_I + \tau Y_F - G$. The private saving and consumption decision is represented by a reduced form exponential consumption equation with wealth effects included via the dependence of consumption (and hence savings) on the interest rate. Each region contributes its total domestic (private plus government) saving, $S_D = S_P + S_G$, to the global pool from which investment is derived.[6]

For each region, the above relations imply the balance of payments identity, which sets the current account surplus equal to the capital account deficit: $X - M = S_P + S_G - I$.[7] From the pool of global savings, investment is allocated across regions and it places demands on capital goods sectors in each region. In the short run considered, however, investment does not add to the installed capital stock. Also, at this length of run, nominal wages are sticky in some regions (the industrialised regions of the US, the EU, Canada and Australia, and those developing countries with heavily regulated labour markets: China and Vietnam) but flexible elsewhere. In the spirit of comparative statics, although price levels do change in response to shocks, agents represented in the model do not expect any continuous inflation and so there is no distinction between the real and nominal interest rates.

In allocating the global savings pool as investment across regions, we have opted for the most flexible approach, implying a high level of global 'capital' mobility.[8] Where controls exist on international capital flows we introduce these explicitly. In the absence of capital controls, the allocation to region j (net investment in that region) depends positively on the expected long-run change in the average rate of return on installed capital, r_j^e, which, in turn, rises when the marginal product of physical capital is expected to increase. This allocation falls when the opportunity cost of financing capital expenditure, the region's real interest rate, r_j, rises. This rate depends, in turn, on a global capital market clearing interest rate, r^w, calculated such that global savings equals global investment: $\Sigma_j S_j^D = \Sigma_j I_j (r_j^e, r_j)$. Here I_j is real gross investment in region j.[9] The region's home interest rate is then $r_j = r^w(1+\pi_j)$ where π_j is a region-specific interest premium thought to be driven by risk factors not incorporated in this analysis. The investment demand equation for region j then takes the form

$$I_j = \delta_j K_j + I_j^N = \delta_j K_j + \beta_j K_j \left(\frac{r_j^e}{r_j}\right)^{\varepsilon_j} = K_j \left[\delta_j + \beta_j \left(\frac{r_j^e}{r_j}\right)^{\varepsilon_j}\right] \tag{1}$$

where K_j is the (exogenous) base year installed capital stock, δ_j is the regional depreciation rate, β_j is a positive constant and ε_j is a positive elasticity. Critically, investment in any region responds positively to changes that are expected to raise the sectoral average of a region's marginal product of physical capital and hence the regional average return on installed capital.[10] Other things being equal, improvements in trans-sectoral efficiency, such as might stem from a trade reform, are thought to raise capital returns permanently and hence they raise r_j^e. If such a shock also causes the rate of unemployment to fall, this raises total labour use and hence the current return on installed physical capital. When the shock is a trade reform, such employment effects are also considered permanent and so they add positively to the expected future return on installed capital, r_j^e.

Investment decisions are assumed to be made by forward-looking agents with access to a long-run version of the model. Thus, the expected change in the (long run) rate of return on installed capital in each region, r_j^e, is exogenous in short-run simulations. It is calculated by first simulating the effects of the same shock but under long-run closure assumptions. These differ from the short-run closure in the following ways: there are no nominal rigidities (no rigidity of nominal wages); larger production and consumption elasticities are used to reflect the additional time for adjustment; physical capital is no longer sector-specific, it redistributes across sectors to equalise rates of return; capital controls are ignored; and in China, irrespective of short-run fiscal policy assumptions, in the long run any loss of government revenue associated with tariff changes is assumed not to be made up via direct (income) tax, with the result that the fiscal deficit expands,so that the ratios of government revenue and expenditure to GDP are endogenous while the average direct tax rate is exogenous.

Note that the short run, comparative-static analysis does not require that the global economy be in a steady state. When shocks are imposed, any change in the counterfactual return on installed capital, r_j^e, need not be the same as the corresponding change in the opportunity cost of capital expenditure, r_j. Most often, in the short run, shocks change income and savings and, therefore, expected returns in directions that differ from corresponding short-run changes in the global interest rate—particularly considering that physical capital is fixed in quantity and sectoral distribution at this length of run. Even in long-run simulations, the global distribution of physical capital at the outset does not equalise rates of return across regions and redistribution through the regional allocation of one year's global savings is insufficient to redress such imbalances.

To include asset markets, region-specific money and homogeneous nominal bonds are introduced. Even though there is no interregional ownership of installed capital in the initial database, regional bonds are traded internationally, making it possible for savers in one region to finance investment in another.[11] Cash in advance constraints cause households to maintain portfolios including both bonds and non-yielding money and the resulting demand for real money balances has the usual reduced form dependence on GDP (transactions demand) and the interest rate. This is equated with the region's real money supply, where purchasing power is measured in terms of its GDP deflator, P^Y. Since all domestic transactions are assumed to use the home region's money, international transactions require currency exchange. For this purpose, a single nominal exchange rate, E_j, is defined for each region. A single key region is identified (here the US) relative to whose currency these nominal rates are defined. For the US, then, $E=1$ and E_j is the number of US dollars per unit of region j's currency. In essence, we are adding to the real model one new equation per region (the LM curve linking the real money supply to GDP and the interest rate) and one new (usually endogenous) variable per region, E_j.[12]

The bilateral rate between region i and region j is then simply the quotient of the two exchange rates with the US, $E_{ij} = E_i/E_j$. Quotients such as this appear in all international transactions. The most straightforward of the international transactions in the original model are trade transactions. There the bilateral exchange rate is simply included in all import price equations, along with cif/fob margins and trade taxes. In the case of savings and investment, the global pool of savings is accumulated in US dollars.

Investment, once allocated to region j, is converted to that region's currency at the rate E_j (US\$ per unit of local currency). The third, and most cryptic, set of international transactions in the original model concerns international transport services. Payments associated with *cif/fob* margins are assumed to be made by the importer in US dollars. The global transport sector then demands inputs from each regional economy and these transactions are converted at the appropriate regional rates.

Without nominal rigidities the model always exhibits money neutrality, both at regional and global levels. Firms respond to changes in nominal product, input and factor prices but a real producer wage is calculated for labour as the quotient of the nominal wage and the GDP deflator, so that $w=W/P^Y$. Thus, money shocks always maintain constant w when nominal rigidities are absent — as expected, money is then neutral. To make possible some rigidity in the setting of the nominal wage, W, a parameter, $\lambda \in (0,1)$ is inserted, such that

$$\frac{W}{W_0} = \Lambda \left(\frac{P^C}{P_0^C} \right)^{\lambda} \tag{2}$$

where W_0 is the initial value of the nominal wage, P_0^C is the corresponding initial value of the consumer price index (CPI), and Λ is a slack constant. Whenever Λ is exogenous and set at unity, the nominal wage carries this relationship to the CPI and the labour market will not clear except in the unlikely event that equation (2) happens to yield a market clearing real wage. The case where the labour market is fully flexible is represented by setting Λ as an endogenous slack variable and thereby rendering (2) ineffective. At the same time, labour demand is forced to equate with exogenous labour supply to reflect the clearing market.

The representation of capital controls

The model assumes that savings are perfectly mobile between regions and that the allocation of investment between them depends on region-specific interest premia and, if they are present, capital controls. In the absence of capital controls, a region's domestic capital market might be represented as in Figure A9.1. Net inflows on the capital account (KA), which comprise the net inflow of foreign savings, S_{NF}, less the net outflow associated with the accumulation of official foreign reserves, ΔR, are perfectly elastic at the global interest rate (this rate is adjusted by the exogenous region-specific risk premium, π).[13] The actual scale of net inflows depends on the net demand for foreign investment, $NFI=I\text{-}S_D$, where the relationship between NFI and r is shifted to the right by an increase in the expected future return on installed capital, r^e, via Equation 1, or by an increase in government spending, G, via its effect on domestic saving. It is shifted to the left by an increase in GDP (Y), via its effect on consumption and tax revenue and hence on domestic savings, S_D. In the figure, net inflows on the capital account are determined by the intersection of the two curves shown. For a balance of payments, these inflows must equate to net outflows on the current account, CA, and prices, and therefore real exchange rates, adjust to ensure that this is the case.

In this analysis, capital controls take the form of a rigid ceiling on net inflows on the capital account. This case is illustrated in Figure A9.2. In this circumstance the link between the home and global interest rates is severed unless net foreign investment falls sufficiently so that the controls cease to bind. To capture this in model simulations, the interest premium, π, is made endogenous while net flows on the capital account, KA, or, equivalently, on the current account, CA, are set as exogenous.

Data and parameters

The regions, primary factors and sectors identified in the analysis are listed in Table A9.1. Considering regions first, we draw on the now well-known GTAP Version 5 global database for 1997, which divides the world into 66 countries and regions. Although this database separates mainland China from Taiwan, it amalgamates Hong Kong with the mainland.[14] Our further aggregation of mainland China with Taiwan does overlook effects that are internal to these regions but such effects are not our focus. Instead, we seek to illustrate the strong interaction between trade reforms and macroeconomic policies, and, particularly, foreign exchange regimes. These interactions are important for all the economies of East Asia and particularly for those with regulated foreign exchange regimes. China is the largest developing economy to maintain, at least *de facto*, fixed US dollar parity and, in this respect, the macroeconomic policy regimes of Hong Kong and Taiwan have been compatible with that of the mainland.

Turning to primary factors, skill is separated from raw labour on occupational grounds, with the 'professional' categories of the International Labour Organization (ILO) classification included as skilled.[15] The structure of factor demand has skill and physical capital as complements. This enables the model to represent the links between skill availability, capital returns and investment that are important in China, which has large skilled and unskilled labour forces that are increasingly mobile between sectors.[16] Finally, the sectoral breakdown we have chosen aggregates the 57 sectors in the database to our more manageable 14, offering the most detail in agricultural and marine products. This is because, amongst China's merchandise trade commitments for WTO accession, a key liberalisation is in the processed food sector, to which these commodities are inputs.

Because the length of run is short, the real part of the short-run model incorporates smaller-than-standard elasticities of substitution in both demand and supply. These are based on a short-run calibration exercise on the Asian crisis, described in Yang and Tyers (2000). For further details of the model, its parameters and its structure, see Yang and Tyers (2000) and Tyers and Yang (2000, 2001).

China's trade policies and reforms

The 2001 pattern of trade and production taxes and subsidies is constructed first from the GTAP Version V global database for 1997. Recent updates to the tariff regime are incorporated, as provided by Ianchovichina and Martin (2001). Of particular importance is the introduction since 1997 of duty drawbacks that are offered to exporting firms on the component of their imports of intermediate goods that is used for export production. The effects of these duty drawbacks are particularly difficult to quantify since it is generally impossible to separate out production for export from production for the domestic market.[17]

For the analysis, the effects of duty drawbacks were approximated by, first, constructing a database comprising inter-industry financial flows after the general tariff reforms for the period 1997-2001. This database emerges from a model simulation in which the only shocks are the documented changes in tariffs by sector. The pattern of these inter-industry flows indicates the magnitude of the expenditures on intermediate inputs and on import tariffs by firms in each industry and the proportions of their respective outputs that are exported. Second, the proportions of expenditures on imported (as distinct from home-produced) intermediate inputs are calculated, along with the average proportions of these that enter export production. Expenditures on tariffs for export production follow for each industry. Finally, this sum is returned through the implementation of equivalent export subsidies (or reduced export taxes).

The application offered here is one in which the manufacturing sector, to which duty drawbacks primarily apply, is aggregated into only 'light manufacturing' and 'other manufacturing'. At this level of aggregation there is a considerable volume of intra-industry trade. A substantial share of the cost of manufactured exports takes the form of expenditure on manufactured intermediate inputs, of which a significant volume is imported (Table A9.2). In these circumstances, the use of export subsidies to proxy duty drawbacks is crude but it offers the following realistic consequences.

- The export industry expands in response to its greater profitability
- The price of the industry's product rises in the home market. Although this effect is not realistic, it has the realistic consequence that there is substitution in favour of imports in intermediate consumption and so the home market share in intermediate inputs falls
- The government is denied the revenue that would have come from the tariffs on intermediates used for export production, in this case by giving it back in the form of export subsidies.

The 2001 export tax rates used are thus modified to take account of the ad valorem equivalent export subsidy rates that are proxies for duty drawbacks. The resulting pattern of equivalent trade taxes and subsidies is listed in Table A9.3. The most substantial effect of the duty drawbacks is in the manufacturing sector with the equivalent export tax rate on light manufacturing falling by about a third and heavy manufacturing receiving an equivalent export subsidy of 1 per cent. In the 1990s 'other crops' and processed food received significant border protection, though at rates that diminished between 1997 and 2001. Moreover, food processing, which is the sector through which most agricultural products flow, still receives considerable protection (Table A9.3).

The equivalent tariff rates following China's accession into the WTO are also summarised in Table A9.3. The associated trade reforms are substantial. The tariff protection in 'other crops', livestock, food processing, light manufacturing and heavy manufacturing are significantly reduced, with the largest reductions being in the food processing and heavy manufacturing sectors. For each product, the database obtained from the WTO website details the decline in tariff rates and the timing of the reductions. To obtain the rates in Table A9.3, the industry classification used in the WTO list of tariff concessions was concorded with the cruder subdivision used in our model and average rates constructed for each sector. The information contained in the database was supplemented by details of the accession tariff rates provided by Ianchovichina and Martin (2001). To represent the behavioural impacts of the changes in equivalent tariff rates as accurately as possible, emphasis was placed on preserving changes in the 'powers of the tariffs' rather than in the rates themselves.[18]

The equivalent Chinese tariff rates of the 1990s vary by country of origin. This means that the application of the same shock to the powers of these equivalent tariffs might have led to negative post-accession rates for some trading partners. The accession shocks to the equivalent bilateral tariff rates were therefore calculated so as to harmonise the post-shock tariff rates across countries of origin. The proportional changes in 'powers of equivalent tariffs' are the same as those implied by the changes in rates detailed in Ianchovichina and Martin (2002: Table 3). For our present purpose, these shocks are the same for both the long run and the short run. As indicated in the WTO database, China is committed to undertaking many of the tariff concessions immediately on accession.[19]

Simulated long-run effects of accession policy reforms

The reasons for examining long-run implications first are twofold. First, the long-run results are useful in their own right, given that they may then be compared with the many other simulations of China's WTO accession reforms. Second, the long-run outlook is required in order that the expectations of investors can be formulated. Recall that they are assumed to take changes in long-run returns on installed capital into account in determining short-run changes in their investment behaviour.

The key elements of the long-run closure, already discussed, in summary, include

- there are no nominal rigidities (no rigidity of nominal wages)

- production and consumption elasticities of substitution are chosen at 'standard' levels to reflect the additional time for adjustment in the long run over the short run (Tyers and Yang 2001)

- physical capital is no longer sector-specific; it redistributes across sectors to equalise rates of return

- capital controls are ignored and

- in China, irrespective of short-run fiscal policy assumptions, in the long run any loss of government revenue associated with tariff changes is assumed to not be made up via direct (income) tax, with the result that the fiscal deficit expands.

The results from the long-run simulation are provided in Table A9.4. They show the expected allocative efficiency gains, reflected here in a rise in GDP, aided by increased returns on installed physical capital that induce greater investment and therefore larger net inflows on the capital account in the long run. The increased average long-run return on installed capital in China is therefore part of investor expectations in the short run and so tends to raise the level of investment in the short run, even if capital controls are maintained, as discussed in the next section. Finally, as discussed earlier, the trade reform causes home consumption to switch away from home-produced goods, the relative prices of home-produced goods to fall, and hence an overall real depreciation occurs. This real depreciation accompanies a rise in Chinese export competitiveness, with overall export volume expanding by 9 per cent.

Of particular interest are the changes in real gross output in each sector of the economy. Although the trade policy regime of 2001 advantaged food processing, 'other crops', fisheries and light manufacturing, apart from the smaller 'beverages' industry, it is the manufacturing sector that is the robust beneficiary of the unilateral trade liberalisation. This somewhat surprising result depends on subtle qualities of China's manufacturing sector in reality and as it is represented in the model. The first of these is its pattern of factor intensities. For the sectors defined in the model, sets of factor proportions show, as expected, that agricultural industries are land-intensive, with beverages being more intensive in capital than the other crops (Table A9.5). Fishing is labour, capital and natural resource intensive and the energy sector is very capital-intensive. Of special relevance in interpreting the effects of unilateral liberalisation, however, are the factor intensities for manufacturing. Note that light manufacturing is highly labour-intensive compared to all the traded goods sectors while heavy, or 'other', manufacturing is one of the most capital-intensive.

The second subtlety is that when manufacturing is aggregated into two types, as in this case, the two sub-industries disguise considerable heterogeneity. One consequence of this is that there is considerable intra-industry trade. Light manufacturing is the most export-oriented of all the sectors—its exports are largest compared to its domestic value added. By contrast, heavy manufacturing is distinctive by the considerable scale of its competing imports. While intra-industry trade is significant in the beverages and other manufacturing sectors, nowhere is it more important than in light manufacturing. Both manufacturing sectors commit approximately half their total costs to inputs in the same product category and about 10 to 15 per cent of those to imports (Table A9.2).

Superficially, trade liberalisation removes the sector's tariff protection and so our intuition, stemming from the standard Heckscher-Ohlin-Samuelson (HOS) trade model, suggests it must contract. But here there are two departures from the HOS model. First, there is extensive intermediate use from the same sector and, second, competing imports, even though they are from the same sector, are differentiated from home products. Under these conditions the tariff reductions on imported intermediates have a direct effect on home industry total cost. Reductions of tariffs on competing, but differentiated, imports have only an indirect effect, the magnitude of which depends on the elasticity of substitution between the two. Indeed, for manufacturing, it turns out that the input cost effect of tariff reductions is considerably greater than that of the loss of protection against competing imports. Cost reductions of similar origin are the reason for similar gains accruing to the domestic transport services sector.

Because the reforms cause the most substantial reductions in protection of China's food processing sector, and therefore lead to long-run contractions in rice and 'other crops' production, they require substantial structural change, including the relocation of employment from agriculture to manufacturing. In the long run, employment in food processing falls by 7 per cent, in rice production by 4 per cent, and in 'other crops' by 2 per cent. As simulated, in the long run, the workers lost from these sectors are re-employed in the energy, manufacturing, and transport and other services sectors.

Simulated short-run effects of trade policy reforms

In the short run, the model used has smaller elasticities of consumption and production. For all the regions represented, the 'standard' closure is as indicated in Table A9.7. Monetary authorities in China and Vietnam are assumed to maintain effective fixed exchange rates against the US$. The other regions identified adopt inflation or CPI targeting. Capital controls are assumed rigid in China and Vietnam, but they are non-existent in the other regions. In the labour markets of China and Vietnam, nominal wages are assumed to be 'sticky'. Full short-run rigidity is assumed in the industrial countries, while nominal wages are assumed to be fully flexible elsewhere in Asia and the developing world. As to fiscal policies (not shown in the table) government spending in all regions is assumed to absorb a fixed proportion of GDP and the rates of direct and indirect tax are constant, so that government deficits vary in response to shocks. Henceforth, this closure is only varied in order to investigate the sensitivity of the effects of trade reforms to the macroeconomic policy environment.

Six macroeconomic regimes were simulated

- The 'standard' closure, with rigid capital controls, a fixed exchange rate and fixed direct and indirect tax rates
- Capital controls and fixed tax rates, monetary policy targets the CPI, and the exchange rate floats.
- The 'standard' closure except that capital controls are removed
- Closure 2 (floating exchange rate), except that capital controls are removed.

- The 'standard' closure except that the direct tax rate adjusts to maintain a government deficit that is fixed as a proportion of GDP.
- Closure 5, except that capital controls are removed. [20]

The short-run shocks are smaller than those administered in the long run, on the assumption that such reforms will be phased in over a period of five years. The results for the first four closures are presented in Table A9.8 and those for the last two are presented in Table A9.9. In general, it is clear that the short-run effects of the trade reform are heavily dependent on the surrounding macroeconomic policy regime. Indeed, the effects range from the contraction alluded to in the introduction through to considerable short-run expansion.

The effects of capital controls and the choice of monetary policy target

In broad terms, the behaviour of the model in the short run with rigid capital controls retained can be represented as in Figure A9.3. The upper diagram represents the domestic capital market and the lower one the domestic market for foreign products. These markets are linked by the requirement that, for a balance of payments, net flows on the capital account must mirror those on the current account. Net demand for foreign products (the downward sloping line in the lower diagram, $NM=M-X$) depends on the relative price of foreign goods. For this purpose we define the real exchange rate as the common-currency ratio of the price of home goods to the price of foreign goods

$$e_R = E\frac{P^Y}{P*} = \frac{P^Y}{P*\!\big/\!E} \tag{3}$$

where, as before, E is the nominal exchange rate in foreign currency per unit of home currency, P^Y is the GDP deflator and $P*$ is the foreign price level. In the numerical model a real effective exchange rate is estimated as the trade-weighted average of the ratio of the home and the foreign GDP deflators. Net imports depend positively on this and negatively on its inverse (the common-currency foreign to home product price ratio). This relationship is shifted to the right by an increase in GDP, Y, or a reduction in protection, τ. The real exchange rate is then determined by the balance of payments requirement that net inflows on the capital account must equal net outflows on the current account, $KA=-CA=NM=M-X$.[21]

The trade liberalisation reduces τ and shifts NM to the right. With tight capital controls, the current account balance cannot change. The shock therefore raises the relative price of foreign goods in the home market and thus depreciates the real exchange rate. If the nominal exchange rate is the target of monetary policy and the home economy is small by comparison with its trading partners then, from (3) a fall in P^Y (a deflation) is required. This must be brought about by a monetary contraction in defence of the exchange rate. To the extent that wages adjust more sluggishly than product prices, the deflation causes the real wage to rise. Were the real depreciation the only consequence of the liberalisation shock, its effects would therefore be contractionary. Fortunately, this need not be the case. The trade reform brings gains in allocative efficiency.[22]

When capital controls remain rigid and the exchange rate fixed, however, these allocative gains are insufficient to offset the contractionary effects of the deflation. This can be seen from the first column of Table A9.8. The real depreciation is substantial and the deflation required is of the order of 2 per cent per year. The production real wage rises by half this and employment falls. Investment demand responds to the expectation of higher real returns to installed capital in the future by shifting outward. The loss of tariff revenue drives the government deficit higher, reducing domestic saving, further reinforcing the outward shift of the NFI curve. But the rigidity of the capital controls causes this to simply push up the interest rate and so real investment falls. Output falls in all sectors except manufacturing and transport. The latter sectors gain in the short run for the same reasons they gain in the long run—cheaper imported inputs. Under these policy circumstances, the overall net gains from trade reform are not robust in the short run.

Now suppose that the monetary policy regime allows the exchange rate to float, targeting the consumer price level instead of the nominal exchange rate. The consumer price is a weighted average of the prices of home-produced goods and the after-tariff home prices of foreign goods in the domestic market

$$\bar{P}^C = Av\left\{P^Y, \frac{P*(1+\tau)}{E}\right\}$$
(4)

where the tariff rate indicated in Equation 4 is an average across imported products. The fall in this tariff rate lowers the domestic price of imported products and therefore shifts demand away from home-produced goods. So there must be a real depreciation (Equation 3). But now the nominal exchange rate can carry some of this adjustment. The question is how much? If monetary policy targets the consumer price level, as expressed above, the primary shock is to the tariff rates. The nominal depreciation cannot be so large as to reverse the effect of this primary shock on the domestic prices of imported goods. The second term in Equation 4 should therefore fall. Thus, to maintain the consumer price target a rise in the GDP price index P^Y is expected, so that monetary policy must be expansionary.

But this is not what is observed in our multi-commodity simulation. In fact, there is a fall in P^Y (a deflation), and this counterintuitive result arises because the two price indices, P^Y and P^C, have significantly different sectoral weightings. The GDP price index has a collective weighting of 55 per cent on 'construction and dwellings', 'light manufacturing' and 'other manufacturing' (Table A9.1) and all three of these product groups have declining prices driven by the tariff reductions.

Consumption, on the other hand, is spread more evenly across commodities, and products that are not subjected to tariff changes weigh more heavily in the consumer price index. Over half of domestic consumption expenditure is allocated to 'other services', 'livestock' and 'other crops'. These three sectors experience a substantial increase in domestic prices whereas both 'light manufacturing' and 'construction' experience a decline. The average price of *home goods* in consumption rises as aggregate demand rises and this increase is sufficient to just offset the fall in the average price of imports in consumption. The result is that a monetary contraction is required to achieve the slight P^Y deflation. The consequence of exchange rate flexibility (with P^C targeting) is a smaller GDP price deflation and hence a smaller real wage increase, a smaller employment slowdown, and therefore a larger net gain from the liberalisation.

In the capital market (upper) part of the diagram in Figure A9.3, this means the GDP-driven tendency of the NFI curve to shift left is larger. This offsets the tendency for investment to rise (which is the same as in the fixed exchange rate case since the same change in return to installed capital is expected). So the net rightward shift in the NFI curve is smaller, as is the rise in the home interest rate; and hence there is a fall in real investment spending. All these results are verified numerically in column 3 of Table A9.8. The key result in the floating rate case is that, although the real production wage still rises, it does so by less than would have occurred had employment been fixed. The gain in allocative efficiency would have yielded a larger real production wage gain. Because this gain is restrained by nominal wage stickiness, there is a rise in employment and an unambiguous increase in GDP in the short run. Therefore, if it can be managed, exchange rate flexibility is superior to a fixed rate regime during a trade reform and where capital controls are tight.

If the capital controls are removed, the corresponding liberalisation shock is as depicted in Figure A9.4. Here, reduced protection also yields a gain in allocative efficiency and hence there is a rise in GDP, reinforcing the rightward shift in the net imports curve. In this case, however, the absence of capital controls allows investment to flow in, responding to the increase in the expected long-run return on installed capital. The increased inflow on the capital account relaxes the balance of payments constraint in the lower diagram and allows a substantial increase in net imports. The net effect on the real exchange rate depends on whether this effect, in raising the net supply of foreign goods, is larger or smaller than the increase in net demand for them due to the tariff reduction and the rise in domestic income. In the case of China, the rise in net demand is dominant and the real exchange rate still depreciates, albeit to a lesser extent than in the presence of capital controls (compare columns 2 and 4 of Table A9.8). Figure A9.4 is drawn correspondingly.

With a real depreciation on the left-hand side of Equation 3, the result is either a nominal depreciation or a domestic deflation, or both, depending on the target of monetary policy. If the nominal exchange rate is fixed, the domestic (P^Y) deflation is larger. With sticky nominal wages, this causes a larger rise in the real production wage and hence a smaller expansion in employment and GDP. Therefore, the floating rate regime with consumer price targeting gives a better short-run outcome when capital controls are relaxed than a fixed exchange rate because there is a smaller deflation and hence greater employment growth. The superiority of the floating rate regime would have seemed even stronger had the target of monetary policy been set, instead, at P^Y. Then there would have been a little (unanticipated) CPI inflation and employment would have expanded further.[23]

To summarise the monetary policy effects: when capital controls are weak or non-existent, the trade liberalisation attracts increased inflows on the capital account to mitigate the real depreciation and associated GDP price deflation that are its inevitable consequences. The real volume of domestic investment rises irrespective of the target of monetary policy, as does the level of GDP. The choice of monetary policy target still matters, however, with CPI targeting offering a smaller GDP price deflation, more modest gains in the real production wage, and better short-run GDP gains.[24]

Fiscal policies

The fiscal impact of the trade reform comes through the associated decline in tariff revenue. Only two fiscal policies were considered. Fiscal policy 1 has no tax revenue switch. Government spending continues at a constant share of GDP and all rates of direct and indirect tax are held constant, except tariffs, which are reduced. The result of the trade liberalisation is therefore an expanded fiscal deficit. This is the fiscal policy applying in the simulations reported in Table A9.8. Fiscal policy 2 has the lost revenue made up via an increase in the direct tax rate, so that the fiscal deficit, government revenue and government spending are all maintained as constant proportions of GDP. The responses to the two fiscal policy regimes are compared in Table A9.9.

As with monetary policies, the ranking assigned to the two depends on the strength of capital controls. In the presence of tight capital controls (that keep net flows on the capital account constant) policy 2 outperforms policy 1 in terms of GDP expansion. This is because the expanded fiscal deficit under policy 1 adds to the demand side of the domestic capital market, pushing up the domestic interest rate and crowding out private investment. Indeed, the home interest rate rises by more than the expected long-run return on installed capital and so the volume of investment falls. Under policy 2, the rise in income taxation reduces pressure on the domestic capital market so that the rise in the home interest rate is smaller, as is, therefore, the fall in real investment. This mitigates, but does not reverse, the contraction in GDP that occurs under policy 1.

In the absence of effective capital controls, the ranking is reversed: policy 1 outperforms policy 2. This is because net inflows on the capital account are now perfectly elastic at the international interest rate (plus an exogenous country risk premium). The added government borrowing therefore draws in additional saving from abroad and does not crowd out new private investment in the short run. Net inflows on the capital account and domestic investment increase substantially: the more so under policy 1. The expanded deficit under policy 1 therefore mitigates the depreciation of the real exchange rate. There is a smaller deflation and, while ever wages adjust more slowly than product prices, this retards real wages more and hence accelerates employment and GDP growth. Thus, the tax mix switch of policy 2 is contractionary by comparison with policy 1 when capital controls are lifted. Yet both give superior results without capital controls than either does in their presence.

Sectoral impacts in the short run

The key determinant of the sectoral mix of changes in the economy is the size of the short-run real depreciation that occurs following the trade reform. When capital controls are tight, this real depreciation is comparatively large. Traded sectors, such as light manufacturing, are advantaged, while non-traded services sectors, such as construction and dwellings, are disadvantaged. When capital controls are ineffective, manufacturing gains are smaller and the non-traded services sectors benefit. Across the board, however, and for reasons that match those given in addressing the long-run simulation results, agriculture and food processing are disadvantaged by the reforms. Interestingly, however, it is only in the case when capital controls are retained and the exchange rate is fixed that almost the entirety of the agricultural sector is hurt. Even where capital controls are retained, a switch to a floating exchange rate would enable the trade reforms to be consistent with further growth in the 'other crops', livestock and fisheries sectors.

As in the long run, a key immediate effect of the reforms is a reduction in protection of China's food processing sector and therefore a contraction in agricultural output. Significant structural change is therefore required in the short run with the movement of employment from agriculture to manufacturing; however, as expected, the scale of the movement is smaller than in the long run. Employment in food processing falls regardless of the macroeconomic policy regime. Under either fiscal regime, the greatest contraction to employment in food processing occurs when capital controls are tight and monetary policy targets the nominal exchange rate. Unlike in the long run, employment in the other agricultural sectors is not necessarily contractionary—the outcome is dependent on the macroeconomic policy regime.

Conclusions

Experiments using a global multi-product, comparative-static, macroeconomic model indicate that the trade reforms to which China has committed as part of its WTO accession yield the well known net gains in the long run. In the short run, however, these gains are not directionally robust to the macroeconomic policy regime. If capital controls are too tight and the quasi-fixed nominal exchange rate regime is retained, the reforms are deflationary. If the labour market is the slowest to adjust, employment growth will slow and the overall reform package will be contractionary. To ensure that the short-run gains are substantial, however, the Chinese government has only to allow sufficient net inflow on the capital account to at least maintain the level of domestic investment. Even if it does not do this, the trade reforms would be expansionary in the short run if a small nominal depreciation were allowed.

The magnitudes of the short-run effects on the rate of economic expansion are quite sensitive to the monetary and fiscal policies adopted. When capital controls are tight, a monetary policy that targets the domestic consumer price level mitigates the GDP price deflation and therefore outperforms one that targets the nominal exchange rate. Moreover, if monetary policy must target the exchange rate and capital controls are tight, the short-run contraction can be mitigated if lost tariff revenue is made up through an increased direct tax rate, rather than left to expand the fiscal deficit and thereby crowd out private investment.

When capital controls are ineffective or the government allows substantial foreign investment, inflows on the capital account are greater and so the real depreciation caused by the reforms is reduced. Although net gains are experienced irrespective of the monetary target, the floating rate alternative with CPI targeting is again superior. Without effective capital controls, however, a fiscal expansion does not crowd out domestic investment. Failing to replace the lost tariff revenue through increased direct taxation is therefore no longer deleterious to short-run growth. Indeed, the associated fiscal expansion augments the growth achieved in this case.

Because the trade reforms to which China is committed remove a comparatively large proportion of the existing protection afforded the food-processing sector, that and some related agricultural activities contract with the reforms, irrespective of the macroeconomic policy environment chosen. The magnitude of the short-run structural change required is, however, reduced if capital controls are relaxed and investment therefore has a larger share of GDP. The fiscal policy response to the loss of import tariff revenue has comparatively little influence over China's economic performance in the short run. In fact, regardless of whether it is the government spending or the government deficit that is held constant, the optimal macro policy environment for China's economy during these reforms is a floating exchange rate regime with no capital controls.

Notes

1 For one line of evolution, see the papers by Ianchovichina and others (Ianchovichina et al. 2000; Ianchovichina and Martin 2001 and 2002 and Walmsley et al. 2001).
2 We do not, however, differentiate rural from urban labour and hence we cannot represent explicitly the hukou system of labour market regulation (Sicular and Zhao 2002).
3 A detailed description of the original model is provided by Hertel (1997).

4   See Liu et al. (1988) for the method adopted.

5   TI includes revenue from taxes on production, consumption, factor use and trade, all of which are accounted for in the original GTAP model and database.

6   Private saving is derived as the difference between disposable income $(Y\text{-}T)$ and consumption expenditure, where real consumption is determined in a Keynesian reduced form equation that takes the form: $C = \gamma r^{\delta}[Y\text{-}T]^{\mu}$, where r is the real interest rate.

7   Note that there is no allowance for interregional capital ownership in the starting equilibrium. At the outset, therefore, there are no factor service flows and the current account is the same as the balance of trade.

8   By which it is meant that households can direct their savings to any region in the world without impediment. Installed physical capital, however, remains immobile even between sectors.

9   Before adding to the global pool, savings in each region is deflated using the regional capital goods price index and converted into US$ at the initial exchange rate. The global investment allocation process, therefore, is made in real volume terms.

10 This investment relation is similar to Tobin's Q in the sense that the numerator depends on expected future returns and the denominator indicates the current cost of capital replacement.

11 Since the initial database (GTAP Version 5) incorporates no 'net income' or factor service component in its current account, the initial equilibria must do likewise. This implies the assumption that, although there are no interregional bond holdings initially, the shocks implemented cause interregional exchanges of bonds and hence a non-zero net income flow in future current accounts not represented.

12 More precisely, since for the US $E=I$, there is one less (usually endogenous) variable. Where nominal exchange rates are to be endogenous and nominal money supplies exogenous, one additional variable must be made endogenous. This could, for example, be balanced by making one price level exogenous, such as by having US monetary policy target the change in the US CPI, PC.

13 The scope of monetary policy includes alterations in the rate at which official foreign reserves are accumulated. When there are no capital controls, however, the perfect capital mobility assumption implies that changes in reserves have no effect on net capital account flows. Where they are important is in the case where capital controls are effective. Because the manipulation of reserves offers only a short-term approach to exchange rate management that is only available if reserves are sufficient in the first place, DR is held exogenous throughout the analysis.

14 Detailed descriptions of the GTAP database's content and sources as they relate to China are available in Gehlhar (2002) who describes the integration of the data for Hong Kong with that of the mainland and discusses the entrepot nature of some of Hong Kong's trade.

15 See Liu et al. (1998) for the method adopted.

16 For further discussion of the role and representation of skill-capital complementarity, see Tyers and Yang (2000).

17 Bach et al. (1996) offer one approximation that requires the construction of a set of equivalent production taxes and subsidies and the rebalancing of the economic database to reflect these. Walmsley et al. (2001) reconstruct their global database to separate out production for exports and domestic sales. An investment of this magnitude is too great for our more illustrative purpose.

18 Consequently, the rates in Table A9.3 tend to reflect the proportional changes in powers of tariffs implied by Ianchovichina and Martin and the magnitudes as detailed in the protocol.

19 To the extent that some of the tariff reductions may be phased in over several years, our analysis will tend to overstate the economic impacts in the short run.

20 For a full enumeration, two more cases were considered: fixed tax rates or fixed government deficit with a floating exchange rate. These cases were excluded to simplify the presentation .The results are available from the authors.

21 The net factor income component of the current account is zero at the outset because that is the assumption embodied in the construction of the original database.

22 To see these at least partially offsetting gains in allocative efficiency, it is necessary to use a multi-commodity, general equilibrium framework such as that used in this paper.

23 It is, at least in part, for this reason that CPI-targeting countries set targets of 2-3 per cent per year. This avoids GDP price deflation following trade reforms or negative external shocks.

24 The trade reform is a positive shock and so it should not be surprising that an open capital account is advantageous. Such openness would, however, risk outflows following negative shocks and it is this risk that justifies the controls in the first place. If the risk of capital flight is to be minimised, controls on the composition of investment may be required. These simulation results simply confirm that such controls should do as little as possible to inhibit the inflow of investment following positive shocks.

References

Bach, C.F., Martin M., and Stevens, J.A., 1996. 'China and the WTO: tariff offers, exemptions and welfare implications', *Weltwirtschaftliches Archiv* 132(3):409-431.

Francois, J. and Spinanger, D., 2001. 'Greater China's accession to the WTO: implications for international trade/production and for Hong Kong', report to the Hong Kong Trade Development Council, December.

Gehlhar, M.J., 2002. 'Hong Kong's re-exports', in Dimaranan, B.V., and McDougall, R.A. (eds), *Global Trade, Assistance and Production: the*

GTAP 5 Data Base, Centre for Global Trade Analysis, Purdue University, West Lafayette. Available online at www.gtap.agecon.purdue.edu/databases/v5/v5_doco.asp.

Gilbert, J. and Wahl, T., 2001. 'Applied general equilibrium assessments of trade liberalisation in China', Department of Economics, Washington State University.

Hertel, T.W. (ed.), *Global Trade Analysis Using the GTAP Model*, New York: Cambridge University Press, 1997.

Ianchovichina, E., Martin, W. and Fukase, E., 2000. 'Modeling the impact of China's accession to the WTO', presented at the 3rd Annual Conference on Global Economic Analysis, Monash University, 27-30 June.

Ianchovichina, E. and Martin, W., 2001. 'Trade liberalisation in China's accession to the WTO', *Journal of Economic Integration*, 16(4):421-45.

——, 2002. *Economic impacts of China's accession to the WTO*, World Bank Policy Research Working Paper No. 3053, World Bank, Washington DC.

Krueger, A.O., 1992. *The Political Economy of Agricultural Pricing Policy: Volume 5, A Synthesis of the Political Economy in Developing Countries*, Johns Hopkins University Press, Baltimore.

Lin, H.C., Chung, L. and Liou, R.W., 2002. 'Taiwan', in Dimaranan, B.V., and McDougall, R.A. (eds), *Global Trade, Assistance and Production: the GTAP 5 Data Base*, Centre for Global Trade Analysis, Purdue University, West Lafayette. Available online at www.gtap.agecon.purdue.edu/databases/v5/v5_doco.asp

Liu, H.C., Van Leeuwen, J.N., Vo, T.T., Tyers, R., and Hertel, T.W., 1998. *Disaggregating labor payments by skill level in GTAP*, Technical Paper No.11, Center for Global Trade Analysis, Department of Agricultural Economics, Purdue University, West Lafayette.

Sicular, T. and Zhao, Y., 2002. 'Labor mobility and China's entry to the WTO', presented at the conference on *WTO* Accession, Policy Reform and Poverty Reduction, Beijing, 26-28 June.

Tyers, R., 2001. 'China after the crisis: the elemental macroeconomics', *Asian Economic Journal*, 15(2):173-99.

Tyers, R. and Rees, L., 2002. *Trade reform and macroeconomic policy in Vietnam*, Working Papers in Economics and Econometrics No. 419. Also presented at the Fifth Annual Conference on Global Economic Analysis, Taipei, June.

Tyers, R. and Yang, Y., 2000. 'Capital-skill complementarity and wage outcomes following technical change in a global model', *Oxford Review of Economic Policy*, 16(3):23–41.

——, 2001. 'Short run global effects of US 'new economy' shocks: the role of capital-skill complementarity', Paper presented at the Fourth Annual Conference on Global Economic Analysis, Purdue University, West Lafayette, 27-29 June.

Walmsley, T., Hertel, T. and Ianchovichina, E., 2001. 'Assessing the impact of China's WTO accession on foreign ownership', GTAP Resource #649, paper presented at the Fourth Annual Conference on Global Economic Analysis, Purdue University, West Lafayette, 27-29 June.

World Bank, 1993. *The East Asian Miracle: economic growth and public policy*, Oxford University Press, Cambridge, UK.

World Trade Organization, 2001. *Accession of the People's Republic of China, decision of 10 November 2001*, World Trade Organization, Geneva.

Yang, Y. and Tyers, R., 2000. *China's post-crisis policy dilemma: multi-sectoral comparative static macroeconomics*, Working Papers in Economics and Econometrics No. 384, Australian National University, June, presented at the Third Annual Conference on Global Economic Analysis, Melbourne, 27-30 June. Available online at http://ecocomm.anu.edu.au/economics/staff/tyers/tyers.html.

Acknowledgments

Funding for the research described comes, in part, from Australian Research Council Large Grant No. A201 and in part from ACIAR Project No.ADP/1998/128. Valuable discussions with Yongzheng Yang, Ron Duncan and Tingsong Jiang are appreciated.

Appendix

Figure A9.1 The domestic capital market without capital controls

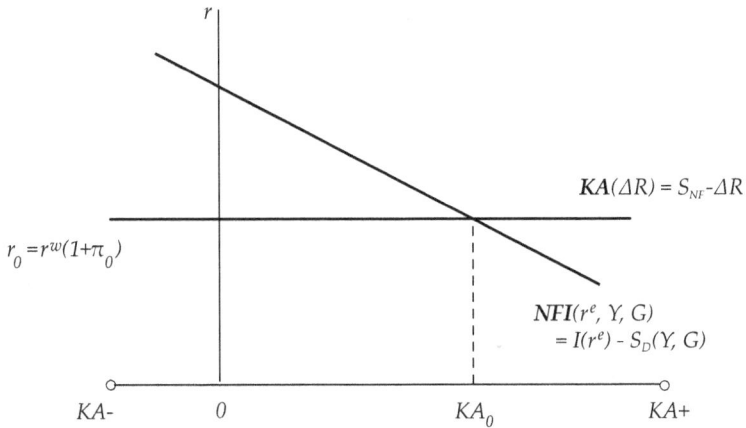

$KA(\Delta R) = S_{NF} - \Delta R$

$r_0 = r^w(1+\pi_0)$

$NFI(r^e, Y, G)$
$= I(r^e) - S_D(Y, G)$

$KA-$ 0 KA_0 $KA+$

Figure A9.2 The domestic capital market with capital controls

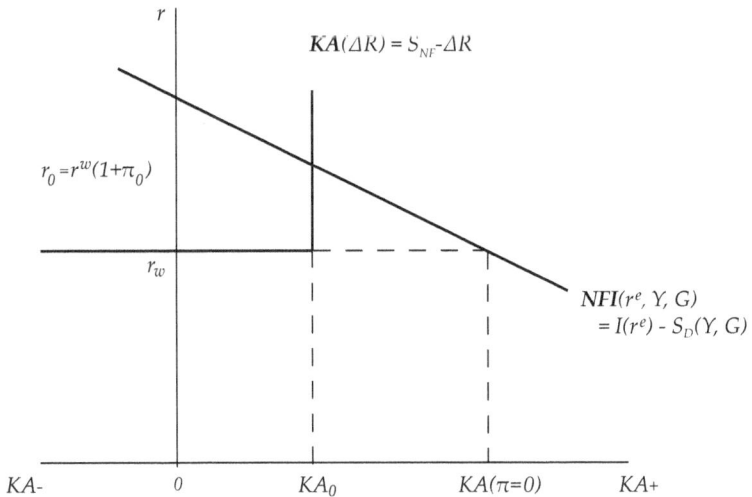

$KA(\Delta R) = S_{NF} - \Delta R$

$r_0 = r^w(1+\pi_0)$

r_w

$NFI(r^e, Y, G)$
$= I(r^e) - S_D(Y, G)$

$KA-$ 0 KA_0 $KA(\pi=0)$ $KA+$

Figure A9.3 Trade reform with capital controls

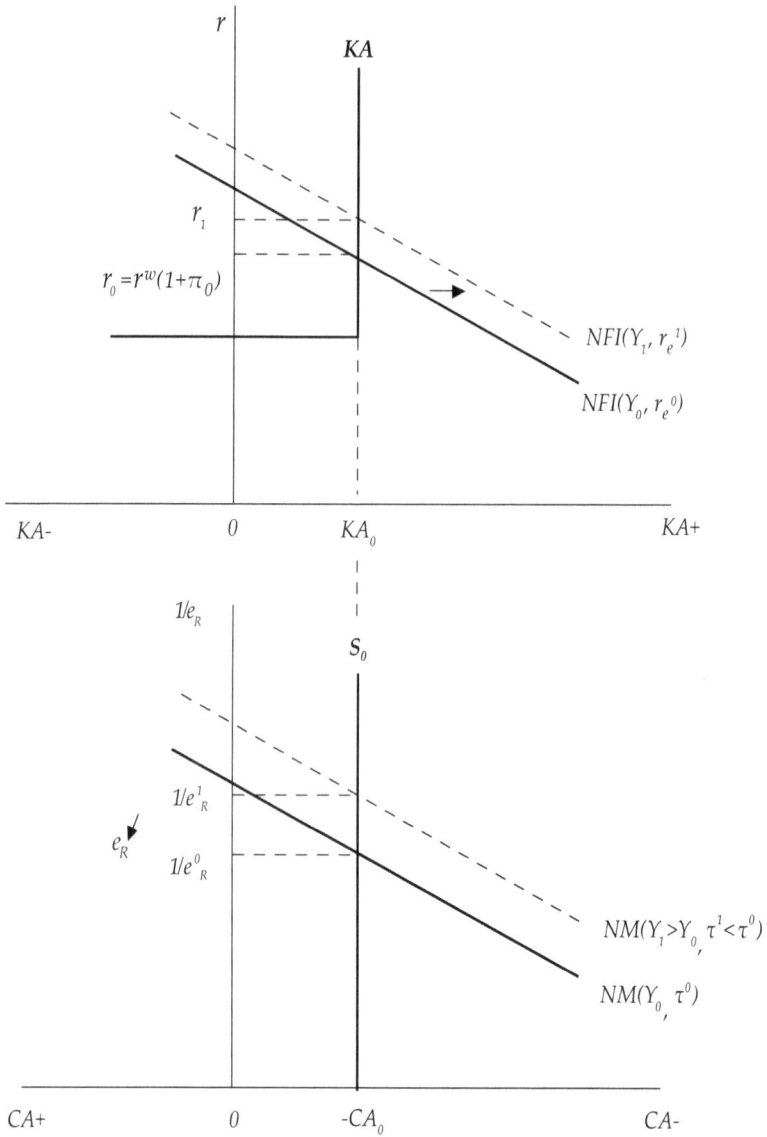

Figure A9.4 Trade reform with no capital controls

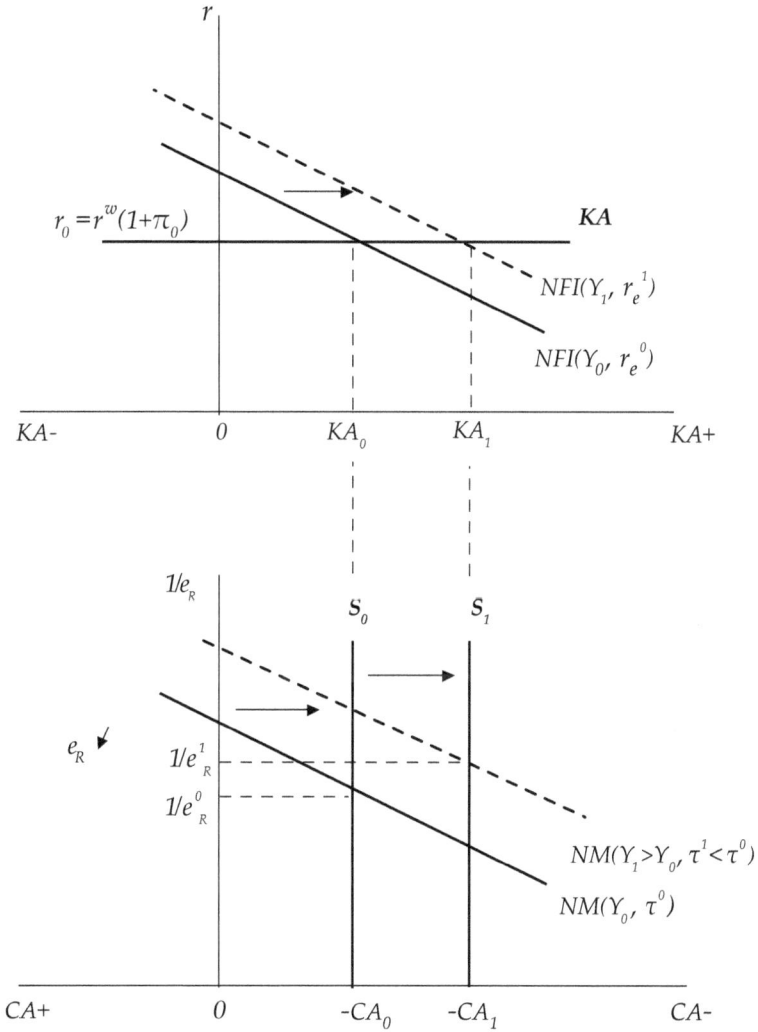

Table A9.1   Model structure

Regions
  1. China, including Hong Kong and Taiwan
  2. Vietnam
  3. Other ASEAN
  4. Japan
  5. Korea
  6. Australia
  7. United States
  8. European Union[a]
  9. Rest of World
Primary factors
  1. Agricultural land
  2. Natural resources
  3. Skill
  4. Labour
  5. Physical capital
Sectors[b]
  1. Paddy rice
  2. Beverages (product 8 OCR, 'crops nec')
  3. Other crops (wheat, other cereal grains, vegetables, fruits, nuts, oil seeds, sugar cane and sugar beet, plant based fibres and forestry)
  4. Livestock products (cattle, sheep, goats, horses, wool, silk-worm cocoons, raw milk, other animal products)
  5. Fish (marine products)
  6. Energy (coal, oil, gas)
  7. Minerals
  8. Processed food (meat of cattle, sheep, goats and horses, other meat products, vegetable oils and fats, dairy products, processed rice, processed sugar, processed beverages and tobacco products)
  9. Light manufacturing (textiles, wearing apparel, leather products and wood products)
  10. Other manufacturing (paper products and publishing, petroleum and coal products, chemicals, rubber and plastic products, other mineral products, ferrous metals, other metals, metal products, motor vehicles and parts, other transport equipment, electronic equipment, other machinery and equipment, other manufactures)
  11. Transport (sea transport, air transport and other transport)
  12. Infrastructure services (electricity, gas manufacturing and distribution, and water)
  13. Construction and dwellings
  14. Other services (retail and wholesale trade, communications, insurance, other financial services, other business services, recreation, other private services, public administration, defence, health and education)

Notes: [a] The European Union of 15.
[b] These are aggregates of the 57 sector GTAP Version 5 database.

Table A9.2 Manufactured inputs as shares of the total cost of production[a]

	Light mfg inputs as a share of production cost			Heavy mfg inputs as a share of production cost		
	Total	Domestic	Imported	Total	Domestic	Imported
Light mfg	44	34	9	11	8	3
Heavy mfg	3	2	0	53	39	14

Note: [a] These are input shares of total value added in each industry, calculated from the 2001 database following the 1997-2001 trade reforms.
Source: The GTAP Version 5 Database, as modified by simulations described in the text.

Table A9.3 Chinese equivalent import tariff and export tax rates[a]

	Equivalent import tariff, per cent		Equivalent export tax[b]
	Pre-accession	Post-accession	(per cent)
Rice	0.04	0.04	-0.13
Beverages	11.00	11.00	-0.11
Other crops	22.00	16.00	-0.25
Livestock	5.00	4.00	-0.02
Food	16.00	6.00	-0.13
Fish	8.00	8.00	-0.19
Minerals	0.40	0.40	-0.32
Energy	3.00	3.00	-0.32
Light manufacturing	14.00	8.00	2.63
Heavy manufacturing	8.00	4.00	-1.01
Transport	0.00	0.00	-0.46
Infrastructure services	0.00	0.00	-0.25
Construction	0.00	0.00	-0.54
Other services	0.00	0.00	-0.28

Notes: [a] All tariff and tax equivalents are ad valorem. They are intended to encompass both tariff and non tariff barriers, though the accounting for non tariff barriers is incomplete.
[b] Negative export tax rates indicate export subsidies. These incorporate the export subsidy equivalents of the duty drawbacks available on imported inputs by exporting firms, calculated as explained in the text.
Sources: The original 1997 numbers were aggregated from the 57 commodity categories in the GTAP Version 5 global database, 2000. They were then modified, as described in the text, based on the work of Ianchovichina, E. and Martin, W., 2002. *Economic impacts of China's accession to the WTO*, World Bank Policy Research Working Paper 3053, World Bank, Washington DC. Finally, the post-accession rates are based on the protocol for China's accession, as obtained from the WTO website at http://www.wto.org

259

Table A9.4   Simulated long-run effects of a unilateral liberalisation of China's 2001 trade policy regime[a]

Change in:	Change
Domestic CPI, P^C, (per cent)	-1.67
Domestic GDP deflator, P^Y (per cent)	-1.90
Price of capital goods, P^K (per cent)	-1.37
Terms of trade (per cent)	-1.25
Real effective exchange rate, e_i^R (per cent)	-1.98
Real exchange rate against USA, e_{ij}^R (per cent)	-1.81
Global interest rate, r^w (per cent)	0.10
Investment premium factor, $(I+\varpi)$ (per cent)	0.00
Home interest rate, r (per cent)	0.10
Current return on installed capital, r (per cent)c	1.30
Real domestic investment, I (per cent)	0.95
Balance of trade, $X-M=-KA=-(I-S_p)$ (US$ bn)	-10.87
Real gross sectoral output (per cent)	
Rice	-3.29
Beverages	2.77
Other crops	-1.25
Livestock	0.03
Food	-5.41
Fish	-0.17
Minerals	0.88
Energy	0.78

Light manufacturing	1.49
Heavy manufacturing	1.08
Transport	1.44
Infrastructure services	0.34
Construction and dwellings	0.73
Other services	0.76
Real GDP, Y	0.41
Unskilled wage and employment (per cent)	
Nominal (unskilled) wage, W	-0.42
Production real wage, $w=W/PY$	1.51
Employment, LD	0.00
Unit factor rewards CPI deflated (per cent)	
Land	-2.65
Unskilled Labour (those employed)	1.27
Skilled Labour	1.38
Physical capital	1.24
Natural Resources	1.20

Notes: [a] All results in this table are based on the adoption of Fiscal policy 1: government spending is held constant as a share of GDP and the revenue lost from tariff reform is not made up in other taxes, so the fiscal deficit expands. Key exogenous variables are highlighted as per the long-run closure discussed in the text.

b E is the nominal exchange rate in US$ per unit of local currency.

c P^c is the domestic consumer price level; the CPI. Note that this is an index of the prices of both home and imported goods.

Source: Model simulations described in the text.

Table A9.5   Factor intensities by industry[a]

	Land	Skilled labour	Unskilled labour	Physical capital	Natural resources
Rice	30	1	58	12	0
Beverages	8	7	36	48	0
Other crops	27	1	60	12	1
Livestock	30	1	58	12	0
Food processing	0	8	38	54	0
Fish	0	0	48	12	40
Minerals	0	6	37	37	20
Energy	0	3	18	42	37
Light manufacturing	0	9	48	43	0
Heavy manufacturing	0	10	42	47	0
Transport	0	16	38	46	0
Infrastructure services	0	11	18	71	0
Construction	0	10	50	39	0
Other services	0	28	30	42	0

Note: [a] These are factor shares of total value added in each industry, calculated from the 2001 database following the 1997-2001 trade reforms.
Source: The GTAP Version 5 Database, as modified by simulations described in the text.

Table A9.6   Trade to value added ratios by industry[a]

	Exports to value added ratio	Competing imports to value added ratio
Rice	0.006	0.000
Beverages	0.665	0.669
Other crops	0.034	0.108
Livestock	0.044	0.068
Food processing	0.177	0.545
Fish	0.057	0.067
Minerals	0.046	0.205
Energy	0.196	0.577
Light manufacturing	1.583	0.677
Heavy manufacturing	0.937	1.097
Transport	0.241	0.190
Infrastructure services	0.015	0.018
Construction	0.012	0.023
Other services	0.105	0.066

Note: [a] These are quotients of the value of exports or imports at world prices and domestic value added in each industry. They are from the 2001 global database (following the trade reforms of 1997-2001).
Source: The GTAP Version 5 Database, as modified by simulations described in the text.

Table A9.7   Short-run closure[a]

Region	Monetary policy target[b]	Labour market closure: nominal wage[c]	Capital controls: capital account net inflow I-SD[d]
China	Nominal exchange rate, E	Sticky (λ=0.5)	Rigid
Vietnam	Nominal exchange rate, E	Sticky (λ=0.5)	Rigid
Other ASEAN	Consumer price level, P^C	Flexible (λ=1)	Flexible
Japan	Consumer price level, P^C	Flexible (λ=1)	Flexible
Korea	Consumer price level, P^C	Flexible (λ=1)	Flexible
Australia	Consumer price level, P^C	Rigid (λ=0)	Flexible
United States	Consumer price level, P^C	Rigid (λ=0)	Flexible
Europe (EU)	Consumer price level, P^C	Rigid (λ=0)	Flexible
Rest of World	Nominal exchange rate, E	Flexible (λ =1)	Flexible

Notes: [a] The expected future return on installed capital is exogenous and determined in a separate long-run solution.
[b] The nominal money supply is endogenous in each case, the corresponding exogenous variable being the listed target.
[c] When the nominal wage is assumed flexible it is endogenous and the corresponding exogenous variable is the employment level. When it is sticky or rigid, equation 2 of Section 2 is activated and the employment level is endogenous.
[d] Capital controls are assumed to maintain a rigid net inflow of foreign investment on the capital account. When $KA = I\text{-}SD$ is made exogenous to represent this, an interest premium opens between the domestic and international capital markets. This premium becomes endogenous. Effectively, the home and foreign capital markets are separated and clear at different interest rates. Where the capital account is flexible (open), this implies that private flows on the capital account are permitted at any level. $KA = I\text{-}SD$ is then endogenous and the home interest premium is exogenous (unchanged by any shock). This means that the home interest rate then moves in proportion to the rate that clears the global savings-investment market.

Table A9.8 Simulated short-run effects of a unilateral liberalisation of China's 2001 trade policy regime: by monetary policy[a]

Change in:	Retaining capital controls, monetary policy targets E^b	Removing capital controls, monetary policy targets E	Retaining capital controls, monetary policy targets P^C	Removing capital controls, monetary policy targets P^C
Nominal exchange rate (US$/•), E_i	0.00	0.00	-1.89	-0.96
Domestic CPI, P^C	-1.74	-0.79	0.00	0.00
Domestic GDP deflator, P^Y	-2.15	-1.03	-0.53	-0.31
Nominal Money supply, M_S	-2.54	-0.83	-0.62	-0.01
Import price index	-0.10	0.06	1.81	1.01
Price of capital goods, P^K	-1.80	-0.89	-0.16	-0.14
Terms of trade	-1.41	-0.85	-1.60	-1.00
Balance of trade, $X-M=-KA=-(I-S_D)$, US$ bn	0.00	-9.50	0.00	-8.60
Real effective exchange rate, e^R_i	-2.30	-1.19	-2.56	-1.43
Real exchange rate against USA, e^R_{ij}	-2.17	-1.03	-2.42	-1.27
Global interest rate, r^w	-0.04	0.08	-0.04	0.07
Investment premium factor, $(1+\pi)$	3.58	0.00	2.92	0.00
Home interest rate, r	3.55	0.08	2.88	0.07
Expected long-run return on capital, r^e	1.30	1.30	1.30	1.30
Current return on installed capital, r^c	0.82	1.99	1.66	2.31
Real domestic investment, I	-1.76	0.98	-1.24	0.98
Nominal domestic investment	-4.24	0.47	-1.91	1.25
Real consumption, C	-1.71	0.12	0.36	1.01
Real private savings, S_p	-0.21	1.52	2.11	2.54
Gov't spending as per cent of GDP (real)	0.00	0.00	0.00	0.00
Power of income tax, $1+\tau$	0.00	0.00	0.00	0.00
Fiscal deficit, per cent change	16.01	18.61	18.33	19.56
Fiscal deficit, change in US$ bn	12.10	14.00	13.80	14.00
Real gross sectoral output				

Rice	-0.43	-0.14	-0.16	-0.03
Beverages	0.47	0.49	0.70	0.61
Other crops	-0.30	0.02	0.02	0.15
Livestock	-0.10	0.49	0.25	0.61
Food	-0.63	-0.37	-0.37	-0.26
Fish	-0.01	0.32	0.24	0.41
Minerals	-0.15	0.30	0.31	0.48
Energy	0.06	0.08	0.20	0.15
Light manufacturing	0.46	0.19	1.03	0.51
Heavy manufacturing	-0.05	0.14	0.52	0.40
Transport	0.23	0.56	0.77	0.80
Infrastructure services	-0.07	0.32	0.38	0.51
Construction and dwellings	-1.48	0.94	-0.97	0.97
Other services	-0.05	0.49	0.44	0.69
Real GDP, Y	-0.09	0.42	0.39	0.61
Unskilled wage and employment				
Nominal (unskilled) wage, W	-0.87	-0.40	0.00	0.00
Production real wage, $w=W/P^Y$	1.31	0.64	0.53	0.31
Employment, L^D	-0.28	0.64	0.59	1.00
Unit factor rewards CPI deflated				
Land	-0.98	1.65	0.75	2.29
Unskilled Labour (those employed)	0.88	0.40	0.00	0.00
Skilled Labour	0.81	0.75	0.33	0.51
Physical capital	0.53	1.33	1.04	1.51
Natural Resources	0.99	2.77	3.24	3.75

Notes: [a] All results in this table are based on the adoption of Fiscal policy 1: government spending is held constant as a share of GDP and the revenue lost from tariff reform is not made up in other taxes, so the fiscal deficit expands.
[b] E is the nominal exchange rate in US$ per unit of local currency.
[c] PC is the domestic consumer price level; the CPI. Note that this is an index of the prices of both home and imported goods.
Source: Model simulations described in the text.

Table A9.9 Simulated short-run effects of a unilateral liberalisation of China's 2001 trade policy regime: by fiscal policy

Change in:	Tight capital controls, monetary policy targets the nominal exchange rate, E		No capital controls, monetary policy targets the nominal exchange rate, E	
	Fiscal Policy 1: increase deficit[a]	Fiscal Policy 2: tax mix switch[b]	Fiscal Policy 1: increase deficit[a]	Fiscal Policy 2: tax mix switch[b]
Nominal exchange rate (US$/•), E_i	0.00	0.00	0.00	0.00
Domestic CPI, p^{C}	-1.74	-1.93	-0.79	-1.36
Domestic GDP deflator, p^{Y}	-2.15	-2.27	-1.03	-1.59
Nominal Money supply, M_S	-2.54	-2.51	-0.83	-1.45
Import price index	-0.10	-0.10	0.06	0.00
Price of capital goods, P_K	-1.80	-1.74	-0.89	-1.18
Terms of trade	-1.41	-1.46	-0.85	-1.12
Balance of trade, $X-M = -KA = -(I-S_I)$, US$ bn	0.00	0.00	-9.5	-5.90
Real effective exchange rate, e_i^R	-2.30	-2.42	-1.19	-1.75
Real exchange rate against USA, e_{ij}^R	-2.17	-2.28	-1.03	-1.59
Global interest rate, r^w	-0.04	-0.04	0.08	0.03
Investment premium factor, $(I+\pi)$	3.58	2.33	0.00	0.00
Home interest rate, r	3.55	2.28	0.08	0.03
Expected long-run return on capital, r^e	1.30	1.24	1.30	1.24
Current return on installed capital, r^c	0.82	0.67	1.99	1.38
Real domestic investment, I	-1.76	-0.82	0.98	0.97
Nominal domestic investment	-4.24	-2.89	0.47	0.18
Real consumption, C	-1.71	-2.43	0.12	-1.37
Real private savings, S_p	-0.21	-2.23	1.52	-1.37
Gov't spending as per cent of GDP (real)	0.00	0.00	0.00	0.00
Power of income tax, $I+\tau$	0.00	1.13	0.00	1.23
Fiscal deficit, per cent change	16.01	0.00	18.61	0.00
Fiscal deficit, change in US$ bn	12.10	0.00	14.00	0.00
Real gross sectoral output				

Rice	-0.43	-0.57	-0.14	-0.41
Beverages	0.47	0.41	0.49	0.42
Other crops	-0.30	-0.45	0.02	-0.26
Livestock	-0.10	-0.32	0.49	0.02
Food	-0.63	-0.80	-0.37	-0.65
Fish	-0.01	-0.19	0.32	0.00
Minerals	-0.15	0.06	0.30	0.36
Energy	0.06	0.09	0.08	0.10
Light manufacturing	0.46	0.49	0.19	0.33
Heavy manufacturing	-0.05	0.11	0.14	0.24
Transport	0.23	0.19	0.56	0.39
Infrastructure services	-0.07	-0.11	0.32	0.12
Construction and dwellings	-1.48	-0.72	0.94	0.85
Other services	-0.05	-0.13	0.49	0.20
Real GDP, Y	-0.09	-0.04	0.42	0.28
Unskilled wage and employment				
Nominal (unskilled) wage, W	-0.87	-0.97	-0.40	-0.68
Production real wage, $w=W/P^Y$	1.31	1.33	0.64	0.92
Employment, L^D	-0.28	-0.19	0.64	0.39
Unit factor rewards CPI deflated				
Land	-0.98	-2.17	1.65	-0.67
Unskilled Labour (those employed)	0.88	0.98	0.40	0.69
Skilled Labour	0.81	0.90	0.75	0.88
Physical capital	0.53	0.67	1.33	1.18
Natural Resources	0.99	1.08	2.77	2.19

Notes: [a] Fiscal policy 1 has government spending held constant as a share of GDP and the revenue lost from tariff reform is not made up in other taxes, so the fiscal deficit expands.
[b] Fiscal policy 2 holds government spending constant as a share of GDP but the rate of indirect tax is allowed to rise so that the fiscal deficit also remains constant as a share of GDP.
Source: Model simulations described in the text.

10 Trade reform, macroeconomic policy and sectoral labour movement in China

Jennifer Chang and Rod Tyers

Although the Chinese economy continues to grow rapidly, since the Asian financial crisis of the late 1990s there is evidence of a slowdown, most prominently in per capita rural income growth. One explanation for this is that the relocation of labour from agriculture to manufacturing and services, essential in any growing developing economy, has been retarded. This could be due to policy disincentives designed to control urban congestion, such as the household registration or *hukou* system (Ianchovichina and Martin 2002a) or information asymmetries and transaction/infrastructural costs (Sicular and Zhao 2002). A further hypothesis attributes comparatively poor performance in China's rural sector to economic reforms; in particular, to trade reform commitments in the leadup to and associated with WTO accession (Anderson et al. 2002).

It is the central hypothesis of this chapter that this comparative decline in rural performance is due, at least in part, to a combination of China's adherence to its *de facto* fixed exchange rate regime and to shocks that have tended to depreciate its real exchange rate. Most significant amongst these shocks was the surge of (largely illegal) outflows on the capital account

First published in Garnaut, R. and Song, L. (eds), 2003. *China: New Engine of World Growth*, Asia Pacific Press, Canberra:231-75.

and the associated private investment slowdown during the Asian financial crisis (Yang and Tyers 2001). Since then, however, numerous trade reforms have been implemented, all of which have tended to encourage Chinese consumption to shift toward foreign goods, so reducing home relative to pre tariff foreign goods prices and hence further depreciating the real exchange rate (Tyers and Rees Chapter 9). By definition, a real depreciation must be accompanied either by a nominal depreciation, a domestic deflation, or a combination of both. The *de facto* peg to the US dollar has therefore necessitated China's deflation. When prices are falling there is downward pressure on wages. Even if wages fall only slightly more slowly than prices, however, other things being equal, employment growth in the wage sectors of the economy can be expected to decline.[1] In the Chinese case this appears as high real wage growth in the modern sector but reduced labour demand growth there and hence a 'bottling up' of workers in the rural sector and reduced rural income per capita.

The relevance of this story to China's comparatively poor recent per capita rural income growth performance is examined here using short and long-run comparative static analysis. The shocks considered are China's WTO accession commitments and the model used is that described in Tyers and Rees (Chapter 9). This model is a development of the model introduced by Yang and Tyers (2000). It is a multi-sector, multi-country comparative static macro model, the microeconomic components of which have their origins in GTAP.[2] All countries have open capital accounts and forward-looking investor behaviour is represented in the short run via expectations formed from long-run simulations. The focus here, however, is on labour relocation and the short run consequences of trade reform shocks for the uptake of labour in the manufacturing and services sectors.[3] We examine the pace of such labour relocation historically and compare this with simulated changes in labour demand following the WTO accession.

We observe a reversal in the rate of relocation of workers from agriculture to manufacturing after 1998 and a glut of workers in the rural sector (Dolven 2003). Our simulations suggest that the retarding effect of the commonly cited system of internal migration restrictions, known as the *hukou* or household registration system (HRS), has been enhanced by China's broader macroeconomic policy regime. Indeed, the exchange rate regime, combined with capital controls, appears to have restricted the flow of workers into manufacturing and services by at least 1 per cent per year and into services by at least 2 per cent per year.

Urban-rural inequality and worker relocation in China

The pace of rural per capita income growth depends on the rate at which surplus labour is generated by agricultural change and the corresponding rate at which workers are accommodated in the manufacturing and services sectors. It is a common view that the economic reform of the early 1980s, which brought de-collectivisation and the household responsibility system, along with associated increases in agricultural labour productivity, created a substantial rural labour surplus.[4] Moreover, prior to that reform, wages were not market-determined and there is evidence of considerable underemployment. This was due, in part, to the seasonal nature of agricultural activity, which meant that many workers were left idle in the off-peak parts of the agricultural cycle, yet unable to take full employment elsewhere (Banister and Taylor 1989).

An exacerbating factor on the supply side in the Chinese case was rapid rural population growth in the pre reform period. Even though the rate of rural population growth slowed in the 1970s and 1980s, the growth of the rural labour supply continued strongly due to the aftermath of the 'baby boom' periods of the 1950s and 1960s.[5] On the demand side, China's pre reform economic strategy promoted relatively capital-intensive heavy industries. These two effects tended to raise underemployment in the rural sector in the post reform period, where, although rural communes were disbanded, all workers were still technically employed.[6] The post reform period therefore carried considerable potential for worker relocation from agriculture into other sectors. The subsequent volume of internal migration would depend, then, on economic incentives in the form of urban–rural income inequality and on both policy-induced and natural barriers to migration.

Urban-rural inequality in China

Income inequality in China exists between urban and rural residents as well as within urban and rural areas. It also has a strong regional pattern, particularly between inland and coastal provinces. Of overall income inequality, Lozada (2002) estimates that 75-80 per cent is due to the urban-rural divide. Official estimates of the trends of rural and urban household incomes are displayed in Figure 10.1. These indicate a general upward trend in income in both areas while urban income experienced comparatively sharp increases in the 1990s. The corresponding proportional difference between urban and rural household incomes is plotted in Figure 10.2. The comparatively rapid growth of rural incomes in the early 1980s, and the associated decline in urban-rural inequality, coincided with the major reforms of that period that first impacted on the rural sector. Since then, however, urban incomes have growth more rapidly, most prominently in the early to mid 1990s leading up to the Asian crisis in 1997-98, and in the post crisis period.

Internal migration

If there were no costs or barriers to internal migration, workers would be expected to respond to the widening income gap between rural and urban areas by migrating until wages were equalised. Indeed, rural labour markets that facilitate this relocation of workers have been essential to economic growth in the more advanced countries (Burgess and Mawson 2003). Although the estimated number of Chinese internal migrants has grown (100 million since the 1980s) these relocations have been insufficient to stem the growing inequality (Dolven 2003). Some of this apparent growth in inequality is due to a rise in skilled employment in urban areas. A more controversial portion is due to reduced urban labour demand associated with real wage growth and to official migration barriers. Together, they appear to have left an expanding labour supply 'bottled up' in the rural sector. The key official barrier has been the HRS.

Introduced in major cities in 1951 and extended to rural areas in 1955, the HRS was intended to deal with the escalating urban influxes of rural migrants at the time of the Great Famine of 1959-61 (Chan and Zhang 1999).

The associated food shortages continued into the 1970s, only ending in the 1980s (Zhao 2000). Each citizen is required to register in their place of regular residence, often their birthplace. Moving from rural to urban areas requires a complex and costly application for a transfer to a local *hukou*. Prior to the reform in the 1980s, this was a huge barrier, as without a proper *hukou* one would not qualify for a government job assignment. Workers who ignored this requirement were denied social security benefits such as housing and other necessities, even food.

Decollectivisation in rural areas allowed income inequality to soften a little and non-farm rural industries to develop. This, and the household responsibility system that accompanied it, had the effect of increasing productivity in the agricultural sector. Thus, while rural incomes grew, so also did the proportion of workers considered redundant.[7] The government began rewarding collective farms and cooperatives for production rather than labour employed, so excess workers could be free to look for other jobs (Seeborg, Jin and Zhu 2000). The marketisation of food also meant that migrants were no longer restrained entirely by the need to obtain a local *hukou* to survive. Combined with the development of special economic zones and the increase in urban private and informal sectors, the cities were on a path of rapid development with growing demand for labour. Yet the HRS created ambiguous incentives for migrants. On the one hand, it constrained people with access to land from migrating, since each was now responsible for their own land, which most saw as a form of social security. On the other hand, individuals with little or no access to land or who had specialised in what were to become low productivity rural activities were now less constrained.

Through the 1980s and 1990s the HRS was gradually weakened by policy reforms and less stringent enforcement. Previously prohibited 'spontaneous' migration to jobs in urban centres was tacitly facilitated and tolerated, conditionally opening urban residency to rural workers and relaxing some of the strict controls. There has therefore been an increase in temporary urban migrants: workers going to cities and towns without official residential status, known as the 'floating population'. Figures from the late 1980s indicate that transients made up over one-fifth of the population of such major cities as Beijing, Shanghai, and Guangzhou (Canadian Immigration and Refugee Board 2002). Many official restrictions remained, however, including the link between *hukou* status and welfare eligibility, the denial of education for migrant children, and state job availability. In 1995, to re-assert control over internal migration, Zhao (2000) notes that tighter controls were imposed over the legality of urban residency and housing subsidies.[8]

As part of China's urbanisation strategy for the 10th Five-Year plan, Central Party Document Number 11 of November 2000 allows a person and his or her immediate family to obtain urban *hukous* if he or she has fixed accommodation and stable work (in a job for more than one year) in the urban area. Urban *hukou* was also offered to those who purchased a local commercial housing unit (to attract outside investment) and to holders of graduate degrees (attracting professionals). The focus of these reforms was still mainly on small towns and small cities, however. Nonetheless, according to China's Committee to Restructure the Economy (SCORES), during 2001 about 600,000 rural residents acquired urban *hukou* in these small urban centres. More recently still, a State Council directive indicated that rural migrants have a legal right to work in cities. It prohibits job discrimination based on residency and orders that urban residency documents are to be provided to any workers who find employment. A further recent directive indicates that businesses should stop delaying wage payments to workers. This is a significant improvement for migrant workers as they often find that they are taken advantage of due to their uncertain legal status.

Official barriers to migration therefore appear to have had an ever diminishing effect and the least effect in the period since the Asian crisis of 1997. Other things equal, then, we would expect an acceleration in the relocation of workers into jobs in the modern sector in that period. This is not what the evidence suggests, however.

Sectoral relocation of workers

The trends in employment by major sector are indicated in Figure 10.3. The early 1990s was a period of rapid industrial expansion during which workers relocated from the rural sector to the industrial and services sectors. Significantly, while the expected long term rise in the share of services in total employment is borne out, industrial employment fell substantially during the Asian crisis period and failed to expand thereafter. Indeed, the farming and forestry sector shows annual declines in employment

273

that peaked in 1993 and continued until 1996. In this period the Chinese economy grew at its greatest historical rates. During the Asian crisis and in the period thereafter, however, declines in the growth rates of activity in manufacturing and services appear to have driven workers back to the rural sector. In the post crisis period this backflow appears to have continued.

Performance since the Asian crisis

Good weather and improved farming incentives brought a resurgence of the rural economy in 1995 and the reform of state-owned industrial enterprises (SOEs) was accelerated, precipitating substantial lay-offs (Meng 1998). These changes changed saving incentives, leading to a sudden increase in the private saving rate (Yang and Tyers 2000, 2001). The contractionary effects of these internal shocks were worsened by the arrival of the Asian financial crisis in 1997, which saw the mostly illegal flight of a substantial part of China's additional private savings, leading to a fall in private investment. At the same time, the slowdowns and currency depreciations in other Asian economies made their exports relatively competitive, so that the dollar value of China's exports grew more slowly than in 1990-96 (Yang 1998).

Our macroeconomic policy story begins with the crisis and China's reaction to it. The shocks of that period, the rise in the domestic saving rate, the increased outflow on the capital account, and the adverse change in the terms of trade depreciated the real exchange rate. This effect can be seen from the following definition of the real bilateral exchange rate, for region i with region j, as the common currency ratio of the price of a basket of region i's goods and the price of a corresponding basket of region j's goods:

$$e_{ij}^R = E_{ij}\left(\frac{P_i^Y}{P_j^Y}\right) = \frac{P_i^Y}{\left(\dfrac{P_j^Y}{E_{ij}}\right)},$$

$$\tag{1}$$

where E_{ij} is i's nominal exchange rate with j, measured as foreign currency units per local unit and the price index used is the GDP price, P^Y.[9] The real exchange rate depreciates when the price of the basket of home goods falls relative to the corresponding basket of foreign goods. The real depreciation in the crisis, combined with the fixed nominal exchange rate of the period, necessitated a monetary policy sufficiently restrictive to bring about a decline in the home price level relative to that in the US—a deflation. This was the first in a series of deflations in the crisis and post crisis periods.

Shocks that cause real depreciations also include the other elements of China's domestic reform program. By raising productivity, these reduce domestic costs relative to foreign costs and hence they tend to reduce relative domestic prices. Similarly, trade reforms divert domestic demand away from home produced goods toward imports and so they also tend to reduce the prices of domestic goods relative to the import prices of foreign goods. In the period 1997—2001 China embarked on substantial trade liberalisation, a major part of which included the introduction of a duty drawback system on the imported inputs of export firms (Tyers and Rees, Chapter 9). Real depreciations due to such shocks are robust to China's macroeconomic policy regime. But it is that regime which distributes the nominal effects of the real depreciations between falls in the home price level, or deflation, on the one hand and depreciation of the nominal exchange rate on the other.

The declines in China's official measures of the domestic price are indicated in Table A10.1.[10] Other things held equal, such deflations are contractionary. This is because, even in the most flexible of industries, wage rates are renegotiated more rarely than product prices are adjusted. Lags in wage adjustment mean that deflation applies a profit squeeze that retards both employment growth and investment in the private sector. Consistent with this, real wage growth in China's modern sector has been extraordinarily high since 1996, as Figure 10.5 shows, suggesting some deterrence of labour demand growth and hence of growth in this sector. Weaker overall performance of the Chinese economy is indicated by its official GDP growth rates (Figure 10.4). These bottomed out in the crisis period at between seven and 8 per cent per year, and have not recovered since.

Since 2001 China has embarked on a still more dramatic set of trade reforms as part of its commitments associated with WTO accession. By themselves, these reforms will cause further real depreciations. Should there be no change in macroeconomic policy, growth and the relocation of workers into China's modern sector will continue to be retarded by deflation. In the remainder of the paper we investigate the relationship between growth, labour relocation and macroeconomic policies in the aftermath of the WTO accession reforms.

Modeling the effects of China's trade policy reforms

Here we simulate the short-run effects of trade reforms in China while accounting for the implications of services trade commitments as well as productivity changes associated with these reforms. As in Tyers and Rees (Chapter 9) and Ianchovichina and Martin (2002a), we make allowance for idiosyncratic trade policies, such as the duty drawbacks on imports used in the manufacture of exported goods. And, as befits a short-run analysis, we also allow for labour market rigidity and associated departures from full employment.

Following Yang and Tyers (2000), we use a comparative static global macroeconomic model, within which the microeconomic (supply) side is adapted from GTAP,[11] a multi-region comparative static model in real variables with price-taking households and all industries comprising identical competitive firms. To this microeconomic base are added independent representations of governments' fiscal regimes, with both direct and indirect taxation, as well as separate assets in each region (currency and bonds) and monetary policies with a range of alternative targets. The details of the model are described in Tyers and Rees (Chapter 9).

Simulated effects of accession policy reforms

A long-run outlook is required in order that the expectations of investors can be formulated for short-run analysis—investors are assumed to take changes in long-run returns on installed capital into account in determining short run changes in their investment behaviour. Long-run results are also of interest in their own right.

The key elements of the long-run closure are as follows
- there are no nominal rigidities (no rigidity of nominal wages)
- production and consumption elasticities of substitution are chosen at 'standard' levels to reflect the additional time for adjustment in the long-run over the short run (Tyers and Yang 2001)

- physical capital is no longer sector-specific; it redistributes across sectors to equalise rates of return
- capital controls are ignored, and
- in China, irrespective of short-run fiscal policy assumptions, in the long-run any loss of government revenue associated with tariff changes is assumed to not be made up via direct (income) taxes, with the result that the fiscal deficit expands.

The key point of difference between our long-run analysis and that of Tyers and Rees (Chapter 9) is that we represent the effects of trade reforms on productivity. There is a substantial literature identifying this association (Chand et al. 1998, Chand 1999 and Stoeckel et al. 1999). These studies use Australian data on the long-run effects of trade reforms to identify elasticities of total factor productivity to protection level by industry. We applied these elasticities to China's intended reforms, albeit with discounts for China's lower starting protection levels in some industries and adjustments to account for services sector reforms (Dee and Hanslow 2000, Verikios and Zhang 2001), to yield the one-off long-run productivity shocks listed in Table A10.4. Although these shocks are applied only in the long run, they are important for short run behaviour (our object here) because they raise the return on installed capital and hence they stimulate investment.

The results from the long-run simulation are provided in Table A10.3. They show the expected allocative efficiency gains, reflected here in a rise in GDP, aided by increased returns on installed physical capital that induce greater investment and therefore larger net inflows on the capital account. Home consumption switches away from home produced goods, the relative prices of home produced goods fall, yielding the predicted real depreciation. The principal downside of the reforms is the long-run shift of activity out of agriculture into manufacturing and services and the associated decline in land rents. Associated with this shift, a substantial relocation of workers from agriculture to the modern sector will therefore be required.

Although the trade policy regime of 2001 advantaged food processing, 'other crops', fisheries and light manufacturing, apart from the smaller 'beverages' industry, it is the manufacturing sectors that are the robust beneficiaries of the unilateral trade liberalisation. This is surprising, given that the protection of the manufacturing sectors is also set to decline. This result arises because both manufacturing sectors commit approximately half their total costs to inputs in the same product category and 10 to 15 per cent of those to imports; and competing imports, even though they are from the same sector, are differentiated from home products (see Tyers and Rees, Chapter 9). Under these conditions the tariff reductions on imported intermediates have a direct effect on home industry total costs. Reductions of tariffs on competing, but differentiated, imports have only an indirect effect, the magnitude of which depends on the elasticity of substitution between the two. Indeed, for manufacturing, it turns out that the input cost effect of tariff reductions is considerably greater than that of the loss of protection against competing imports. Cost reductions of similar origin are the reason for similar gains accruing to the domestic transport services sector.

The reforms result in the most substantial reductions in protection to China's food processing sector and therefore lead to long-run contractions in that sector and in the local supply of its inputs (especially rice and 'other crops'). The more income elastic and lightly protected agricultural sectors, the 'beverages' group and livestock, expand. Labour is assumed to be perfectly mobile between sectors, so that our results indicate the labour movement needed in order to achieve the maximum gain from the reforms. In the long run, employment in food processing falls by 7 per cent, in rice production by 4 per cent and in 'other crops' by 2 per cent. Workers lost from these sectors are re-employed primarily in manufacturing and services.

When reform-driven productivity improvements are included in the long-run analysis, the economic effects of the reforms are greatly amplified, even though the assumed productivity changes are one-off and modest. The GDP increase is almost ten times larger, domestic investment is four times larger, as is the return on installed capital that will drive short term investment. Increases in sectoral expansions are largest in manufacturing and services with productivity growth in services having widespread effects through their role as intermediates. The contraction in the agricultural sector is much reduced, with the beverages, livestock and fisheries sectors now showing robust expansion.

Simulated short-run effects

In the short run, the standard closure is as indicated in Table A10.4. Monetary authorities in China, Vietnam and the Rest of the World are assumed to maintain fixed exchange rates against the US$. The other regions identified adopt price level (CPI) targeting. Capital controls are assumed to be rigid in China and Vietnam, but they are non-existent in the other regions. In the labour markets of China and Vietnam, nominal wages are assumed to be sticky. Full short run rigidity is assumed in the industrial countries, while nominal wages are assumed to be fully flexible elsewhere in Asia and the developing world. As to fiscal policies (not shown in the table), government spending in all regions is assumed to absorb a fixed proportion of GDP and the rates of direct and indirect tax are constant, so that government deficits vary in response to shocks.

Three macroeconomic policy regimes are considered.

- The 'standard': rigid capital controls, a fixed exchange rate and fixed direct and indirect tax rates.

- No capital controls, a fixed exchange rate and fixed direct and indirect taxes.

- No capital controls, a floating exchange rate with monetary policy targeting the GDP price, and fixed direct and indirect tax rates (Table A10.4).

In addition, two different assumptions are made about investment and the services capital stock. In a pessimistic scenario, investors do not see the long-run benefits of productivity gains that would accompany trade reform, and there is no immediate effect from reforms in services trade. In an optimistic counterpart, investors are motivated by the long-run return on installed capital listed in the second column of Table A10.3. Moreover, drawing on the conclusions from studies by Dee and Hanslow (2000) and Verikios and Zhang (2001), service trade reforms are assumed to result in the short run bolstering of the capital stock in finance and communications indicated in Table A10.2. Not surprisingly, the short-run effects of China's WTO commitments prove to be heavily dependent on its macroeconomic policy regime and these associated productivity and capital stock changes.

Indeed, the effects range from the contraction alluded to in the introduction through to a substantial short run expansion.

The effects of capital controls and the choice of the monetary policy target

In broad terms, the behaviour of the model in the short run with rigid capital controls retained can be represented as in Figure 10.6. The upper diagram represents the domestic capital market and the lower one the domestic market for foreign products. These markets are linked by the requirement that, for a balance of payments, net flows on the capital account must mirror those on the current account. Net demand for foreign products (the downward sloping line in the lower diagram, NM=M-X) depends on the relative price of foreign goods. For this purpose we define the real exchange rate as in Equation 1 as the common currency ratio of the price of home goods to the price of foreign goods. Net imports depend positively on this real exchange rate and negatively on its inverse (the common currency, foreign to home product price ratio). This excess demand curve is shifted to the right by an increase in GDP, Y, or a reduction in protection, τ. The real exchange rate is then determined by the balance of payments requirement that net inflows on the capital account must equal net outflows on the current account, KA=-CA=NM=M-X.[12]

The trade liberalisation reduces τ and shifts NM to the right. With tight capital controls, the current account balance cannot change. The shock therefore raises the relative price of foreign goods in the home market and depreciates the real exchange rate. If the nominal exchange rate is the target of monetary policy and the home economy is small by comparison with its trading partners (P* is unaffected) then a fall in P^Y (a deflation) is required. This must be brought about by a monetary contraction in defence of the exchange rate. To the extent that wages adjust more sluggishly than product prices, the deflation causes the real wage to rise. Were the real depreciation the only consequence of the liberalisation shock, its effects would be contractionary. Fortunately, this need not be the case. The trade reform brings gains in allocative efficiency.[13]

However, when capital controls remain rigid and the exchange rate is fixed, these allocative gains are insufficient to offset the contractionary effects of the deflation. This can be seen from the first column of Table A10.5. The real depreciation is substantial and the deflation required is of the order of 2 per cent per year. The production real wage rises by half this and employment falls. Investment demand responds to the expectation of higher real returns to installed capital in the future by shifting outward (Figure 10.6). The loss of tariff revenue drives the government deficit higher, reducing domestic saving, and further reinforcing the outward shift of the net foreign investment demand (NFI) curve. But the rigidity of the capital controls causes this to simply push up the domestic interest rate and so real investment falls. Output falls in all sectors except beverages, energy, manufacturing and transport. Manufacturing gains in the short run for the same reasons it gains in the long run—cheaper imported inputs. Therefore, under these policy circumstances the net gains from trade reform are not robust in the short run—at least when pessimistic assumptions are made about productivity effects and services reform.

If the capital controls are removed, the corresponding liberalisation shock is as depicted in Figure 10.7 (the results are shown in column 2 of Table A10.5). Here, reduced protection also yields a gain in allocative efficiency but this time it is large enough to generate a net gain in GDP, reinforcing the rightward shift in the net imports curve in the lower diagram. In this case, however, the absence of capital controls allows investment to flow in, responding to the increase in the expected long-run return on installed capital. The increased inflow on the capital account relaxes the balance of payments constraint in the lower diagram and allows a shift toward net imports. The net effect on the real exchange rate depends on whether the capital account shift, which raises the net supply of foreign goods, is larger or smaller than the increase in their net demand due to the tariff reduction and the rise in domestic income. In this case, the increase in net demand is dominant and the real exchange rate still depreciates, albeit to a lesser extent than in the presence of capital controls. Thus, when capital controls are weak or non-existent, trade liberalisation attracts increased inflows on the capital account in order to mitigate the real depreciation and associated GDP price deflation that are its inevitable consequences.

In the third column of Table A10.4 the target of monetary policy is the GDP price, so that the nominal exchange rate is allowed to depreciate. This removes the deflation that must accompany a fixed exchange rate and hence reduces the rise in the production real wage due to the reforms. The GDP gain is therefore almost doubled and now only the processed food sector contracts in the short run. Interestingly, the additional investment and greater employment generated with the policy regimes of columns 2 and 3 ensure that real land rents rise in the short run.

Finally, the three right hand columns of Table A10.4 indicate the short-run effects of the WTO accession reforms under the same three policy regimes but with the more optimistic ancillary effects of those reforms (Table A10.2) included. There are two key differences. First, investors are motivated by the effects of increased productivity in response to the reforms in the long run and, second, services reforms see a short run increment to the capital stocks in the financial and communications sectors. These changes cause substantial increases in investment and, in association, they cause larger net inflows on the capital account; they also boost the construction sector and reduce the cost of service inputs to other sectors. Overall, expansions are consistently larger in these cases as is the demand for the sectoral relocation of workers. These are due, in part, to the increases in total employment that occur because of sticky nominal wages. The upward movement in nominal wages is slower than that in labour productivity, expanding aggregate labour demand by up to 2 per cent.

Short run sectoral impacts

The key determinants of the sectoral mix of changes in the economy are the tariff reductions; these reduce product prices in affected sectors and the size of the resulting short run real depreciation, which reduces non traded (largely services) prices relative to traded goods prices. When capital controls are tight the real depreciation is comparatively large. Traded sectors, such as light manufacturing, are advantaged, while non traded services sectors, such as construction and dwellings, are disadvantaged. On the other hand, when capital controls are ineffective, manufacturing gains are smaller and the non traded services sectors gain. Processed food suffers because of the decline in its protection and other agricultural industries contract as that sector demands fewer local inputs. When the macroeconomic policy regime is expansionary, however, only the processed food sector contracts. And, when the optimistic ancillary effects are included, the decline in the processed food sector becomes trivially small, ensuring gains to the agricultural sector as a whole. This result is quite important since, sensing losses due to reduced protection, the farm sector opposes the reforms. Yet, if the reforms were embraced along with expansionary macroeconomic policy, the farm sector would be a net gainer.

Worker relocation demand in the short run

The sectors defined for the purpose of our simulations are here aggregated for ease of comparison with the classification of employment by China's National Bureau of Statistics. In facilitating this comparison, it has occasionally been necessary to aggregate sectors in both classifications. The result is the seven sectoral groupings in Table A10.6. Also listed are the maximum and average annual changes in employment by sector, drawn from official statistics since 1978. The detailed official record of annual changes in employment by sector is provided in the Appendix. These changes include some of extraordinary magnitude, most especially the growth in employment in the services sectors in 1984. We discount these as due to changes in measurement in that year. There have, however, also been some extraordinary employment growth periods since then, including the service employment expansions of 1993 and 1994. Of course, the services sectors were considerably smaller then than they are now, and so the numbers of workers relocated to achieve those employment growth numbers were smaller than would now be required. Even so, these statistics suggest that China's capacity for the rapid sectoral relocation of workers has been considerable.

In the same table we show the range of simulated short run worker relocation demands associated with China's WTO accession reforms. In interpreting these it must be borne in mind that employment in China is endogenous in these simulations. From Table A10.5, depending on the macroeconomic policy settings, it either contracts a quarter of a per cent or expands by almost 2 per cent. This opens the possibility that, at least in the short run, rural activity can expand at sufficient pace to retain its workers. When we combine the most expansionary policy scenario with the ancillary effects of productivity and services reform (Table A10.2), the expansion in overall employment permits employment in each sector to increase, as indicated in the final column of Table A10.6. Of course, there remains considerable redistribution of employment in these results, since the services sectors expand their labour use much more rapidly than does agriculture. Since we focus on the short run, we take no explicit account of technical change in Chinese agriculture. In other countries at the same stage of development this change has been labour saving, enabling the agricultural sector to shed workers at a considerable rate (Anderson et al. 2002).

Our contention that the macroeconomic policy environment is important in determining the pace of worker relocation demand is borne out in these results. In all sectors there is a stark contrast between employment growth under tight capital controls and a fixed exchange rate regime on the one hand and an investment policy that renders the capital controls ineffective, combined with a flexible exchange rate, on the other. When the ancillary shocks to productivity and services capital are included, this gulf widens further. In the case of this most optimistic of scenarios, employment growth exceeds the average since 1990 in all but two sectors and the excess is largest in construction and dwellings and transport services. These strong worker relocation demands nonetheless fall short of the maxima achieved in a single year in all sectors, even since 1990.

Yet the WTO accession reforms are but a small part of the pantheon of China's overall reform program, the bulk of which is growth enhancing. We not only exclude the trend of technology and associated organisational changes, as mentioned previously, but we take no account of the ongoing financial sector reforms and the continuing transformation of urban activity from the public to the private sector. Even in the case of China's WTO accession, we take no account of commitments by China's trading partners to reduce protection against its labour-intensive exports. These changes will improve China's terms of trade and stimulate its growth. Considering such omissions, worker relocation demand in a more expansionary policy environment could well approach, or even exceed, the high rates of change observed in the early 1990s.

Finally, our simulations rest on the assumption that production and skilled workers in one sector are perfectly transformable into corresponding workers in other sectors. Because some workers have narrowly sector-specific education and training and because sectoral relocation has high transaction costs, particularly for rural families, the actual transformability of production workers will remain imperfect (Sicular and Zhao 2002). Our results therefore place upper bounds on the economic performance results from the WTO accession reforms and, more importantly for our purpose, upper bounds on worker relocation demands in the short run.

Conclusions

Our chronicle of changes in economic performance, income distribution and internal migration in China suggests a recent slowing of employment growth in the modern sector and the 'bottling up' of labour in rural activities, widening the income gap between urban and rural workers. This has happened in spite of what our review suggests is a considerable relaxation of the worker registration, or *hukou*, system that has constrained internal migration in the past.

To examine the hypothesis that the slowdown in economic growth, and the pace of worker relocation in particular, is due at least in part to a restrictive macroeconomic policy regime, we adapt a comparative static multi-product, multi-region macroeconomic model. We use the model to compare the economic effects of a key element of China's current economic reform program, namely its commitments associated with its accession to the WTO, under a variety of macroeconomic policy regimes. These range from very restrictive capital controls, combined with a fixed exchange rate, to a regime with no capital controls (or the equivalent in FDI flexibility) and a flexible exchange rate. The results suggest these regimes make a difference of at least 1 per cent per year in GDP growth and at least 2 per cent per year in employment growth in the economy's modern sector. They therefore support our hypothesis, at least to the extent that the macroeconomic policy regime has contributed to a slowdown in the pace of expansion and worker relocation. Indeed, with an expansionary macroeconomic policy and optimistic assumptions about productivity effects associated with the WTO accession reforms and inter-sectoral worker transformability, simulated worker relocation demands from these reforms alone exceed the average of China's recent experience.

While we are confident about our conclusion that China's macroeconomic policy regime has reduced economic performance relative to its theoretical potential, it does not follow that we advocate the immediate elimination of capital controls and the adoption of a floating exchange rate. The latter has been advocated recently by government representatives in Japan and the US, including such significant players as the Chairman of the US Federal Reserve.[14] We are more inclined to caution on the issue of the exchange rate, recognising that a float would be premature considering China's underdeveloped financial sector, its partially reformed banking industry and its still vulnerable state-owned enterprises.[15] The fact that more flexibility would enhance China's growth should be taken as advocating the acceleration of the reform processes needed before a floating rate regime can be implemented.[16]

Notes

1   The contractionary effects of deflation do not end there (Bordo and Redish 2003).
2   See Hertel (1997) for the original specification.
3   Ianchovichina and Martin (2001, 2002b) also examine the effects of WTO accession on labour relocation demand and they offer an explicit analysis of the HRS. Their analysis is strictly long run, however, ignoring the short run contractionary effects of macroeconomic policy emphasised here.
4   See Banister and Taylor (1989); Chai and Chai (1997); Hui (1989); Seeborg, Jin and Zhu (2000); and Multinational Monitor.
5   See Banister and Taylor (1989); Hui (1989); Seeborg, Jin and Zhu (2000).
6   See Banister and Taylor (1989); Hong Kong Liaison Office (IHLO). China's unemployment statistics measure only the urban unemployed.
7   In 1984, it was estimated that about 40 per cent of workers in the countryside were redundant. See Chai and Chai (1997).
8   The effect of these restrictions was to tend to leave marginal urban residents underqualified for their jobs while migrants from rural areas tend to be overqualified. Meng and Zhang (2001) estimate these rates of under and over qualification at 22 per cent for urban residents and 6 per cent for migrants from rural areas.

9   Often quoted is the real effective exchange rate, which is a weighted average of bilateral real rates

$$e_i^R = \sum_j E_j \left(\frac{P_i^Y}{P_j^Y} \right) \left(\frac{X_j + M_j}{X_i + M_i} \right)$$

where Xi and Mi are region i's total values of exports and imports, respectively.

10 The analysis by Yang and Tyers (2000, 2001) suggests that the magnitudes of China's crisis period deflation might be understated by official statistics.

11 A detailed description of the original model is provided by Hertel (1997).

12 The net factor income component of the current account is zero at the outset because that is the assumption embodied in the construction of the original database.

13 To see these at least partially offsetting gains in allocative efficiency, it is necessary to use a multi-commodity, general equilibrium framework such as used in this paper.

14 Interestingly, this pressure has tended to be for a revaluation of the RMB, which would benefit both Japan and the US in the short run. The shocks we examine yield real depreciations, however. Were these shocks dominant, we believe the fundamentals are more likely to support a devaluation of the RMB. A revaluation would be justified only in the event of Chinese investment demand taking on 'bubble' characteristics, capital controls notwithstanding.

15 See the discussion on this point by Edwards (2003).

16 See both Roberts and Tyers (2003) and Edwards and Levy-Yeyati (2003).

References

Anderson, K., Huang, J. and Ianchovichina, E., 2002. 'Long-run impact of China's WTO accession on farm–non-farm income inequality and rural poverty', Paper presented at World Bank Conference on China's WTO Accession, Policy Reform and Poverty Alleviation, Beijing, 28-29 June.

Banister, J. and Taylor, J.R., 1989. 'China: surplus labour and migration', *Asia-Pacific Population Journal*, 4(4):3-20.

Bordo, M.D. and Redish, A., 2003. *Is deflation depressing? Evidence from the classical gold standard*, National Bureau of Economic Research Working Paper No. W9520, Cambridge, Massachusetts.

Burgess, S. and Mawson, D., 2003. *Aggregate growth and the efficiency of labour reallocation*, Discussion Paper No. 3848, Centre for Economic Policy Research, London.

287

Canadian Immigration and Refugee Board, accessed 9 December 2002. *China: Internal migration and the floating population* http://www.irb. gc.ca/en/Researchpub/research/publications/chn11_e.htm.

Chai, J.C.H. and Chai, B.K., 1997. 'China's floating population and implications', *International Journal of Social Economics*, 24(7-8-9):1,038-51.

Chan, K.W. and Zhang, L., 1999. 'The Hukou system and the rural-urban migration in China: processes and changes', *The China Quarterly*, Dec 0(160):818-55.

Chand, S., 1999. 'Trade liberalisation and productivity growth: time-series evidence from Australian manufacturing', *The Economic Record*, 75(228):28-36.

Chand, S., McCalman. P. and Gretton, P., 1998. 'Trade liberalisation and manufacturing industry productivity growth', Paper in Productivity Commission and Australian National University, Microeconomic Reform and Productivity Growth, Workshop Proceedings, AusInfo, Canberra:239-81.

Dee, P. and Hanslow, K., 2000. *Multilateral liberalisation of services trade*, Productivity Commission of Australia, Canberra.

Dolven, B., 2003. 'Take our workers, please', *Far Eastern Economic Review*, 27 (February):24-26.

Edwards, S., 2003. 'China should not rush to float its currency', *Financial Times*, 3 August.

—— and Levy-Yeyati, E., 2003. *Flexible exchange rates as shock absorbers*, NBER Working Paper No.W9867, Cambridge, Massachusetts.

Garnaut, R., 1999. 'China after the East Asian crisis', *China Update 1999 Conference Papers*, National Centre for Development Studies, Australian National University, Canberra.

Gehlhar, M.J., 2002. 'Hong Kong's re-exports', Dimaranan, B.V., McDougall, R.A., (eds), *Global Trade, Assistance and Production: the GTAP 5 Data Base*, Center for Global Trade Analysis, Purdue University, West Lafayette, www.gtap.agecon.purdue.edu/databases/v5/v5_doco.asp.

Hertel, T.W. (ed.), 1997. *Global Trade Analysis Using the GTAP Model*, Cambridge University Press, New York.

Hong Kong Liaison Office (IHLO), accessed 4 December 2002. China and the WTO, update and analysis, May. Available online at http://www. ihlo.org/item3/item3h-3.htm.

Hui, D., 1989. 'Rural labour force transition and patterns of urbanisation in China', *Asia Pacific Population Journal*, 4(3):41-51.

Ianchovichina, E. and Martin, W., 2001. 'Trade liberalisation in China's accession to the WTO', *Journal of Economic Integration* 16(4):421-45.

——, 2002a. *Economic impacts of China's accession to the WTO*, World Bank Policy Research Working Paper No. 3053, World Bank, Washington DC.

——, 2002b. 'Evaluating accession to WTO by China and Chinese Taipei', Paper presented at the Fifth Annual Conference on Global Economic Analysis, Taipei, June.

Lee, JW. and Rhee, C., 1998. 'Social impacts of the Asian crisis: policy challenges and lessons', Paper prepared for the United Nations Development Programme Human Development Report Office, New York.

Lin, H.C., Chung, L., and Liou, R.W., 2002. 'Taiwan', Chapter 11F in Dimaranan, B.V., McDougall, R.A. (eds), *Global Trade, Assistance and Production: the GTAP 5 Data Base*, Center for Global Trade Analysis, Purdue University, West Lafayette. Available online at www.gtap.agecon. purdue.edu/databases/v5/v5_doco.asp.

Lozada, C., 2002. 'Globalisation reduces inequality in China', *The NBER Digest*, National Bureau of Economic Research, Cambridge, Massachusetts.

Meng, X., 1998. 'Recent development in China's labour market', *China Update 1998 Conference Papers*, National Centre for Development Studies, Australian National University, Canberra.

Meng, X. and Zhang, J., 2001. 'The two-tier labour market in urban China: occupational segregation and wage differentials between urban residents and rural migrants in Shanghai', *Journal of Comparative Economics*, 29(3):485-504.

Multinational Monitor, 2002. 'Sewing up the Chinese market: the effect of WTO entry on the Chinese rural sector'. Available online at www. multinationalmonitor.org/mm2000/00may/weil.html (accessed 4 December 2002).

National Bureau of Statistics, 2001, 2002. *China Statistical Yearbook*, China Statistics Press, Beijing.

Roberts, I. and Tyers, R., 2003. 'China's exchange rate policy: the case for greater flexibility', *Asian Economic Journal*, 17(2):157-86.

Seeborg, M.C., Jin, Z., and Zhu, Y., 2000. 'The new rural-urban labour mobility in China: causes and implications', *The Journal of Socio-Economics*, 29:39–56.

Sicular, T. and Zhao, Y., 2002. 'Labor mobility and China's entry to the WTO', Paper presented at the conference on WTO Accession, Policy Reform and Poverty Reduction, Beijing, 26-28 June.

Stoeckel, A., Tang, K.K., and McKibbin, W., 1999. 'The gains from trade liberalisation with endogenous productivity and risk premium effects', Technical Paper prepared for seminar: Reasons versus Emotion: Requirements for a Successful WTO Round, Seattle, 2 December.

Tyers, R. and Rees, L., 2000. 'Capital-skill complementarity and wage outcomes following technical change in a global model', *Oxford Review of Economic Policy*, 16(3):23–41.

——, 2001. 'Short run global effects of US 'new economy' shocks: the role of capital-skill complementarity', Paper presented at the Fourth Annual Conference on Global Economic Analysis, Purdue University, West Lafayette, 27-29 June. http://ecocomm.anu.edu.au/people/rod.tyers

Verikios, G. and Zhang, X., 2001. 'The economic effects of removing barriers to trade in telecommunications and financial services', Paper presented at the Fourth Annual Conference on Global Economic Analysis, Purdue University, West Lafayette, 27-29 June.

Wang, Z., Zhai, F., and Xu, D., 2002. 'China', in B.V. Dimaranan and R.A. McDougall (eds), *Global Trade, Assistance and Production: the GTAP 5 Data Base*, Center for Global Trade Analysis, Purdue University, West Lafayette, www.gtap.agecon.purdue.edu/databases/v5/v5_doco.asp.

Yang, Y., 1998. 'China in the middle of the East Asian crisis—export growth and the exchange rate', *China Update 1998 Conference Papers*, National Centre for Development Studies, The Australian National University, Canberra.

Yang, Y. and Tyers, R., 2000. *China's post-crisis policy dilemma: multi-sectoral comparative static macroeconomics*, Working Papers in Economics and Econometrics No. 384, The Australian National University, Canberra.

_____, 2001. 'The Asian crisis and economic change in China'. *Japanese Economic Review*, 52(4):491–520.

Zhao, Y., 2000. 'Rural to urban labour migration in China: past and present', in L. A. West and Y. Zhao (eds), *Rural Labour Flows in China*, University of California Press, Berkeley:15-33.

Appendix

Table A10.1 Official price level changes since the Asian crisis

	Consumer prices	Retail prices	Ex factory prices	GDP deflator
1997	2.8	0.8	-0.3	0.9
1998	-0.8	-2.6	-4.1	-2.6
1999	-1.4	-3.0	-2.4	-2.4
2000	0.4	-1.5	2.8	1.0
2001	0.7	-0.8	-1.3	0.0

Source: National Bureau of Statistics, 2002. *China Statistical Yearbook*, China Statistics Press, Beijing:Tables 3-3 and 9-1.

Table A10.2 Ancillary effects of WTO trade reforms[a]

	Long-run total factor productivity rises associated with reduced protection,[a] per cent	Short-run effect of reduced services protection on capital accumulation in financial and communications,[c] per cent/year
Rice	1.0	
Beverage	1.0	
Other crops	1.0	
Livestock	1.0	
Food	2.0	
Fish	0	
Minerals	0	
Energy	0	
Light manufacturing	1.5	
Heavy manufacturing	3.0	
Transport services	1.0	
Infrastructural services (electricity, gas, water)	1.0	
Construction and dwellings	2.0	
Other (including financial and communications)	5.0	2.0

Notes: [a] These are supplementary exogenous shocks applied, where indicated, along with the tariff reductions. They incorporate the findings of other research, particularly on the effects of services reform.
[b] Estimates based on the results from research by Chand et al. (1998), Chand (1999) and Stoeckel et al. (1999).
[c] Estimated one year effect, based on the results from research by Dee and Hanslow (2000) and Verikios and Zhang (2001).
Sources: Chand, S., 1999. 'Trade liberalisation and productivity growth: time-series evidence from Australian manufacturing', *The Economic Record*, 75(228):28–36; Chand, S., McCalman. P. and Gretton, P., 1998. 'Trade liberalisation and manufacturing industry productivity growth', Paper in Productivity Commission and Australian National University, Microeconomic Reform and Productivity Growth, Workshop Proceedings, AusInfo, Canberra:239–281; Stoeckel, A., Tang, K.K., and McKibbin, W., 1999. 'The gains from trade liberalisation with endogenous productivity and risk premium effects', Technical Paper prepared for seminar: Reasons versus Emotion: Requirements for a Successful WTO Round, Seattle, 2 December 1999; Dee, P. and Hanslow, K., 2000. *Multilateral liberalisation of services trade*, Productivity Commission of Australia, March; Verikios, G. and Zhang, X., 2001. 'The economic effects of removing barriers to trade in telecommunications and financial services', Paper presented at the Fourth Annual Conference on Global Economic Analysis, Purdue University, West Lafayette, 27-29 June.

Table A10.3 Simulated long-run effects of a unilateral liberalisation of China's 2001 trade policy regime[a]

Change in:	No ancillary effects on productivity or services capital	With ancillary effects on productivity and services capital [b]
Terms of trade (per cent)	-1.25	-1.52
Real effective exchange rate, e^R (per cent)	-1.98	-2.56
Real exchange rate against USA, e_{ij} (per cent)	-1.81	-2.37
Global interest rate, r_w (per cent)	0.10	0.05
Investment premium factor, (1+π) (per cent)	0.00	0.00
Home interest rate, r (per cent)	0.10	0.05
Return on installed capital, r_c (per cent)	1.30	4.98
Real domestic investment, I (per cent)	0.95	3.85
Balance of trade, X-M = - KA = -(I-S) (US$ bn)	-10.87	-56.51
Real gross sectoral output (per cent)		
Rice	-3.29	-2.54
Beverage	2.77	3.33
Other crops	-1.25	-0.36
Livestock	0.03	1.32
Food	-5.41	-4.32
Fish	-0.17	0.35
Minerals	0.88	2.38
Energy	0.78	0.95
Light manufacturing	1.49	2.26
Heavy manufacturing	1.08	4.60
Transport	1.44	2.87
Infrastructure services	0.34	2.25
Construction and dwellings	0.73	3.46
Other services	0.76	4.96
Real GDP, Y	0.41	3.31
Unskilled wage and employment (per cent)		
Nominal (unskilled) wage, W_Y	-0.42	1.85
Production real wage, $w=W/P_b$	1.51	4.53
Employment, L	0.00	0.00
Unit factor rewards CPI deflated (per cent)		
Land	-2.65	-1.39
Unskilled labour (those employed)	1.27	4.08
Skilled labour	1.38	4.22
Physical capital	1.24	4.04
Natural resources	1.20	2.65

Notes: [a] All results in this table are based on the adoption of fiscal policy 1: government spending is held constant as a share of GDP and the revenue lost from tariff reform is not made up in other taxes, so the fiscal deficit expands. Key exogenous variables are highlighted as per the long-run closure discussed in the text.
[b] For these additional shocks, see Table A10.4.
Source: Model simulations described in the text.

Table A10.4 Short-run closure[a]

Region	Monetary policy target[b]	Labour market closure: nominal wage[c]	Capital controls: capital account net inflow I-S$_D$[d]
China (1)	Nominal exchange rate, E	Sticky (λ=0.5)	Rigid
China (2)	Nominal exchange rate, E	Sticky (λ=0.5)	Flexible
China (3)	GDP price, P^Y	Sticky (λ=0.5)	Flexible
Vietnam	Nominal exchange rate, E	Sticky (λ=0.5)	Rigid
Other ASEAN	Consumer price level, P^C	Flexible (λ=1)	Flexible
Japan	Consumer price level, P^C	Sticky (λ=0.5)	Flexible
Korea	Consumer price level, P^C	Flexible (λ=1)	Flexible
Australia	Consumer price level, P^C	Sticky (λ=0.5)	Flexible
United States	Consumer price level, P^C	Sticky (λ=0.5)	Flexible
Europe (EU)	Consumer price level, P^C	Rigid (λ=0)	Flexible
Rest of World	Nominal exchange rate, E	Flexible (λ=1)	Flexible

Notes: [a] The expected future return on installed capital is exogenous in the short run, determined in a separate long-run solution. There are three macroeconomic policy regimes for China, with (1) the most restrictive and (3) the most expansionary.
[b] The nominal money supply is endogenous in each case, the corresponding exogenous variable being the listed target.
[c] When the nominal wage is assumed flexible it is endogenous and the corresponding exogenous variable is the employment level. When it is sticky or rigid, Equation 2 is activated and the employment level is endogenous.
[d] Capital controls are assumed to maintain a rigid net inflow of foreign investment on the capital account. When KA = I-SD is made exogenous to represent this, an interest premium opens between the domestic and international capital markets. This premium becomes endogenous. Effectively, the home and foreign capital markets are separated and clear at different interest rates. Where the capital account is flexible (open), this implies that private flows on the capital account are permitted at any level. KA = I-SD is then endogenous and the home interest premium is exogenous (unchanged by any shock). This means that the home interest rate then moves in proportion to the rate that clears the global savings-investment market.

Table A10.5 Simulated short-run effects of a unilateral liberalisation of China's 2001 trade policy regime: by macroeconomic policy regime[a]

Change in:	No productivity and services reform effects			With productivity and service reform effects		
	Capital controls, monetary policy targets E^b	No capital controls, monetary policy targets E	No capital controls, monetary policy targets P^Y	Capital controls, monetary policy targets E^b	No capital controls, monetary policy targets E	No capital controls, monetary policy targets P^Y
Nominal exch rate (US$/•), E_i	0.00	0.00	-1.36	0.00	0.00	-1.07
Domestic CPI,[b] P^C	-1.73	-0.79	0.34	-2.12	-0.68	0.20
Domestic GDP deflator, P^Y	-2.15	-1.03	0.00	-2.51	-0.80	0.00
Price of capital goods, P^I	-1.80	-0.89	0.18	-1.81	-0.42	0.42
Terms of trade	-1.41	-0.85	-1.06	-1.48	-0.62	-0.79
Real effective exchange rate, e_i^R	-2.30	-1.19	-1.53	-2.70	-1.02	-1.29
Real exchange rate vs USA, e_{ij}^R	-2.16	-1.03	-1.37	-2.53	-0.80	-1.07
Global interest rate, r^w	-0.04	0.08	0.06	-0.01	0.17	0.16
Interest premium factor, $(1+\pi)$	3.58	0.00	0.00	5.54	0.00	0.00
Home interest rate, r	3.54	0.08	0.06	5.53	0.17	0.16
Exp long-run capital return, r^c[c]	1.30	1.30	1.30	4.98	4.98	4.98
Short run capital return, r^c	0.83	1.99	2.44	-0.57	1.19	1.54
Real domestic investment, I	-1.76	0.98	0.99	-0.43	3.78	3.79
Balance of trade, X–M, US$m	0.00	-9,520	-8,250	0.00	-14,600	-13,600
Real gross sectoral output						
Rice	-0.43	-0.14	0.01	-0.39	0.04	0.16
Beverages	0.47	0.49	0.66	0.52	0.55	0.68
Other crops	-0.30	0.02	0.20	-0.26	0.23	0.38
Livestock	-0.10	0.49	0.67	-0.04	0.86	1.00

Food	-0.63	-0.37	-0.22	-0.62	-0.23	-0.11
Fish	-0.01	0.32	0.45	0.00	0.50	0.60
Minerals	-0.15	0.30	0.56	0.33	1.01	1.21
Energy	0.06	0.08	0.18	0.15	0.17	0.25
Light manufacturing	0.46	0.19	0.64	0.85	0.45	0.80
Heavy manufacturing	-0.05	0.14	0.52	0.46	0.74	1.04
Transport	0.23	0.56	0.90	0.60	1.11	1.38
Infrastructure services	-0.07	0.32	0.59	0.23	0.83	1.04
Construction and dwellings	-1.48	0.94	0.99	-0.31	3.40	3.44
Other services	-0.05	0.49	0.77	0.58	1.41	1.63
Real GDP, Y	-0.09	0.42	0.70	0.41	1.19	1.41
Unskilled wage and employment						
Nominal (unskilled) wage, W	-0.87	-0.40	0.17	-1.06	-0.34	0.10
Production real wage, $w=W/P^Y$	1.31	0.64	0.17	1.49	0.47	0.10
Employment, L^D	-0.28	0.65	1.15	0.11	1.53	1.92
Factor rewards CPI deflated						
Land	-0.98	1.65	2.56	-0.38	3.70	4.42
Production labour	0.88	0.40	-0.17	1.08	0.34	-0.10
Skilled labour	0.81	0.75	0.41	1.60	1.51	1.23
Capital	0.53	1.33	1.59	-0.10	1.12	1.32
Natural resources	1.00	2.77	4.17	3.03	5.84	6.98

Notes: [a] Variables fixed as exogenous are shown in bold letters. All results in this table are based on the adoption of a fiscal policy wherein government spending is held constant as a share of GDP and the revenue lost from tariff reform is not made up in other taxes, so the fiscal deficit expands.

[b] P^C is the domestic consumer price level; the CPI. Note that this is an index of the prices of both home and imported goods.

[c] The expected long-run return on capital motivates investment in these simulations. It is exogenous and its values are from Table A10.5.

Source: Model simulations described in the text.

Table A10.6 Comparing simulated trade reform-driven labour relocation demand with historical experience[a]

	Agriculture and fishing	Minerals	Manufacturing	Electricity, gas and water	Construction and dwellings	Transport services	Other services, incl finance & communication
Official record of historical annual employment changes, per cent							
1978–2001							
Maximum +ve (year)	3.6 (1982)	4.8 (1990)	8.2 (1986)	11.6 (1993)	28.6 (1984)	19.9 (1984)	16.8 (1984)
Maximum .ve (year)	-2.4 (1993)	-16.9 (1998)	-13.5 (1998)	-0.4 (2000)	-3.3 (1989)	-3.0 (1998)	-
Average[b]	0.68	-0.50	1.92	4.44	6.74	4.54	5.95
1990–2001							
Maximum +ve (year)	2.5 (1991)	3.8 (1993)	3.4 (1994)	11.6 (1993)	14.8 (1993)	10.4 (1994)	10.9 (1994)
Maximum .ve (year)	-2.4 (1993)	-16.9 (1998)	-13.5 (1998)	-0.4 (2000)	-3.3 (1998)	-3.0 (1998)	-
Average[b]	-0.3	-3.84	-0.48	3.81	4.02	2.47	5.13
Simulated WTO accession effects: annual employment changes, per cent							
Restrictive macro policy[c]							
Skilled workers	0.12	-0.13	0.02	-0.03	-0.9	0.21	-0.02
Production workers	-0.33	-0.37	0.02	0.01	-2.75	0.52	-0.14
Expansionary macro policy[d]							
Skilled workers	0.33	0.47	0.13	0.57	0.36	0.48	0.51
Production workers	0.62	1.46	1.12	1.92	1.90	2.18	2.07
Expansionary policy + ancillary shocks[e]							
Skilled workers	0.34	1.02	0.17	0.88	1.53	0.53	1.23
Production workers	0.98	3.21	2.11	3.26	6.61	3.42	1.49

Notes: [a] The sectoral aggregation used in this table concords the model sectors defined in Table A10.2 with those defined in the official statistics. Agriculture and fishing includes model sectors rice, beverages, other crops, livestock products and fish and the official sector farming, forestry, animal husbandry and fisheries; Minerals includes the corresponding sector and, from official statistics, mining and quarrying; electricity, gas and water corresponds with the official sector and it includes the model sectors energy and infrastructure services; construction and dwellings is the official sector construction and real estate; transport services corresponds to the model sector and includes official sectors transport, storage, post and telecommunications services; other services is the corresponding model sector and includes the remaining sectors from the official list.

[b] Arithmetic average of annual proportional changes.

[c] Rigid capital controls are maintained, monetary policy targets the exchange rate, there are no exogenous productivity effects, nor any boost in the capital stock in finance and communications (per Table A10.4).

[d] Capital controls are removed and monetary policy targets the GDP price. These results coincide with the third column of Table A10.7.

[e] To the expansionary policy environment is added the effects of ancillary productivity and services capital shocks, the results coinciding with the final column of Table A10.7.

Source: National Bureau of Statistics, 2001 and 2002. *China Statistical Yearbook*, China Statistics Press, Beijing; model simulations described in the text.

Table A10.7 Chinese labour relocation: the official record since 1978 (employment change, per cent/year)

	Agriculture and fishing	Minerals	Manufacturing	Electricity, gas and water	Construction and dwellings	Transport services	Other services, incl finance & communication
1978-79	1.12	2.76	3.45	4.67	7.34	4.13	6.16
1979-80	1.70	4.03	6.94	5.36	8.42	3.07	7.52
1980-81	2.25	4.45	3.78	5.93	3.50	4.84	7.95
1981-82	3.63	2.61	3.38	2.40	10.69	4.03	2.19
1982-83	0.95	1.47	2.83	2.34	11.78	6.61	8.87
1983-84	-0.91	1.19	8.01	2.29	28.58	19.87	16.83
1984-85	0.85	3.65	5.45	5.97	22.11	13.99	7.04
1985-86	0.40	1.76	8.19	7.04	9.80	7.58	5.01
1986-87	1.31	1.24	4.24	7.89	6.55	5.60	6.84
1987-88	1.85	1.59	3.51	7.93	4.54	4.68	5.95
1988-89	3.03	1.20	-1.21	1.69	-3.28	0.07	2.28
1989-90	2.68	4.75	0.90	6.67	0.73	2.89	3.82
1990-91	2.46	2.61	2.49	5.73	2.51	3.26	4.71
1991-92	-0.46	-0.77	3.02	5.91	7.27	3.53	7.74
1992-93	-2.38	3.79	2.08	11.63	14.81	0.84	9.51
1993-94	-1.71	-1.82	3.42	2.50	4.69	10.43	10.93
1994-95	-1.10	1.86	1.98	4.88	4.29	4.18	6.96
1995-96	-0.33	-3.22	-0.41	5.81	2.65	3.66	3.49
1996-97	0.56	-3.77	-1.55	3.66	1.26	2.43	5.35
1997-98	0.41	-16.94	-13.45	0.00	-3.25	-3.01	1.24
1998-99	0.79	-7.49	-2.52	0.71	2.54	1.10	0.14
1999-00	-0.41	-10.49	-0.81	-0.35	4.10	0.35	4.24
2000-01	-1.14	-6.03	0.50	1.41	3.40	0.39	2.10
Average	0.68	-0.50	1.92	4.44	6.74	4.54	5.95

Source: National Bureau of Statistics, 2002. *China Statistical Yearbook*, China Statistics Press, Beijing:Table 5-5.

Figure 10A.1 Per capita annual income of urban and rural households, 1978–2001

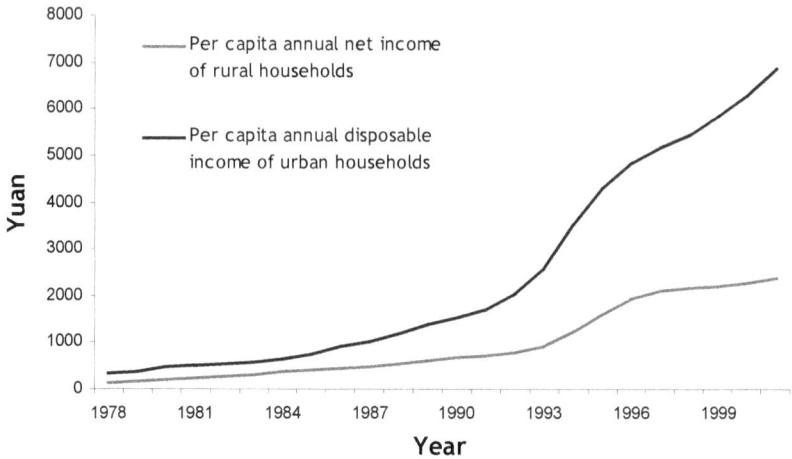

Source: National Bureau of Statistics, 2002. *China Statistical Yearbook*, China Statistics Press, Beijing:Table 10-3.

Figure 10A.2 Gap between urban and rural incomes, 1978–2001

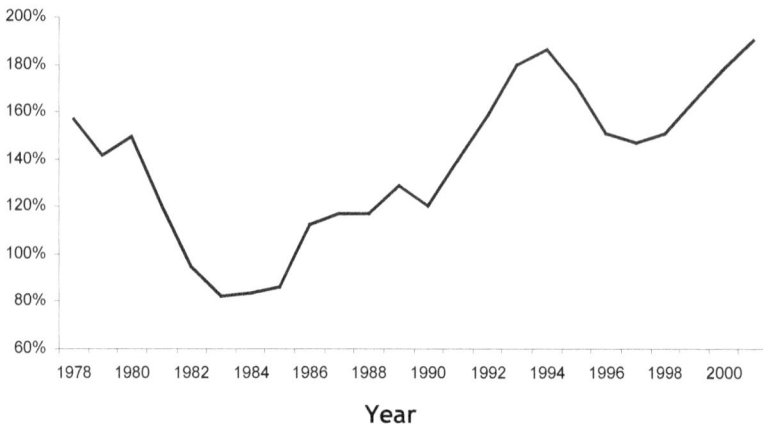

Source: National Bureau of Statistics, 2002. *China Statistical Yearbook*, China Statistics Press, Beijing:Table 10-3.

Figure 10A.3 Employment by industry group, 1990–2001

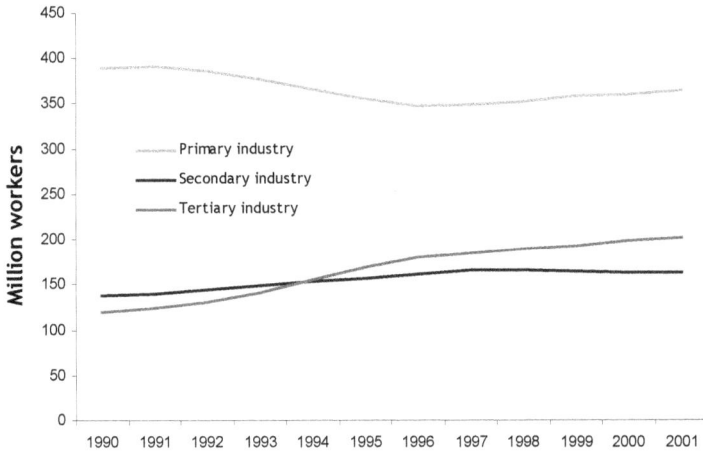

Source: National Bureau of Statistics, 2002. *China Statistical Yearbook*, China Statistics Press, Beijing:Table 5-2. Primary industry: farming, forestry, animal husbandry and fishery; secondary industry: mining, manufacturing, electricity, water, gas, construction; tertiary: all other industries.

Figure 10A.4 Official growth rate of real GDP, 1990–2001 (per cent/year)

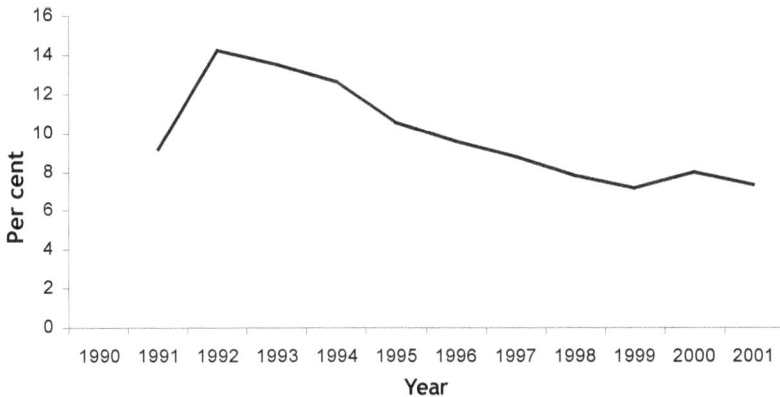

Source: National Bureau of Statistics, 2002. *China Statistical Yearbook*, China Statistics Press, Beijing:Table 3-3.

Figure 10A.5   Official growth rate of average real manufacturing
wage,1996–2000 (per cent/year)

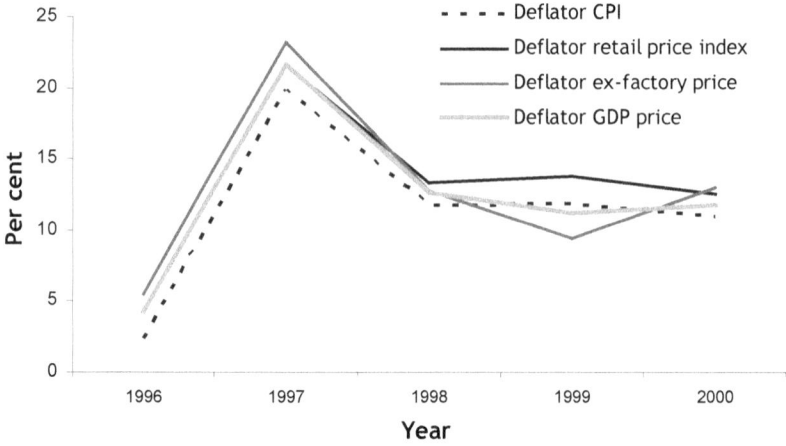

Source: National Bureau of Statistics, 2002. *China Statistical Yearbook*, China Statistics Press, Beijing:Table 5-22 and 9-1.

Figure 10A.6   Trade reform with capital controls

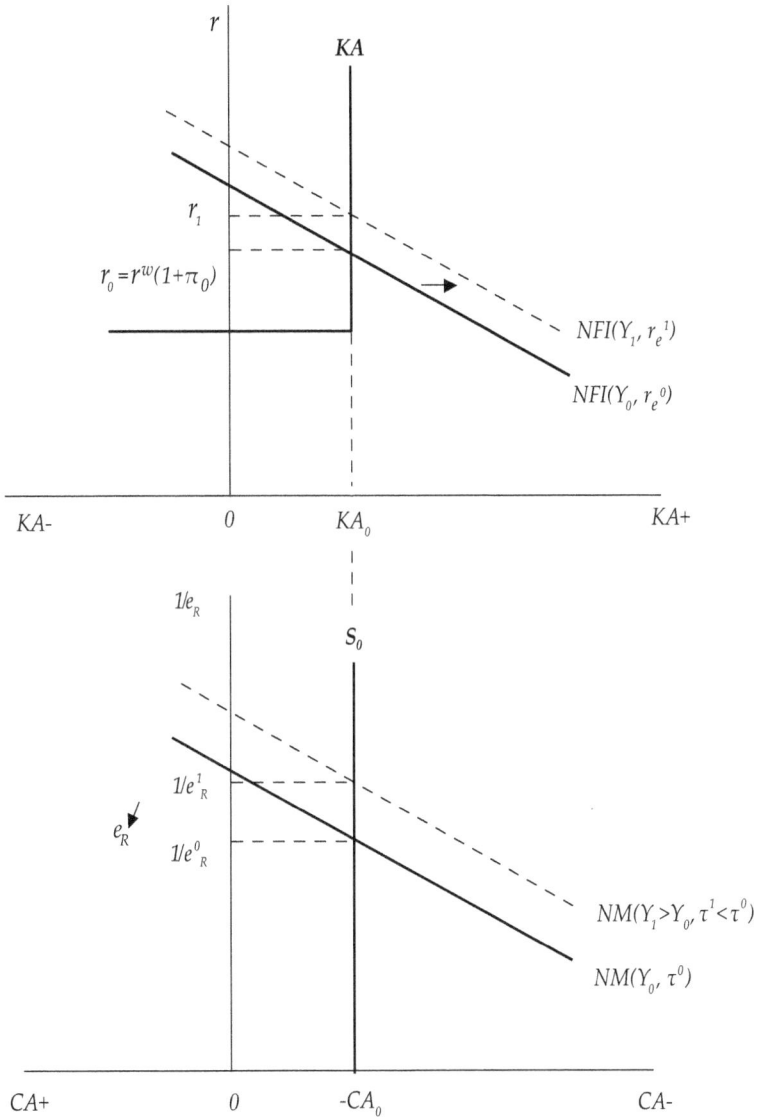

Figure 10A.7   Trade reform with no capital controls

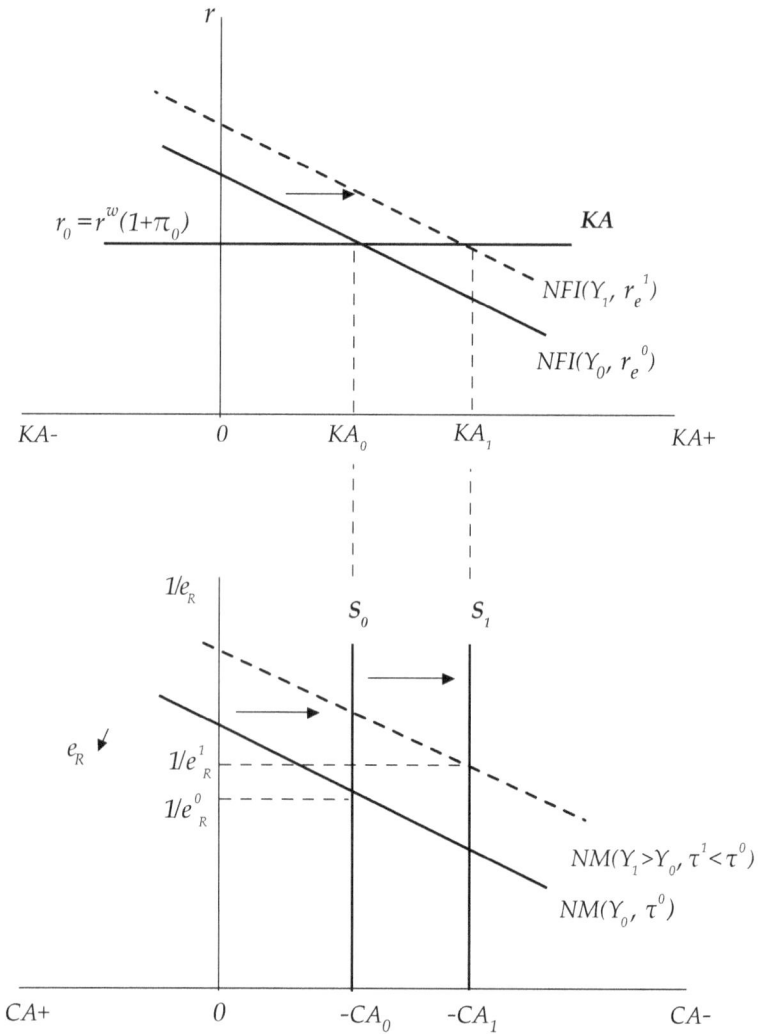

11 China's agricultural trade following its WTO accession

Chunlai Chen

Since its entry into the WTO in December 2001, China's economy has grown rapidly. The average annual growth rate of China's real GDP was more than 9.8 per cent during 2002-05.[1] China's foreign trade has been expanding even more rapidly than its economic growth. The total value of China's foreign trade has increased from US$457 billion in 2001 to US$1263 billion in 2005—an annual growth rate of 28.6 per cent, as compared with 9.4 per cent during the 1990s.[2] Undoubtedly, China's economy has benefited from its more open international trade regime resulting from accession to the WTO.[3]

The impact of China's entry into the WTO on its agricultural sector has been the major concern of the Chinese government and has been the hottest topic among policy makers and academics in and outside China (for example, Anderson 1997; Cheng 1997; Development Research Centre 1998; Huang 1998; Huang and Chen 1999; Wang 1997). In general, experts argued that based on China's resource endowments and comparative advantage, after entry into the WTO China's land-intensive farming sector would shrink but its labour-intensive horticultural sector, its animal husbandry sector and its processed agricultural product sector would expand. As a result, China would import more land-intensive agricultural products, such as grains and vegetable oils, and export more labour-intensive agricultural products, such as vegetables and fruits, animal products and processed agricultural products.

What has happened to China's agricultural trade since its accession to the WTO? Have there been any changes in the pattern of China's agricultural trade and in its revealed comparative advantages in agriculture? If so, what factors have driven these changes? This chapter examines these questions.

Classification of agricultural commodities and sources of data

In analysing agricultural trade, the first step is to identify the coverage of agricultural commodities in international trade. Here, the classification of agricultural commodities in international trade is based on the Harmonised System (HS) of Trade Classification 1992. Table 11.1 presents the product coverage used and the product coverage in the Uruguay Round Agreement on Agriculture (URAA). The product coverage in this chapter and in the URAA is very similar. The differences are that here fish and fish products are included, but HS Code 2905.43 (mannitol), HS Code 2905.44 (sorbitol), HS Heading 33.01 (essential oils), HS Headings 35.01 to 35.05 (albuminoidal substances, modified starches, glues), HS Code 3809.10 (finishing agents) and HS Code 3823.60 (sorbitol n.e.p.) are excluded. The main reasons for these inclusions and exclusions are that fish and fish products are very important agricultural products in China's trade while the trade values of those excluded are negligible.

The agricultural trade data for the period 1992 to 2004 are from the United National Statistics Division, Commodity Trade Statistics Database (COMTRADE). Data for 2005 are from the China Customs Statistical Monthly Report.[4] All the values of agricultural trade data presented here are at 2000 constant US$ prices.

For the purpose of analysing changes in the pattern of China's agricultural trade, the data have been grouped in two ways. First, the data are divided into five categories based on the nature of the commodities

- cereals, edible vegetable oilseeds and vegetable oils
- horticultural products
- animal products (including fish)
- processed agricultural products (including processed fish products)
- raw materials for textiles.

Second, the data are grouped into two categories based on the factor intensity of production

- land-intensive agricultural products
- labour-intensive agricultural products.

Agricultural trade data for the period 1992-2005 categorised in this way are presented in the appendix in Tables A11.1 and A11.2.

Table 11.1   Comparison of agricultural product coverage

Product coverage in this study	Product coverage in the URAA
HS Chapters 1 to 24, plus HS Headings 41.01 to 41.03 (hides and skins) HS Heading 43.01 (raw fur skins) HS Headings 50.01 to 50.03 (raw silk and silk waste) HS Headings 51.01 to 51.03 (wool and animal hair) HS Headings 52.01 to 52.03 (raw cotton, waste and cotton carded or combed) HS Heading 53.01 (raw flax) HS Heading 53.02 (raw hemp)	HS Chapters 1 to 24 less fish and fish products, plus HS Code 2905.43 (mannitol) HS Code 2905.44 (sorbitol) HS Heading 33.01 (essential oils) HS Headings 35.01 to 35.05 (albuminoidal substances, modified starches, glues) HS Code 3809.10 (finishing agents) HS Code 3823.60 (sorbitol n.e.p.) HS Headings 41.01 to 41.03 (hides and skins) HS Heading 43.01 (raw fur skins) HS Headings 50.01 to 50.03 (raw silk and silk waste) HS Headings 51.01 to 51.03 (wool and animal hair) HS Headings 52.01 to 52.03 (raw cotton, waste and cotton carded or combed) HS Heading 53.01 (raw flax) HS Heading 53.02 (raw hemp)

Sources: The Uruguay Round Agreement on Agriculture and author's classification.

307

Trends in China's agricultural trade following WTO accession

Aggregate trends in agricultural trade

Between 1992 and 2001, China's agricultural trade experienced large fluctuations but did not grow (Figure 11.1). Since its entry into the WTO, the value of China's agricultural trade has increased dramatically to reach US$50.44 billion in 2005—an increase of 90 per cent over the 2001 figure.

The trends in exports and imports were similar to that of total agricultural trade during the period 1992 to 2001. However, following entry into the WTO, agricultural imports have increased more rapidly than agricultural exports. From 2002 to 2005, the annual growth rate of agricultural imports was 31.5 per cent, while that of agricultural exports was 11.7 per cent. As a result, in 2004 and 2005 agricultural imports exceeded agricultural exports

Figure 11.1   China's agricultural trade, 1992-2005 (at constant 2000 US$ prices)

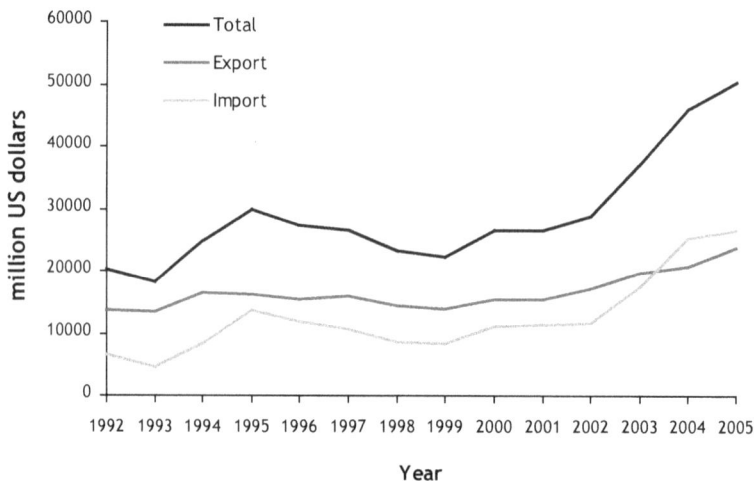

Sources: Data from 1992 to 2004 are from United Nations Statistics Division, *Commodity Trade Statistics Database*, COMTRADE. http://unstats.un.org/unsd/comtrade/default.aspx. Data for 2005 are from China General Administration of Customs (various issues, 2005). *Zhongguo Haiguan Tongji Yuebao* [China Customs Statistical Monthly Report], Zhongguo Haiguan Chubanshe, Beijing.

and China has experienced agricultural trade deficits for the first time since the early 1990s. It is expected that the higher growth of agricultural imports will continue.

Trends in agricultural trade, by commodity grouping

Agricultural exports, by commodity grouping

Figure 11.2 presents China's agricultural exports based on commodity groupings for the period 1992 to 2005. China's agricultural exports are dominated by processed agricultural products, followed by animal products and horticultural products. The export values of cereals, vegetable oilseeds and vegetable oils and, in particular, raw materials for textiles are small.

Figure 11.2   China's agricultural exports by category, 1992-2005 (at constant 2000 US$ prices)

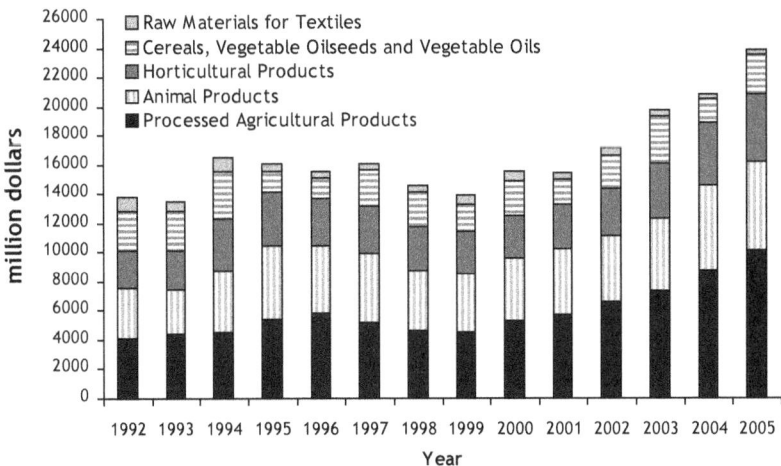

Sources: Data from 1992 to 2004 are from United Nations Statistics Division, *Commodity Trade Statistics Database*, COMTRADE. http://unstats.un.org/unsd/comtrade/default.aspx. Data for 2005 are from China General Administration of Customs (various issues, 2005). *Zhongguo Haiguan Tongji Yuebao* [China Customs Statistical Monthly Report], Zhongguo Haiguan Chubanshe, Beijing.

Processed agricultural products

The group of processed agricultural products has been the largest commodity group in China's agricultural exports. Following entry into the WTO, the export value of processed agricultural products has risen sharply, from US$5.74 billion in 2001 to US$11.12 billion in 2005—an annual average growth rate of 15.3 per cent. As a result, its share in China's total agricultural exports increased from 32.9 per cent in the period of 1992-2001 to 40.2 per cent in the period of 2002-05.

Within this group, exports have been dominated by two product categories, namely the preparations of meat and fish, and the preparations of vegetables and fruits. The export value of the preparations of meat and fish increased from US$2 billion in 2001 to US$3.9 billion in 2005. The export value of the preparations of vegetables and fruits has increased from US$1.5 billion in 2001 to US$2.8 billion in 2005, an increase of 88 per cent. Their combined share in the export value of processed agricultural products has increased from 48.8 per cent in the period of 1992-2001 to 63 per cent in the period of 2002-05. Moreover, their combined share in China's total agricultural exports increased from 16 per cent in the earlier period to 25.3 per cent in 2002-05.

Animal products

The group of animal products has been the second largest commodity group in China's agricultural exports. The exporting of animal products has also been increasing steadily since China's accession, up from US$4.5 billion in 2001 to US$6 billion in 2005—an average annual growth rate of 7.5 per cent. However, because of the larger share and faster growth of the exports of processed agricultural products, the export share of animal products in China's total agricultural exports has declined marginally.

In this group, aquatic products have been the most important component, followed by meats, products of animal origin, and live animals. After 2001, exports of aquatic products increased very fast. Consequently, the share of aquatic products in total exports of animal products increased from 48 per cent during 1992-2001 to 62.5 per cent during 2002-05. As a result, the shares of other animal products have declined.

Horticultural products

The group of horticultural products has been the third largest commodity group in China's agricultural exports. Exports of horticultural products have increased relatively quickly, from US$3 billion in 2001 to US$4.7 billion in 2005, at an average annual growth rate of 11.6 per cent, which is similar to the annual growth rate of total agricultural exports. As a result, the share of horticultural products in China's total agricultural exports has remained at around 20 per cent.

Within this group, vegetables are the most important commodities, followed by fruits and the product categories of tea, coffee, mate and spices. After 2001 exports of vegetables and fruits increased more rapidly than other commodities in this group and the share of vegetables in horticultural exports has increased from 52.4 per cent during 1992-2001 to 55.3 per cent during 2002-05. At the same time the share of fruits in total horticultural exports has increased from 14.7 per cent to 18.8 per cent.

Cereals, edible vegetable oilseeds and vegetable oils

The 'cereals, edible vegetable oilseeds and vegetable oils' group ranks fourth in China's agricultural exports. The annual export value of cereals, edible vegetable oilseeds and vegetable oils has fluctuated extensively over the period 1992 to 2005. Following China's entry into the WTO, the importance of this commodity group has declined, falling from 14.8 per cent during 1992-2001 to 12.2 per cent during 2002-05.

Within the group, corn has been the single most important export commodity. Annual exports of corn averaged US$0.8 billion in 1992-2001, increased to US$1.1 billion in 2002, and to US$1.7 billion in 2003. In 2004, because of several economic and policy factors, including changes in the relationship between domestic prices and world prices,[5] and the reduction in corn export quotas (Gale 2005), China's corn exports declined sharply to US$0.3 billion. However, corn exports increased again in 2005, reaching almost US$1 billion. On average, corn exports have accounted for around 40 per cent of the total exports of this group during 2002-05.

Raw materials for textiles

Finally, China's exports of the commodity group of raw materials for textiles have been small. Both the export value and the export share of this commodity group in China's total agricultural exports have declined substantially since its WTO accession. The export value has declined from around US$0.6 billion annually over 1992-2001 to around US$0.35 billion annually in 2002-05. As a result, the export share of this commodity group in China's total agricultural exports has fallen from 4 per cent in 1992-2001 to 1.75 per cent in 2002-05.

Within this group, silk has been the largest export commodity for a long time. However, silk exports have been declining since the mid 1990s. The export values of cotton and raw hides have declined substantially since 2001.

Agricultural imports, by commodity grouping

China's imports of agricultural products are overwhelmingly dominated by cereals, vegetable oilseeds and vegetable oils, followed closely by raw materials for textiles (Figure 11.3). The imports of animal products, processed agricultural products, and horticultural products are relatively low but have been rising rapidly, especially since 2003.

Cereals, edible vegetable oilseeds and vegetable oils

Cereals, edible vegetable oilseeds and vegetable oils is the largest group in China's agricultural imports. But imports of these commodities experience large fluctuations (Figure 11.3). Imports of this commodity group averaged around US$3.9 billion in 1992-01 but increased sharply to US$8.9 billion in 2002-05. As a result, the share of the group in China's total agricultural imports increased from 41.6 per cent in 1992-2001 to 43.9 per cent in 2002-05.

Cereals dominated the imports of this commodity group from 1992 to 1996. However, from 1997 to 2003, imports of cereals declined, reaching their lowest level of US$0.4 billion in 2003. In 2004, imports jumped to US$2 billion and then fell to US$1.2 billion in 2005. Wheat has dominated

Figure 11.3   China's agricultural imports by category, 1992-2005 (at constant 2000 US$ prices)

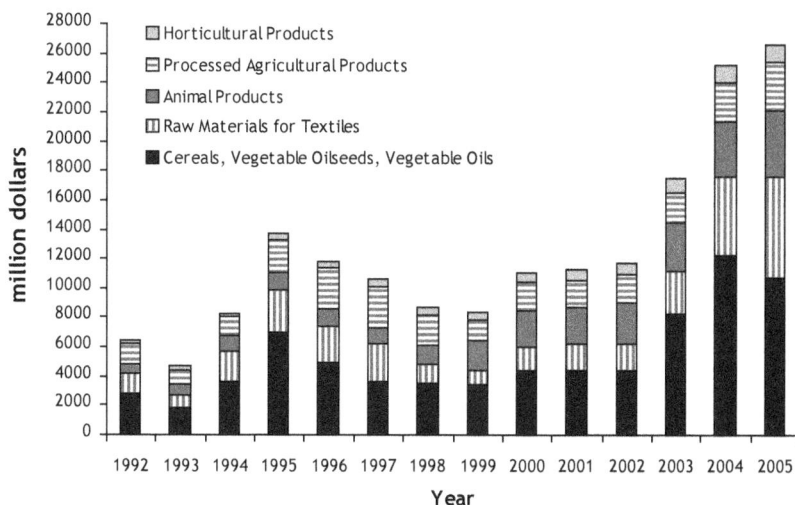

Sources: Data from 1992 to 2004 are from United Nations Statistics Division, *Commodity Trade Statistics Database*, COMTRADE. http://unstats.un.org/unsd/comtrade/default.aspx. Data for 2005 are from China General Administration of Customs (various issues, 2005). *Zhongguo Haiguan Tongji Yuebao* [China Customs Statistical Monthly Report], Zhongguo Haiguan Chubanshe, Beijing.

imports of cereals. Wheat import rose to above US$2 billion in 1995-96 because of a sharp increase in China's domestic grain prices in 1994. However, from 1997 to 2003 wheat imports declined to reach their lowest level of US$0.07 billion. The decline in wheat imports during this period was mainly caused by consecutive bumper domestic harvests from 1996 to 1999. After 2000, grain production in China declined and government had to use state grain reserves to fill the gap between demand and supply. In late 2003, China's domestic grain prices began to increase sharply. In 2004 the Chinese government implemented a series of policies, including subsidies to farmers and gradually abolishing agricultural taxes,[6] with the aim of increasing grain production and farmers' incomes. At the same time, China began to increase

313

wheat imports. China imported US$1.5 billion of wheat in 2004 and US$0.7 billion in 2005, mainly to replenish state grain reserves.

In the early 1990s, China imported a limited amount of edible vegetable oilseeds. However, since 1997, imports of edible vegetable oilseeds have increased rapidly, from US$1 billion to US$3.2 billion in 2001. Following entry into the WTO, imports of edible vegetable oilseeds have been increasing even more rapidly, jumping to US$5.3 billion in 2003, to US$6.7 billion in 2004, and to US$7.3 billion in 2005. The share of edible vegetable oilseeds in total imports of cereals, edible vegetable oilseeds and vegetable oils reached 61.2 per cent during 2002–05. Within the edible vegetable oilseeds group, soybean has been overwhelmingly the largest import, accounting for 95 per cent of total imports during 2002-05.

Edible vegetable oils are the next important commodities in the imports of cereals, edible vegetable oilseeds and vegetable oils. China imported US$1.3 billion of edible vegetable oils annually during 1992-2001. Imports of these commodities increased sharply during 2002–05, up to US$2.4 billion annually. Soybean oil and palm oil are the most important commodities in this group. Their combined share was 88 per cent of total imports of edible vegetable oils from 2002 to 2005.

Raw materials for textiles

Raw materials for textiles is the second large group in China's agricultural imports. Since 2003 imports of raw materials for textiles have increased sharply from US$2.9 billion, to US$5.4 billion in 2004 and US$6.9 billion in 2005. As a result, the share of raw materials for textiles in China's total agricultural imports increased from 19.1 per cent in 1992-2001 to 21.1 per cent in 2002-05. The dramatic increase in the import of raw materials for textiles during 2003 to 2005 was mainly driven by the large expansion of China's textile industry as the Agreement on Textiles and Clothing (ATC) was phased out at the end of 2004 and the import quotas on textiles and clothing were to be abolished from 1 January 2005.

Within this group, wool is a very important import commodity. From 1992 to 2001, imports of wool were fairly stable with an average annual import value around US$0.65 billion. However, in 2004 and 2005, imports increased to above US$1 billion. The share of wool in the total imports of this group averaged 35.8 per cent from 1992 to 2001. However, its share declined to 21.2 per cent during 2002 to 2005 due to the large increases in imports of cotton and raw hides and skins.

Cotton is also an important commodity in this group. Cotton imports averaged around US$1.3 billion from 1994 to 1997, but declined to less than US$0.2 billion from 1998 to 2002. From 2003 to 2005, cotton imports increased significantly to reach US$2.9 billion. As a result, cotton became the largest import commodity in this group, accounting for 41.4 per cent of total imports of the group over the period 2002 to 2005.

Raw hides and skins are also important commodities in this group. Imports of raw hides and skins have been increasing steadily since 1992. However, since 2003, imports have increased sharply, up from US$0.9 billion in 2003 to US$1.4 billion in 2004 and with a further jump to US$2.8 billion in 2005. As a result, the share in total imports of raw materials for textiles increased to 33.3 per cent during 2002 to 2005.

Animal products

Animal products rank third in China's agricultural imports. The import value of animal products increased gradually from 1992 to 1998 then increased rapidly during 1999-2001. After entry into the WTO, imports of animal products increased even faster, rising from US$2.75 billion in 2002 to US$4.5 billion in 2005. As a result, the share of animal products in China's total agricultural imports increased from 14.9 per cent in the period of 1992-2001 to 17.6 per cent in the period 2002-05.

Within the group, fish and other aquatic products are the most important commodities. Imports of fish and other aquatic products have increased particularly quickly since entry into the WTO, rising to US$2.6 billion in 2005, nearly double that of 2001. Hence, the share of fish and other aquatic products in China's total imports of animal products increased to 55.7 per cent during 2002-05.[7]

Meat and dairy products are the next most important commodities in this group. In the 1990s China imported very limited amounts of these products. However, since 1999 and particularly since 2001, imports of meat and dairy products have increased quickly. The import value of meat increased from US$0.26 billion annually in 1992-2001 to US$0.57 billion annually in 2002-05, while the import value of dairy products increased from US$0.12 billion annually in 1992-2001 to US$0.35 billion annually in 2002-05. The imports of other animal products have also shown an increasing trend since 2001.

Processed agricultural products

The group of processed agricultural products holds fourth place in China's total agricultural imports. Imports of processed agricultural products have been very variable during 1992 to 2005. They increased quickly in the mid 1990s, reaching their highest level of US$2.8 billion in 1997. They declined to US$1.4 billion in 1999. In the early 2000s, imports recovered to around US$2 billion, and then rose rapidly to US$2.7 billion in 2004 and US$3.2 billion in 2005. However, because of the larger increase in the imports of other agricultural products, particularly edible vegetable oilseeds, vegetable oils, and raw materials for textiles, the import share of processed agricultural products in China's total imports of agricultural products has declined from 19.8 per cent in 1992-2001 to 12.4 per cent in 2002-05.

The imports of processed agricultural products have been dominated by imports of animal feed (residues of the food industry and feedstuffs). China imported a large quantity of animal feed during 1996 to 1998. The import value of animal feed reached a historical high of US$1.9 billion in 1997. During 1999 to 2004, imports of animal feed have been US$0.7-0.8 billion. In 2005, imports of animal feed increased to US$1.2 billion. Another important import commodity in this group is sugar. China imported large volumes of sugar in the mid 1990s, reaching a high point of US$1.1 billion in 1995. Since then sugar imports have declined and are around US$0.3 billion. Imports of tobacco, miscellaneous edible preparations, and beverages and spirits have been small, but show a slightly increasing trend after 2002.

Horticultural products

Horticultural products have been the smallest component in China's agricultural imports, accounting for around 5 per cent of total agricultural imports during the period 1992 to 2005. From 1992 to 2001, imports of horticultural products increased gradually from US$0.25 billion to US$0.74 billion. After 2002, they increased quickly to US$1.26 billion in 2005.

Fruits and vegetables dominate the imports of horticultural products. Import of fruits and vegetables surged during 2003 to 2005. This was mainly because of the implementation of the 'early-harvest' program (EHP), which is part of the China-ASEAN Free Trade Area Framework Agreement signed by China and the ASEAN in 2002. Under the EHP, which was implemented on 1 January 2004, the two sides immediately cut tariffs on about 600 agricultural imports to between 2 per cent and 15 per cent, and agreed to scrap these tariffs in 2006. Thailand has taken the lead among the ASEAN members in initiating this free trade accord as it has phased out all import tariffs on 188 fruits and vegetables imports from China starting in October 2003 (*China Daily*, 9 August 2004). China's import of fruits from Thailand

Figure 11.4   Share of china's agricultural trade in total trade, 1992-2005 (per cent)

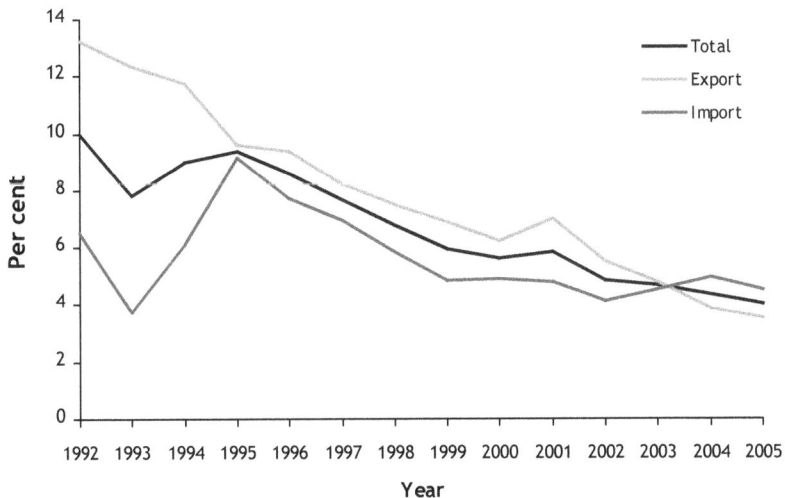

Sources: Data from 1992 to 2004 are from United Nations Statistics Division, *Commodity Trade Statistics Database*, COMTRADE. http://unstats.un.org/unsd/comtrade/default.aspx. Data for 2005 are from China General Administration of Customs (various issues, 2005). *Zhongguo Haiguan Tongji Yuebao* [China Customs Statistical Monthly Report], Zhongguo Haiguan Chubanshe, Beijing.

increased sharply from US$77.7 million in 2003 to US$165 million in 2004, while its imports of vegetables from Thailand increased from US$141.4 million in 2003 to US$249.5 million in 2004.

Changes in the pattern of China's agricultural trade after WTO accession

Despite the recent rapid increases in absolute values of agricultural trade, its importance in China's total trade has been declining. As shown in Figure 11.4, the share of agricultural trade in China's total trade declined from 10 per cent in 1992 to 5.8 per cent in 2001. After China's entry into the WTO, this decline has become even more pronounced. The share of agricultural trade was only 4 per cent in 2005, and the share of agricultural exports in China's total exports declined even faster—from 7 per cent in 2001 to 3.5 per cent in 2005.

According to international trade theory, a country's pattern of trade with the rest of the world is determined by its comparative advantage, and its comparative advantage is determined by its resource endowments. In the case of China, the characteristics of China's resource endowments with respect to agricultural production are that it is scarce in land resources but abundant in labour. China's per capita arable land is 0.11 hectares, only 43 per cent of the world average, and its per capita pasture land is 0.3 hectares, only 33 per cent of the world average. However, China has abundant labour—1.3 billion population—with nearly 70 per cent living in rural areas, and half of the labour force is in the agricultural sector. Based on its resource endowments, China should have a comparative advantage in labour-intensive agriculture and a comparative disadvantage in land-intensive agriculture. Further, China should have a bias towards exporting labour-intensive agricultural products and importing more land-intensive agricultural products.

Given that the WTO accession commitments should have led to China producing and trading more in accord with its comparative advantage, an interesting question is whether there has been any change in the pattern of China's agricultural trade since its accession? To answer this question, we compare China's agricultural trade patterns for the two periods 1992-2001 and 2002-05, both by commodity group and by factor intensity of production.

Changes in the pattern of China's agricultural trade, by commodity group

Figure 11.5 shows the composition of China's agricultural exports by commodity group for the two periods 1992-2001 and 2002-05, illustrating how the dominance of processed agricultural products has increased. Between the two periods, its export share increased by 7.3 percentage points, while the shares of all other agricultural commodity groups declined.

Figure 11.6 shows the composition of China's agricultural imports by commodity group for the same two periods. China's imports of agricultural products are dominated by the group of cereals, vegetable oilseeds and

---

Figure 11.5   Shares of China's agricultural exports by commodity groupings, 1992-2001 and 2002-2005 (per cent)

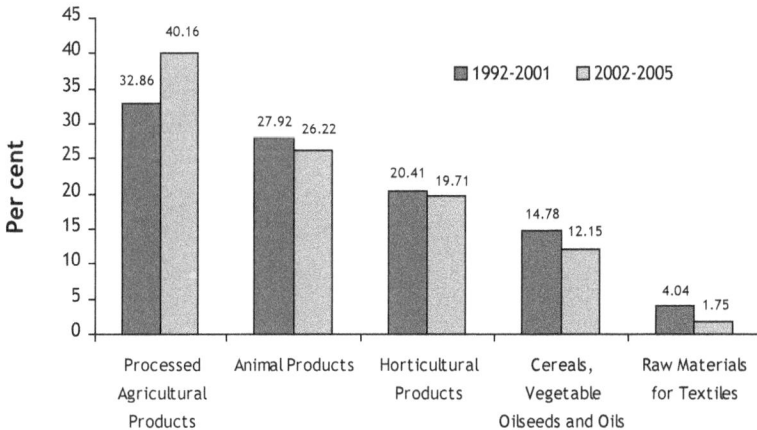

Sources: Author's calculation. Data from 1992 to 2004 are from United Nations Statistics Division, *Commodity Trade Statistics Database*, COMTRADE. http://unstats.un.org/unsd/comtrade/default.aspx. Data for 2005 are from China General Administration of Customs (various issues, 2005). *Zhongguo Haiguan Tongji Yuebao* [China Customs Statistical Monthly Report], Zhongguo Haiguan Chubanshe, Beijing.

Figure 11.6   Share of China's agricultural imports by commodity grouping, 1992-2001 and 2002-2005 (per cent)

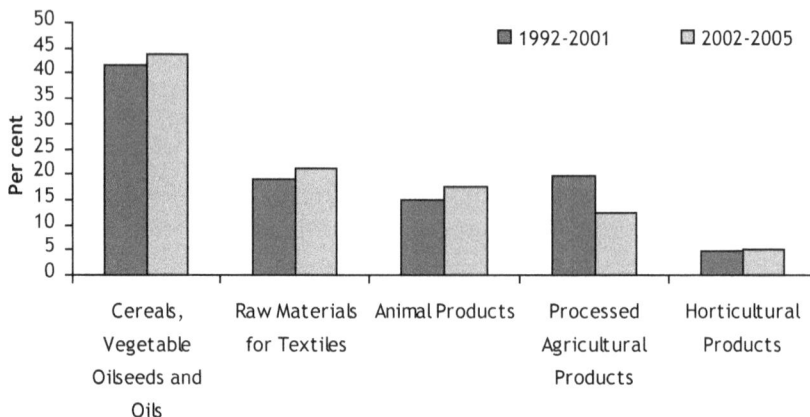

Sources: Author's calculation. Data from 1992 to 2004 are from United Nations Statistics Division, *Commodity Trade Statistics Database*, COMTRADE. http://unstats.un.org/unsd/comtrade/default.aspx. Data for 2005 are from China General Administration of Customs (various issues, 2005). *Zhongguo Haiguan Tongji Yuebao* [China Customs Statistical Monthly Report], Zhongguo Haiguan Chubanshe, Beijing.

vegetable oils. Raw materials for textiles and animal products are also important. All increased their shares since 2001 at the expense of processed agricultural products.

Changes in the pattern of China's agricultural trade, by factor intensity of production

To see if there have been any changes in the pattern of agricultural trade in terms of factor intensity, agricultural trade was grouped into labour-intensive products and land-intensive products. The labour-intensive products include processed agricultural products, animal products, horticultural products and silk, while the land-intensive products include cereals, vegetable oilseeds and vegetable oils, and raw materials (excluding silk) for textiles.

320

Figure 11.7 shows the composition of China's agricultural exports by factor intensity of production. It can be seen that China's agricultural exports are overwhelmingly dominated by exports of labour-intensive products and that this domination has increased since 2001.

Figure 11.8 shows that China's agricultural imports are biased towards land-intensive products, accounting for 60.6 per cent and 65 per cent of China's total agricultural imports during the two periods of 1992–2001 and 2002–05, respectively.

Since its accession into the WTO, China's agricultural trade has been moving in line with its comparative advantage and is now more consistent with its resource endowments of relative scarcity of land resources and relative abundance of labour.

Figure 11.7   Shares of China's agricultural exports by factor intensity of production, 1992–2001 and 2002–2005

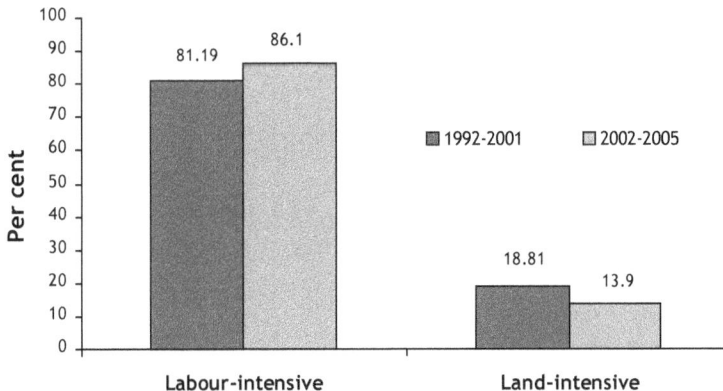

Sources: Author's calculation. Data from 1992 to 2004 are from United Nations Statistics Division, *Commodity Trade Statistics Database*, COMTRADE. http://unstats.un.org/unsd/comtrade/default.aspx. Data for 2005 are from China General Administration of Customs (various issues, 2005). *Zhongguo Haiguan Tongji Yuebao* [China Customs Statistical Monthly Report], Zhongguo Haiguan Chubanshe, Beijing.

Figure 11.8   Share of China's agricultural imports by factor intensity of production, 1992-2001 and 2002-2005 (per cent)

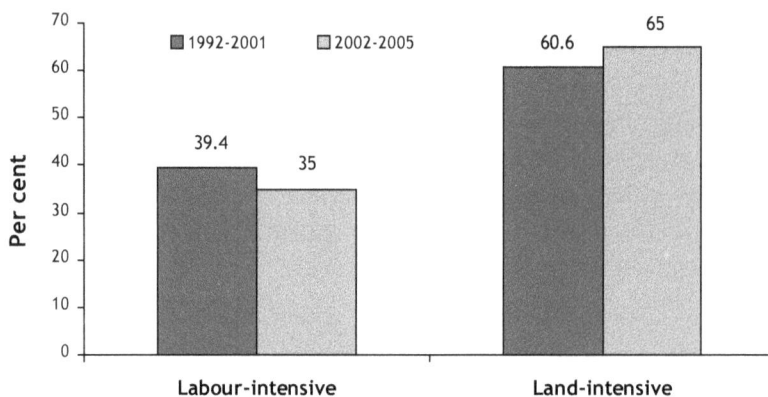

Sources: Author's calculation. Data from 1992 to 2004 are from United Nations Statistics Division, *Commodity Trade Statistics Database*, COMTRADE. http://unstats.un.org/unsd/comtrade/default.aspx. Data for 2005 are from China General Administration of Customs (various issues, 2005). *Zhongguo Haiguan Tongji Yuebao* [China Customs Statistical Monthly Report], Zhongguo Haiguan Chubanshe, Beijing.

Changes in revealed comparative advantage in China's agriculture following WTO accession

The above section showed that China's agricultural trade has moved in line with its resource endowments, exporting mainly labour-intensive products and importing mainly land-intensive products. This trade pattern has been strengthened since WTO accession. However, have there been any changes in China's comparative advantage in agriculture as revealed by its international agricultural trade performance?

It is difficult if not impossible to measure a country's comparative advantage directly. The most common indirect approach is the principle of revealed comparative advantage (RCA) proposed by Balassa (1965). It is argued that since its trade is generated by a country's underlying comparative advantage, data on exports and imports can be used to infer the underlying pattern of comparative advantage. This idea has given rise to various RCA indicators. One of these indicators is the net export ratio (NER_{ij}), which is defined as

$$RCA\ (NER_{ij}) = (X_{ij} - M_{ij}) / (X_{ij} + M_{ij})$$

where X_{ij} are the exports of good i by country j and M_{ij} are the imports of good i into country j.

The rationale behind the index is that countries are revealed as having a comparative advantage in a particular good if they export more of it than they import. However, to simply consider net exports might be misleading where, for example, we compare a large and a small country. For this reason net exports are divided by total trade (exports plus imports). Net export ratios have a minimum value of -1 (the country only imports the good concerned) and a maximum value of +1 (the country only exports the good). Positive values are taken to reveal a comparative advantage and negative values are taken to reveal a comparative disadvantage.

However, RCA indices have one major flaw. The principle of revealed comparative advantage presumes that observed trade flows are generated by underlying comparative advantages and disadvantages. However, observed trade flows are not just created by underlying economic forces but are often significantly affected by government policies. Because of the higher levels of government intervention in agriculture, this problem has been potentially more serious for trade in agricultural products than in manufactured goods. As far as China is concerned, the WTO accession commitments have led to trade being less affected by government intervention. Therefore, RCAs for the period since 2001 are more likely to show true comparative advantage than previously.

Revealed comparative advantage in China's agriculture

Table A11.3 presents China's RCA indices calculated by using the measure of net export ratio for agricultural products for the period of 1992 to 2005.

With respect to individual commodities, in the group of cereals, vegetable oilseeds and vegetable oils, with the exception of some years China has a revealed comparative advantage in corn, rice, peanuts, other oilseeds and miscellaneous grains. However, China has a revealed comparative disadvantage in wheat, soybean, rapeseeds, and all vegetable oils.

In the group of horticultural products, China has a revealed comparative advantage in all horticultural products, except vegetable plaiting materials.

In the group of animal products, China has a revealed comparative advantage in live animals (including pigs and poultry), beef, pork, fish and aquatic products, and products of animal origin, but has a revealed comparative disadvantage in mutton, dairy products and animal fats.

For the group of processed agricultural products, China has a revealed comparative advantage in products of the milling industry, preparations of meat, fish and aquatic products, preparations of cereals, preparations of vegetables and fruits, miscellaneous edible preparations, beverages and spirits, and tobacco products, while it has a revealed comparative disadvantage in sugar and sugar confectionary, cocoa and cocoa preparations, and residues from food industry and animal feed.

In the group of raw materials for textiles, China has a revealed comparative advantage in silk, but has a revealed comparative disadvantage in raw hides and skins, wool, cotton, and other vegetable textile fibres.

In terms of commodity groups, China has a revealed comparative advantage in horticultural products, in processed agricultural products, and in animal products. But has a revealed comparative disadvantage in cereals, vegetable oilseeds and vegetable oils, and in raw materials for textiles.

In terms of the factor intensity of production, China has a revealed comparative advantage in labour-intensive agricultural products, but has a revealed comparative disadvantage in land-intensive agricultural products.

These patterns of China's revealed comparative advantage and dis-advantage are consistent with the country's resource endowments.

Changes in the revealed comparative advantage of agriculture

China's revealed comparative advantage in agriculture has been on a declining trend, especially after 2002 (Figure 11.9). The values of China's revealed comparative advantage indices for all agricultural products declined from around 0.4 in the early 1990s to around 0.2 in 2002, and then fell into negative territory in 2004 and 2005. In other words, since 2004 China's agriculture as a whole has lost its comparative advantage to non agricultural activities. In fact, agriculture may well have not had a comparative advantage prior to 2002 but the removal of government protection through the WTO accession has made this clear. Certainly, individual agricultural industries and commodities have a comparative advantage, as seen above. Also, the regional dimension of China's agricultural comparative advantage is not examined here and comparative advantage could vary widely throughout the country.

Figures 11.10 and 11.11 show in terms of commodity groups and factor intensities where agriculture has comparative advantage and where it does not. Across commodity groups the RCA indices have been declining, except for processed agricultural products. In fact, the comparative advantage of

Figure 11.9   China's revealed comparative advantage indices (NER) of all agricultural products, 1992-2005

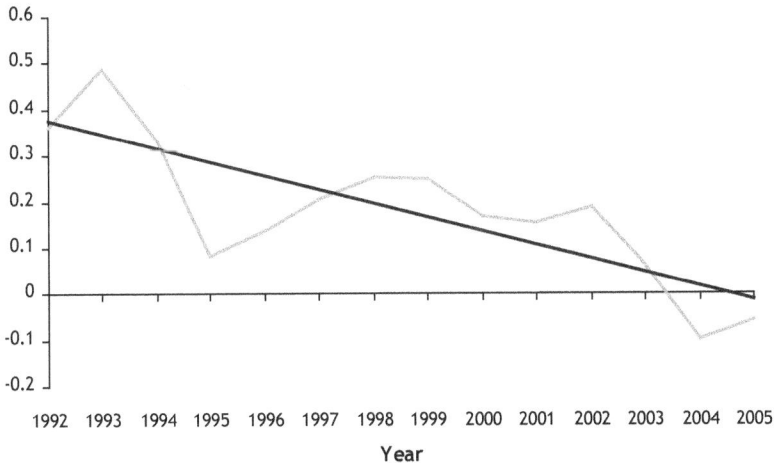

Sources: Author's calculation. Data from 1992 to 2004 are from United Nations Statistics Division, *Commodity Trade Statistics Database*, COMTRADE. http://unstats.un.org/unsd/comtrade/default.aspx. Data for 2005 are from China General Administration of Customs (various issues, 2005). *Zhongguo Haiguan Tongji Yuebao* [China Customs Statistical Monthly Report], Zhongguo Haiguan Chubanshe, Beijing.

Figure 11.10 China's revealed comparative advantage indices (NER) of agricultural products by commodity group, 1992-2005

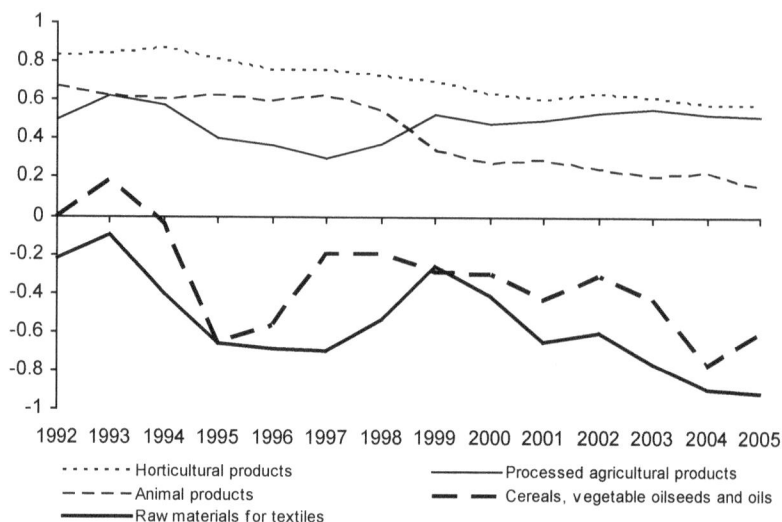

Sources: Author's calculation. Data from 1992 to 2004 are from United Nations Statistics Division, *Commodity Trade Statistics Database*, COMTRADE. http://unstats.un.org/unsd/comtrade/default.aspx. Data for 2005 are from China General Administration of Customs (various issues, 2005). *Zhongguo Haiguan Tongji Yuebao* [China Customs Statistical Monthly Report], Zhongguo Haiguan Chubanshe, Beijing.

agricultural processing, which is a labour-intensive activity, has increased slightly. Though declining, the RCAs for horticultural products and animal products indicate that these activities still have comparative advantage. However, the RCA for animal products has moved close to zero. The RCAs for the land-intensive agricultural products, cereals, vegetable oilseeds and vegetable oils and raw materials for textiles, have declined rapidly, particularly since 2003, and have become significantly negative. These are the activities from which we will most likely see resource flows and structural adjustment.

Figure 11.11 illustrates China's comparative advantage in labour-intensive activities and its comparative disadvantage in land-intensive activities.

While maintaining its comparative advantage, the RCA for labour-intensive agricultural products suggests a weakening, due most likely to competition from labour-intensive non agricultural activities.

Factors driving the changes in revealed comparative advantage in China's agriculture

What are the reasons for the changes in China's revealed comparative advantage in agriculture? Empirical studies have shown that during the process of economic growth a country's comparative advantage in agriculture declines, and for those countries where arable land is scarce,

Figure 11.11 China's revealed comparative advantage indices (NER) of agricultural products by factor intensity of production, 1992–2005

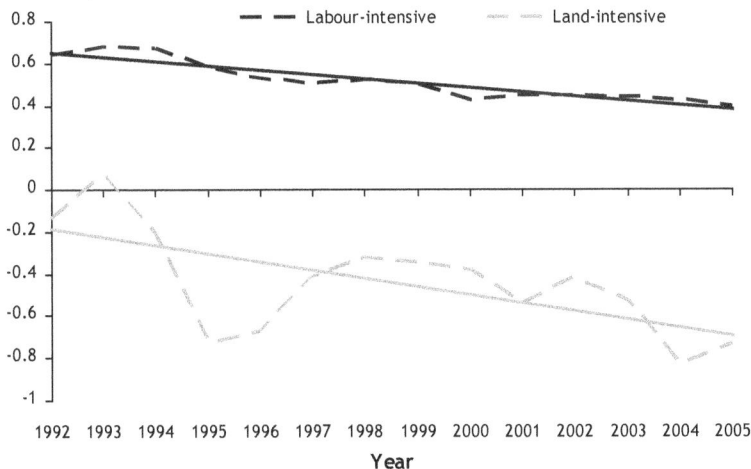

Sources: Author's calculation. Data from 1992 to 2004 are from United Nations Statistics Division, *Commodity Trade Statistics Database*, COMTRADE. http://unstats.un.org/unsd/comtrade/default.aspx. Data for 2005 are from China General Administration of Customs (various issues, 2005). *Zhongguo Haiguan Tongji Yuebao* [China Customs Statistical Monthly Report], Zhongguo Haiguan Chubanshe, Beijing.

the comparative advantage in agriculture tends to decline more rapidly (Anderson 1990). Undoubtedly, the changes in China's RCAs for agriculture during the period 1992 to 2005 have mainly been the result of the fast economic growth and the resulting dramatic structural changes. These structural changes appear to have become even more rapid following entry into the WTO and the removal of agricultural protection.

Economic factors

Since entry into the WTO in 2001, China's economy has been growing at a rapid average annual growth rate of around 9.8 per cent. This rapid economic growth has led to changes in the structure of the economy, with the growth of manufacturing and services sectors much faster than the growth of the agricultural sector. The share of agricultural GDP in total GDP has declined from 15 per cent in 2001 to 13.8 per cent in 2004 (calculated from various issues of the SSB).

The structure of the agricultural economy has also been changing. Although the farming sector remains the most important agricultural sector, its share has declined from 55.2 per cent in 2001 to 50 per cent in 2004. However, the animal husbandry and fishery sectors have been growing rapidly and the share of these sectors has increased from 41.2 per cent in 2001 to 46 per cent in 2004 (calculated from various issues of SSB).

With the rapid economic growth, especially after China's entry into the WTO, it is likely that the comparative advantage of China's agricultural sector has been declining, and in particular that the comparative advantage of China's farming sector has been declining. This changing pattern of comparative advantage is consistent with China's resource endowments. It is also an indication of the improvement in resource allocation among China's economic sectors.

China's remarkable industrial growth has also played a large part in driving up agricultural imports. Over 30 per cent of the growth in China's agricultural imports in 2004 came from raw materials used in production of non food manufactured products: such as cotton, wool, animal hides, and rubber. In particular, growing textile production is generating demand for cotton and wool that is beyond China's production capacity. China's exports of apparel and footwear grew at double digit rates during 2004, and its domestic retail sales of apparel, shoes, and textiles rose 18.7 per cent. Chinese yarn production grew 13.9 per cent and cloth production grew 18.8 per cent during 2004 (Gale 2005).

The continued increase in per capita income in China has led to not only a rise in food consumption but also a change in the structure of food consumption. Since the late 1990s, China has sharply increased imports of vegetable oilseeds (mainly soybeans) and vegetable oils (mainly soybean oil and palm oil). Soybeans are crushed to produce vegetable oil for human consumption and animal feed to help the rapid growth in animal production. Driven by consumer and food industry demands, China has also rapidly increased imports of meats, fish, milk, cheese, wines, and fruits since the early 2000s.

Trade barriers

Apart from the economic factors discussed above, other factors could also affect China's revealed comparative advantage in agriculture. RCA indices are not only created by underlying economic forces but are often significantly distorted by government policies. This problem has been more serious for trade in agricultural products. Admittedly, after the establishment of the WTO and the implementation of the Uruguay Round Agreement on Agriculture (URAA), some liberalisation of trade in agriculture has taken place. However, significant trade barriers remain. In particular, the developed countries have increasingly resorted to sanitary and phytosanitary (SPS) measures for animal and plant health and technical barriers to trade (TBT) to block agricultural imports, especially from developing countries; these actions have seriously affected the developing countries' exports of agricultural products in which they have a comparative advantage.

Chinese farmers and exporters anticipated a large, positive impact on exports of agricultural products following accession to the WTO, especially for labour-intensive agricultural products such as vegetables, fruits, animal products, and aquatic products. In fact, these products have been hardest hit by the need to meet significant SPS standards, which has prevented substantial growth in these exports.

According to official Chinese sources, SPS and TBT actions have resulted in huge direct losses for agricultural exports. The indirect losses are even larger. In 2001, about US$7 billion worth of Chinese exports were affected by SPS and TBT actions. In early 2002, the EU banned imports of Chinese animal-derived food, seafood and aquatic products, resulting in a 70 per cent slump in China's aquatic product exports during the second half of that year (MOFCOM 2005). Also, according to an investigation by China's Ministry of Commerce (MOFCOM), in 2002 about 90 per cent of China's exporters of foodstuffs, domestic produce, and animal by-products were affected by foreign TBTs and suffered losses totalling US$9 billion (*China Daily* 2003).

Although the WTO's SPS Agreement requires members to ensure that SPS measures are based on sufficient scientific evidence, there are some well-founded concerns that countries may abuse SPS measures and use them as trade barriers. Because of very low production and labour costs, some agricultural products exported from China are very competitive in world markets. Consequently, importing countries may look to restrict imports from China by setting relatively high standards or strict inspections in order to protect domestic markets.

Conclusions

Entry into the WTO has boosted China's agricultural trade, especially its agricultural imports. The pattern of China's agricultural trade appears consistent with its resource endowments. Following entry into the WTO, changes in the pattern of agricultural trade indicate that China is moving closer to its comparative advantage.

China has a comparative advantage in labour-intensive agricultural products and a comparative disadvantage in land-intensive agricultural products. Since entry into the WTO, its comparative advantage in labour-intensive agricultural products has been declining, especially in animal and horticultural products, and China's agriculture as a whole has lost comparative advantage since 2004.

Fast economic growth and the associated structural changes have played significant roles in driving the changes in comparative advantage in China's agriculture. However, the application of TBT and SPS measures by importing countries may have also contributed to the rapid decline in China's revealed comparative advantage in labour-intensive animal and horticultural products. Because of China's low production costs, some agricultural products exported from China are very competitive in world markets. Consequently, importing countries may look to restrict imports from China by setting relatively high SPS standards or may impose strict inspections in order to protect domestic markets.

China itself should first increase and strengthen SPS domestic standards to meet the international standards in order to increase its exports of animal and horticultural products, especially to developed countries' markets. As China is likely to face more and more SPS disputes, the government needs to initiate bilateral negotiations to counter unfair trade restrictions and discrimination and could use the WTO to coordinate and resolve trade disputes. As a member of the WTO, China can now participate in the negotiation and establishment of international regulations and standards to obtain a more equal position for its agricultural exports.

Notes

1 In January 2006 China revised its GDP growth rate for the period of 1979-2004. The revised growth rates for the three years 2002 to 2004 were 9.1 per cent, 10.0 per cent and 10.1 per cent respectively. The GDP growth rate in 2005 was 9.9 per cent.
2 All the trade values are at 2000 constant US$ prices.
3 China's average tariff level has been reduced to 9.9 per cent in 2005 as against 15.6 per cent in 2000. In 2005 the average tariff on industrial products was 9.3 per cent as against 14.8 per cent in 2000; for agricultural products the change was from 23.2 per cent in 2000 to 15.3 per cent.
4 Data for 2005 are preliminary statistics subject to final revision.
5 China's domestic grain prices increased sharply in the last quarter of 2003. From September to December 2003, rice prices increased by 27 per cent, wheat prices by 37 per cent, and corn prices by 14 per cent.
6 China abolished agricultural taxes at the beginning of 2006.
7 According to a Chinese customs official, a large amount of the imported fish was processed for export, which contributed to the rapid increase in exports of processed fish and aquatic products.

References

Anderson, K., 1990. *Changing Comparative Advantages in China*, Organisation for Economic Cooperation and Development (OECD), Paris.

Anderson, K., 1997. 'On the complexities of China's WTO accession', *The World Economy*, 20(6):749-72.

Balassa, B., 1965. 'Trade liberalisation and "revealed comparative advantage"', *The Manchester School of Economic and Social Studies*, 33:99-123.

Cheng, G., 1997. *Jiaru Shimao Zuzhi dui Woguo Nongye Yingxiang de Fenxi* [Studies on the Impact of Joining the WTO on China's Agriculture], *Brief Research Report*, Institute of Agricultural Economics of Chinese Academy of Agricultural Sciences, No. 6, Beijing.

China Daily, 2003. 'Ministry sounds warning on trade barriers against China', 11 June.

China Daily, 2004. 'China-ASEAN agricultural trade on fast track', 9 August.

China General Administration of Customs (various issues), 2005. *Zhongguo Haiguan Tongji Yuebao* [China Customs Statistical Monthly Report], Zhongguo Haiguan Chubanshe, Beijing.

Development Research Centre, 1998. *The global and domestic impact of China joining the World Trade Organisation*, Project Report, Development Research Centre, State Council, Beijing.

Gale, F., 2005. *China's agricultural imports boomed during 2003-04*, United Sates Department of Agriculture, Electronic Outlook Report from the Economic Research Service, WRS-05-04, www.ers.usda.gov

Huang, J., 1998. 'The impact of joining the WTO on China's grain market', *International Trade*, 20:10-13.

Huang, J. and Chen, C., 1999. *Effects of Trade Liberalisation on Agriculture in China*, Working Paper, United Nations ESCAP CGPRT Centre.

Ministry of Commerce of China, 2005. *Challenges for Agricultural Trade*, 27 February, http://english.mofcom.gov.cn

State Statistical Bureau (SSB), various issues, *Zhongguo Tongji Nianjian* [China Statistical Yearbook], Zhongguo Tongji Chubanshe, Beijing.

United Nations Statistics Division, *Commodity Trade Statistics Database*, COMTRADE. http://unstats.un.org/unsd/comtrade/default.aspx

Wang, Z., 1997. *The impact of China and Taiwan joining the World Trade Organisation on US and world agricultural trade: a computable general equilibrium analysis*, Technical Bulletin Number 1858, US Department of Agriculture, Washington, DC.

Appendix

Table 11A.1 China's Agricultural Exports by Commodity Group, 1992–2005 (US$ million at 2000 constant prices)

	1992	1993	1994	1995	1996	1997	1998	1999	2000	2001	2002	2003	2004	2005
Total Agricultural Exports	13,778.52	13,530.96	16,483.80	16,095.57	15,530.50	16,054.78	14,528.82	13,904.08	15,524.50	15,405.71	17,108.66	19,715.37	20,777.77	23,808.82
Cereals, vegetable oilseeds and vegetable oils	2,709.83	2,635.12	3,336.60	14,13.01	1,368.27	2,477.60	2,382.56	1,873.40	2,345.10	1,746.44	2,327.66	3,299.59	1,595.50	2,664.70
Cereals (10)	1,822.43	1,755.87	1,779.80	85.88	205.30	1,262.67	1,578.24	1,173.33	1,643.00	1,005.58	1,579.60	2,423.52	674.99	1,254.51
Wheat (1001)	0.37	9.54	11.62	1.70	0	0.11	1.48	0.21	0.20	45.71	67.01	248.06	102.20	0
Corn (1005)	1,456.72	1,375.61	1,096.69	14.69	32.94	921.52	562.00	465.20	1052.00	608.79	1,117.21	1,654.06	295.65	983.69
Paddy, Rice (1006)	267.53	301.59	598.30	18.08	122.96	284.29	976.10	674.02	561.00	319.96	363.79	463.36	212.05	206.36
Vegetable Oilseeds	727.13	645.61	993.29	830.58	776.96	556.24	512.77	585.73	605.70	646.72	671.38	793.70	830.61	1,228.59
Soybean (1201)	196.36	121.59	257.91	113.01	72.46	78.31	66.55	64.09	64.00	74.88	73.71	81.44	132.10	150.62
Peanuts (1202)	233.17	233.64	365.95	290.42	278.86	149.12	164.80	200.55	232.00	255.77	252.74	297.67	272.31	283.85
Rapeseeds (1205)	13.87	22.65	3.49	0.57	1.32	0	0.53	0	0.40	0	0.67	0.94	0.09	0.09
Other oilseeds (1203-1204, 1206-1209, 1212)	283.73	267.73	365.95	426.59	424.32	328.81	280.89	321.09	309.30	316.07	344.26	413.65	426.11	794.03
Vegetable Oils	160.28	233.64	563.51	496.54	386.01	658.69	291.56	144.33	96.40	94.14	76.68	82.38	89.89	181.60
Soybean oil (1507)	4.42	14.30	56.93	54.24	94.42	393.71	142.61	35.15	17.00	22.37	20.68	6.08	12.13	35.61
Palm oil (1511)	23.69	78.67	274.17	222.62	110.88	71.88	23.24	0.62	0.30	0	3.06	0	0.18	0
Rapeseeds oil (1514)	33.87	35.76	146.38	155.95	125.16	98.70	50.71	18.61	24.00	23.34	10.05	3.74	3.73	145.99
Other vegetable oils (1508-1510, 1512-1513, 1515)	98.30	104.90	86.03	63.73	55.55	94.41	75.00	59.96	55.10	48.43	42.89	72.55	73.84	0
Horticultural products	2,645.28	2,732.63	3,475.95	3,646.66	3,266.48	3,243.57	2,996.55	2,881.53	2,861.30	3,028.41	3,247.94	3,745.36	4,360.84	4,695.42
Live trees and other plants (06)	18.41	22.65	27.88	31.64	32.94	34.33	31.69	32.05	32.00	34.03	41.17	45.87	58.62	68.48
Edible vegetables (07)	1,292.27	1,351.77	1,843.69	1,935.77	1,692.91	1,623.13	1,566.62	1,570.26	1,545.00	1,698.01	1,802.66	2,040.66	2,313.16	2,710.94
Edible Fruits (08)	347.30	411.25	480.96	542.42	506.12	497.77	459.53	439.35	417.00	423.04	531.32	703.93	835.45	947.84

Coffee, tea, mate and species (09)	568.21	554.30	527.43	525.47	540.15	593.25	549.32	505.51	506.00	527.10	528.45	584.11	788.22	823.77
Vegetable plaiting materials (14)	62.59	53.64	62.73	68.93	57.09	53.64	47.54	41.35	43.00	41.82	42.12	43.06	39.66	43.74
Other vegetable products (1210-1214, 13)	356.51	339.02	533.24	542.42	437.28	441.45	341.84	292.97	318.30	304.40	302.23	327.72	325.74	100.65
Animal products	3,435.49	3,067.12	4,224.06	5,090.40	4,625.74	4,677.35	4,113.57	3,974.22	4,388.60	4,512.66	4,555.37	4,917.16	5,800.07	6,022.16
Live Animals (01)	587.84	540.00	543.70	568.41	533.56	510.45	465.87	398.00	385.00	334.55	329.32	306.10	301.12	292.01
Pig (0103)	357.12	324.23	313.67	315.28	322.77	323.98	307.41	245.00	232.00	214.93	205.83	202.19	219.53	201.49
Poultry (0105)	114.13	107.28	121.98	141.26	132.84	122.30	101.41	102.34	104.00	75.86	78.50	62.72	30.36	33.22
Meat and edible meat offal (02)	456.53	413.64	736.55	1,154.91	1,192.28	1,040.60	887.37	714.33	753.00	817.89	636.63	604.71	644.36	659.75
Beef (0201, 0202)	47.86	33.38	36.01	37.86	55.99	57.93	77.12	26.88	23.00	32.09	18.29	14.04	27.62	36.86
Pork (0203)	93.27	75.10	148.70	276.86	236.04	209.19	191.21	69.26	67.00	132.26	201.04	251.81	418.91	360.71
Mutton (0204)	2.95	3.58	2.79	3.39	2.20	2.90	4.23	4.14	6.00	4.38	7.47	19.66	38.12	30.00
Poultry (0207)	204.95	220.53	441.46	701.76	758.62	659.76	552.49	556.17	587.00	580.59	383.89	299.55	132.01	98.99
Fish and aquatic products (03)	1,676.39	1,494.82	2,109.73	2,358.40	1,908.09	2,126.26	1,833.89	2,012.75	2,270.00	2,519.79	2,750.41	3,121.83	3,697.33	3,863.76
Dairy products, Eggs and natural honey (04)	195.13	170.46	181.23	183.07	214.08	177.01	184.87	169.54	188.00	186.72	185.72	207.81	213.51	237.51
Dairy products (0401-0406)	27.74	31.71	25.56	30.06	32.94	43.98	42.26	44.45	50.00	37.93	52.46	44.00	51.33	50.00
Product of animal origin not elsewhere specified (05)	508.07	436.29	634.31	803.46	744.35	750.95	686.65	648.17	760.00	633.11	626.57	693.64	889.32	898.48
Animal fats (1501-1506, 1516-1518, 1520-1521)	11.54	11.92	18.53	22.15	33.38	71.88	54.93	31.43	32.60	20.62	26.71	37.07	54.43	70.65
Processed agricultural products	4,101.38	4,381.94	4,549.40	5,342.85	5,811.00	5,194.43	4,638.60	4,555.82	5,259.00	5,735.41	6,527.09	7,311.73	8,735.53	1,0122.25
Products of milling industry (11)	57.68	76.29	91.78	114.13	239.33	199.54	111.98	82.70	93.00	104.06	112.97	132.92	155.26	177.63

	1992	1993	1994	1995	1996	1997	1998	1999	2000	2001	2002	2003	2004	2005
Preparations of meet, fish and aquatic products (16)	488.44	717.61	906.16	1,261.13	1,612.76	1,489.03	1,291.96	1,432.80	1,883.00	1,989.77	2,227.71	2,507.76	3,180.69	3,876.35
Sugars and sugar confectionery (17)	824.69	780.79	421.71	264.43	333.75	208.12	193.32	144.73	173.00	151.71	217.31	183.47	229.92	371.61
Cocoa and cocoa preparations (18)	44.18	53.64	40.66	46.33	53.80	60.08	46.48	41.35	29.00	26.74	34.46	51.48	63.45	97.59
Preparations of cereals, flour (19)	157.09	145.43	192.85	238.44	258.00	290.73	276.77	299.79	360.00	401.65	434.63	493.32	595.13	674.77
Preparations of vegetables and fruits (20)	844.33	814.16	947.99	1,223.84	1,149.46	1,121.06	1,089.14	1,164.02	1,315.00	1,455.86	1,682.03	2,029.43	2,350.35	2,748.85
Miscellaneous edible preparations (21)	136.22	145.43	185.88	244.09	275.56	326.13	347.55	348.38	359.00	388.03	441.33	509.23	559.03	635.42
Beverages, spirits and vinegar (22)	403.76	356.42	439.14	441.85	435.85	498.85	470.09	472.43	493.00	556.28	571.53	582.24	677.36	638.26
Residues from food industry and animal feeds (23)	603.79	529.27	526.27	379.70	380.96	295.02	200.71	222.26	252.00	285.92	390.59	360.39	456.10	424.60
Tobacco and manufactured tobacco substitutes (24)	541.21	762.91	796.96	1128.92	1071.52	705.89	610.59	347.35	302.00	375.39	414.52	461.49	468.23	477.17
Agricultural Products as Raw Materials for Textiles	886.53	714.15	897.80	602.65	459.02	461.83	397.52	619.12	670.50	382.78	450.62	387.54	285.84	304.29
Raw hides and skins, leather, fur skins and articles	52.28	42.08	72.38	57.29	35.35	45.38	26.73	12.72	11.30	10.60	11.87	8.61	8.15	7.94
Raw hides and skins and articles (4101-4103)	43.08	28.01	67.96	53.56	29.75	30.47	16.80	9.20	6.20	6.62	7.18	4.31	1.13	1.10
Raw fur skins (4301)	9.20	14.07	4.41	3.73	5.60	14.91	9.93	3.51	5.10	4.08	4.69	4.31	7.02	6.84
Silk	398.85	293.24	470.51	400.04	343.63	335.78	265.15	291.21	331.70	273.76	256.18	223.63	215.24	237.19
Cocoon (5001)	28.23	28.61	65.06	27.12	31.84	12.87	6.34	3.82	5.70	3.40	1.53	0.84	0.91	0.89

Raw silk (5002)	342.40	225.30	343.88	340.14	289.84	295.02	225.01	247.07	272.00	236.32	232.63	198.45	194.73	217.20
Waste silk (5003)	28.23	39.34	61.57	32.77	21.96	27.89	33.80	40.32	54.00	34.04	22.02	24.34	19.60	19.10
Wool, fine or coarse animal hair	162.73	105.50	159.16	85.09	60.71	69.73	38.24	15.82	12.70	15.75	14.84	26.49	43.67	47.78
Uncarded wool (5101)	12.27	13.11	8.60	14.69	30.19	45.06	25.35	9.30	10.00	12.62	11.97	23.40	40.20	44.41
Uncarded fine or coarse animal hair (5102)	149.72	90.59	149.05	67.80	29.09	22.53	10.56	4.76	1.40	1.85	0.77	0.37	0.55	0.53
Waste wool and fine or coarse animal hair (5103)	0.74	1.79	1.51	2.60	1.43	2.15	2.32	1.76	1.30	1.26	2.11	2.71	2.92	2.84
Cotton	266.68	265.82	186.58	55.15	15.37	5.26	62.33	294.83	307.90	79.65	165.43	126.46	15.77	8.45
Uncarded cotton (5201)	258.94	227.68	173.10	53.11	13.72	3.54	59.16	292.56	306.00	77.80	162.75	124.50	14.31	7.03
Waste cotton (5202)	6.14	30.99	9.29	2.03	1.65	1.07	3.17	1.55	1.20	1.36	2.11	1.22	1.19	1.15
Carded cotton (5203)	1.60	7.15	4.18	0	0	0.64	0	0.72	0.70	0.49	0.57	0.75	0.27	0.27
Other vegetable textile fibres (5301-5302)	6.00	7.51	9.18	5.09	3.95	5.69	5.07	4.55	6.90	3.01	2.30	2.34	3.01	2.93

Table 11A.2 China's Agricultural Imports by Commodity Group, 1992-2005 (US$ million at 2000 constant price)

	1992	1993	1994	1995	1996	1997	1998	1999	2000	2001	2002	2003	2004	2005
Total Agricultural Imports	6,462.93	4,671.29	8,252.71	13,684.17	11,784.04	10,574.66	8,685.94	8,334.66	11,048.10	11,293.18	11,705.20	17,463.26	25,258.46	26,636.15
Cereals, vegetable oilseeds and vegetable oils	2,741.74	1,799.38	3,615.59	6,964.24	4,933.58	3,650.26	3,561.82	3384.04	4,359.00	4,372.82	4,422.20	8,243.13	12,245.78	10,699.22
Cereals (10)	2,058.06	1,188.46	1,488.20	4,046.69	2,805.04	956.92	735.25	513.78	574.00	590.32	461.43	415.62	2,021.33	1,237.93
Wheat (1001)	1,845.75	994.16	1,116.44	2,288.34	2,074.96	394.79	294.73	88.90	147.00	117.67	98.61	72.08	1495.49	686.44
Corn (1005)	0	0.24	0.23	922.12	80.14	0.21	33.80	8.27	0.40	4.67	1.53	0.37	0.73	1.31
Paddy, Rice (1006)	47.86	41.72	163.81	490.44	313.99	150.19	126.77	80.63	113.00	96.28	76.59	90.80	229.28	177.14
Vegetable oilseeds	88.36	66.75	101.89	165.44	404.78	1,006.38	1,364.96	1,649.38	3,029.80	3,207.45	2,609.88	5,266.68	6,677.36	7,248.27
Soybean (1201)	35.59	30.99	15.92	84.75	351.32	904.36	850.39	920.05	2,270.00	2,732.77	2,377.05	5,070.75	6,362.66	6,908.95
Peanuts (1202)	0.37	0.48	2.67	0.34	0.22	2.68	2.11	0.41	0.20	0	0.38	0.09	0.73	0
Rapeseeds (1205)	0.12	0	45.31	29.38	0.11	17.16	424.67	649.21	658.00	363.72	140.73	44.00	122.53	152.80
Other oilseeds (1203-1204, 1206-1209, 1212)	52.28	35.28	37.99	50.97	53.14	82.18	87.79	79.70	101.60	110.96	91.71	151.83	191.45	186.52
Vegetable oils	595.33	544.17	2,025.51	2,752.11	1,723.76	1,686.96	1,461.62	1,220.88	755.20	575.05	1,350.89	2,560.84	3,547.09	2,213.02
Soybean oil (1507)	122.72	45.30	756.30	1,157.17	838.77	734.86	552.49	435.22	126.00	22.85	390.59	950.12	1,412.16	806.26
Palm oil (1511)	284.72	379.07	801.61	977.49	578.57	653.33	623.27	617.16	456.00	413.32	812.77	1,350.77	1,702.89	1,065.17
Rapeseeds oil (1514)	109.22	87.02	370.60	466.71	205.30	211.34	184.87	39.28	28.00	18.96	37.62	79.57	198.56	92.55
Other vegetable oils (1508-1510, 1512-1513, 1515)	78.67	32.78	97.01	150.75	101.11	87.43	100.99	129.22	145.20	119.91	109.90	180.38	233.48	249.04
Horticultural products	245.44	233.64	239.67	384.89	451.55	463.34	487.78	514.30	653.20	741.93	726.71	906.78	1,186.07	1,263.16
Live trees and other plants (06)	7.36	11.92	6.97	6.78	5.49	8.58	11.62	17.57	21.00	21.40	31.59	42.12	46.86	61.00
Edible vegetables (07)	46.63	29.80	17.42	88.14	84.55	79.39	75.00	85.80	82.00	204.23	185.72	226.53	369.04	465.13
Edible Fruits (08)	50.32	53.64	76.68	94.92	216.28	252.10	255.65	266.71	368.00	356.91	361.87	464.30	564.23	583.96
Coffee, tea, mate and species (09)	29.45	16.69	11.62	16.95	30.74	10.73	21.13	19.64	23.00	20.42	22.02	26.21	29.54	37.00

Vegetable plaiting materials (14)	38.04	36.95	46.47	101.70	43.91	35.40	45.42	49.62	83.00	63.21	42.41	67.40	80.13	60.20
Other vegetable products (1210-1214, 13)	73.63	84.63	80.51	76.39	70.59	77.13	75.95	74.95	76.20	75.76	83.10	80.22	96.27	55.87
Animal products	693.26	725.36	1,044.99	1,189.49	1,183.39	1,096.92	1,224.78	1,972.43	2,546.80	2,470.87	2,754.72	3,274.69	3,694.69	4,509.74
Live Animals (01)	24.54	22.65	26.72	40.68	51.60	43.98	58.10	68.23	52.00	34.04	50.74	109.52	200.47	96.73
Pig (0103)	0	0	0.58	2.26	1.87	2.15	3.17	4.14	4.00	1.85	1.94	1.87	2.55	na
Poultry (0105)	20.49	13.11	12.20	16.95	16.69	15.02	10.56	11.37	11.00	9.73	12.45	14.04	11.76	na
Meat and edible meat offal (02)	69.95	81.06	98.75	107.35	172.36	159.85	151.06	515.85	637.00	581.56	600.25	709.55	433.77	521.12
Beef (0201, 0202)	4.91	6.44	6.12	4.52	4.39	3.22	5.28	6.20	7.00	5.84	12.25	11.23	9.12	na
Pork (0203)	0	0.08	0.23	1.13	1.10	2.15	7.39	24.81	58.00	40.85	78.50	85.18	49.69	na
Mutton (0204)	0.37	0.36	0.70	0.79	1.76	2.57	5.28	8.27	14.00	19.45	25.85	36.51	38.84	na
Poultry (0207)	61.36	69.14	83.65	90.40	153.70	139.46	114.09	423.85	481.00	431.80	407.82	432.47	140.12	na
Fish and aquatic products (03)	396.39	433.90	666.84	676.90	655.42	583.60	703.55	912.82	1,212.00	1,294.42	1,498.22	1,745.79	2,133.19	2,557.08
Dairy products, Eggs and natural honey (04)	84.68	67.95	98.75	72.32	62.58	72.95	94.02	169.54	218.00	212.98	260.39	327.63	408.51	410.66
Dairy products (0401-0406)	75.84	59.14	89.80	65.99	59.17	67.69	89.79	164.37	214.90	210.58	256.95	324.07	405.14	na
Product of animal origin not elsewhere specified (05)	50.32	48.87	56.93	80.23	104.30	123.37	107.75	121.98	160.00	168.25	183.42	205.00	229.10	196.67
Animal fats (1501-1506, 1516-1518, 1520-1521)	67.37	70.93	97.01	212.00	137.12	113.18	110.29	184.01	267.80	179.62	161.69	177.20	289.63	727.50
Processed agricultural products	1,387.99	1,035.88	1,248.88	2,257.83	2,723.80	2,823.57	2,114.89	1,413.16	1,874.00	1,910.02	1,986.94	2,097.76	2,733.61	3,236.90
Products of milling industry (11)	46.63	33.38	42.98	81.36	77.95	72.95	58.10	81.67	64.00	78.77	90.95	127.31	172.03	164.96
Preparations of meet, fish and aquatic products (16)	8.59	5.96	9.29	13.56	8.78	8.58	7.39	12.41	12.00	13.62	18.67	25.27	23.89	25.53

	1992	1993	1994	1995	1996	1997	1998	1999	2000	2001	2002	2003	2004	2005
Sugars and sugar confectionery (17)	335.03	158.54	518.14	1056.59	469.89	268.20	180.64	188.15	177.00	365.67	268.05	202.19	306.50	400.93
Cocoa and cocoa preparations (18)	44.18	46.49	56.93	66.67	64.77	76.17	67.61	55.82	71.00	77.80	76.59	108.59	123.71	157.65
Preparations of cereals, flour (19)	19.64	28.61	37.18	25.99	18.66	18.24	15.85	49.62	71.00	90.44	142.64	138.54	177.68	213.13
Preparations of vegetables and fruits (20)	17.18	30.99	20.91	16.95	17.57	19.31	25.35	44.45	60.00	82.66	105.31	125.43	129.46	139.06
Miscellaneous edible preparations (21)	57.68	72.71	79.00	74.58	93.32	93.33	87.68	123.02	147.00	177.00	171.36	292.06	433.22	271.23
Beverages, spirits and vinegar (22)	42.95	54.83	42.98	41.81	46.11	72.95	79.23	127.15	161.00	141.99	141.69	174.11	238.03	363.76
Residues from food industry and animal feeds (23)	565.75	365.96	404.29	474.62	1425.03	1921.36	1481.06	639.90	907.00	621.44	739.06	616.88	862.89	1159.76
Tobacco and manufactured tobacco substitutes (24)	250.35	238.41	37.16	405.69	501.72	272.49	111.98	90.97	204.00	260.63	232.63	287.38	266.20	340.89
Agricultural products as raw materials for textiles	1,394.50	8,77.03	2,103.58	2,887.72	2,491.71	2,540.57	1,299.67	1,050.72	1,615.10	1,797.54	1,814.62	2,940.89	5,398.31	6,927.13
Raw hides and skins, leather, fur skins and articles	158.80	206.82	292.76	444.22	396.77	429.33	418.75	412.27	627.80	819.54	739.25	929.34	1,268.85	2,755.22
Raw hides and skins (4101–4103)	102.47	118.01	240.48	396.65	353.51	381.91	366.57	365.23	564.00	753.70	679.71	845.28	1,137.11	2,626.88
Raw fur skins (4301)	56.33	88.81	52.28	47.57	43.26	47.42	52.19	47.04	63.80	65.84	59.55	84.06	131.73	128.34
Silk	3.56	2.03	10.69	18.42	7.58	21.24	13.20	13.13	14.20	9.82	6.80	9.08	14.22	13.86
Cocoon (5001)	2.58	0.60	3.95	6.44	1.43	4.29	2.11	1.76	1.70	3.31	0.86	0.47	0.46	0.44
Raw silk (5002)	0.25	1.07	1.05	0.68	0.99	0.86	0.53	2.07	6.50	2.63	2.11	2.53	2.10	2.04
Waste silk (5003)	0.74	0.36	5.69	11.30	5.16	16.09	10.56	9.30	6.00	3.89	3.83	6.08	11.67	11.37

Wool, fine or coarse animal hair	670.56	623.79	691.12	768.77	668.71	558.60	446.22	473.98	779.60	786.08	785.68	727.90	1,009.57	1,105.24
Uncarded wool (5101)	650.43	606.75	654.06	711.93	643.35	532.10	436.29	454.86	745.00	768.29	780.23	704.87	981.58	1,077.94
Uncarded fine or coarse animal hair (5102)	17.18	14.30	32.53	51.98	18.66	23.60	8.45	18.61	33.00	16.24	4.98	20.78	24.98	24.34
Waste wool and fine or coarse animal hair (5103)	2.95	2.74	4.53	4.86	6.70	2.90	1.48	0.52	1.60	1.56	0.48	2.25	3.01	2.93
Cotton	557.04	28.61	1,068.81	1,607.15	1,388.25	1,497.61	379.24	85.80	86.10	81.40	183.23	1,111.32	2,913.48	2,865.61
Uncarded cotton (5201)	527.71	19.07	1,022.34	1,557.20	1,314.14	1,427.88	350.72	69.26	74.00	69.05	172.32	1,088.66	2,886.77	2,839.59
Waste cotton (5202)	4.17	3.58	4.65	4.75	4.39	2.15	2.11	3.10	4.00	7.68	4.50	17.04	17.77	17.32
Carded cotton (5203)	25.16	5.96	41.82	45.20	69.71	67.59	26.41	13.44	8.10	4.67	6.41	5.62	8.93	8.70
Other vegetable textile fibres (5301-5302)	4.54	15.78	40.20	49.16	30.41	33.79	42.26	65.54	107.40	100.69	99.66	163.25	192.19	187.24

341

Table 11A.3 China's revealed comparative advantage indices (NER) 1992–2005

	1992	1993	1994	1995	1996	1997	1998	1999	2000	2001	2002	2003	2004	2005
All agricultural products	0.36	0.49	0.33	0.08	0.14	0.21	0.25	0.25	0.17	0.15	0.19	0.06	-0.10	-0.06
By factor intensity of production														
Labour-intensive agricultural products	0.64	0.68	0.67	0.58	0.53	0.51	0.52	0.50	0.43	0.45	0.45	0.44	0.43	0.40
Land-intensive agricultural products	-0.13	0.07	-0.21	-0.72	-0.67	-0.41	-0.32	-0.34	-0.38	-0.54	-0.42	-0.53	-0.83	-0.73
By commodity group														
Cereals, vegetable oilseeds and vegetable oils	-0.01	0.19	-0.04	-0.66	-0.57	-0.19	-0.20	-0.29	-0.30	-0.43	-0.31	-0.43	-0.77	-0.60
Cereals (10)	-0.06	0.19	0.09	-0.96	-0.86	0.14	0.36	0.39	0.48	0.26	0.55	0.71	-0.50	0.01
Wheat (1001)	-1.00	-0.98	-0.98	-1.00	-1.00	-1.00	-0.99	-1.00	-1.00	-0.44	-0.19	0.55	-0.87	-1.00
Corn (1005)	1.00	1.00	1.00	-0.97	-0.42	1.00	0.89	0.97	1.00	0.98	1.00	1.00	1.00	1.00
Paddy, Rice (1006)	0.70	0.76	0.57	-0.93	-0.44	0.31	0.77	0.79	0.66	0.54	0.65	0.67	-0.04	0.08
Vegetable oilseeds	0.78	0.81	0.81	0.67	0.31	-0.29	-0.45	-0.48	-0.67	-0.66	-0.59	-0.74	-0.78	-0.71
Soybean (1201)	0.69	0.59	0.88	0.14	-0.66	-0.84	-0.85	-0.87	-0.95	-0.95	-0.94	-0.97	-0.96	-0.96
Peanuts (1202)	1.00	1.00	0.99	1.00	1.00	0.96	0.97	1.00	1.00	1.00	1.00	1.00	0.99	1.00
Rapeseeds (1205)	0.98	1.00	-0.86	-0.96	0.85	-1.00	-1.00	-1.00	-1.00	-1.00	-0.99	-0.96	-1.00	-1.00
Other oilseeds (1203–1204, 1206–1209, 1212)	0.69	0.77	0.81	0.79	0.78	0.60	0.52	0.60	0.51	0.48	0.58	0.46	0.38	0.62
Vegetable oils	-0.58	-0.40	-0.56	-0.69	-0.63	-0.44	-0.67	-0.83	-0.77	-0.72	-0.89	-0.94	-0.95	-0.85
Soybean oil (1507)	-0.93	-0.52	-0.86	-0.91	-0.80	-0.30	-0.59	-0.85	-0.76	-0.01	-0.90	-0.99	-0.98	-0.92
Palm oil (1511)	-0.85	-0.66	-0.49	-0.63	-0.68	-0.80	-0.93	-1.00	-1.00	-1.00	-0.99	-1.00	-1.00	-1.00
Rapeseeds oil (1514)	-0.53	-0.42	-0.43	-0.50	-0.24	-0.36	-0.57	-0.36	-0.08	0.10	-0.58	-0.91	-0.96	0.22
Other vegetable oils (1508–1510, 1512–1513, 1515)	0.11	0.52	-0.06	-0.41	-0.29	0.04	-0.15	-0.37	-0.45	-0.42	-0.44	-0.43	-0.52	-1.00

	0.83	0.84	0.87	0.81	0.76	0.75	0.72	0.70	0.63	0.61	0.63	0.61	0.57	0.58
Horticultural products	0.83	0.84	0.87	0.81	0.76	0.75	0.72	0.70	0.63	0.61	0.63	0.61	0.57	0.58
Live trees and other plants (06)	0.43	0.31	0.60	0.65	0.71	0.60	0.46	0.29	0.21	0.23	0.13	0.04	0.11	0.06
Edible vegetables (07)	0.93	0.96	0.98	0.91	0.90	0.91	0.91	0.90	0.90	0.79	0.81	0.80	0.72	0.71
Edible fruits (08)	0.75	0.77	0.73	0.70	0.40	0.33	0.29	0.24	0.06	0.08	0.19	0.21	0.19	0.24
Coffee, tea, mate and species (09)	0.90	0.94	0.96	0.94	0.89	0.96	0.93	0.93	0.91	0.93	0.92	0.91	0.93	0.91
Vegetable plaiting materials (14)	0.24	0.18	0.15	-0.19	0.13	0.20	0.02	-0.09	-0.32	-0.20	.000	-0.22	-0.34	-0.16
Other vegetable products (1210--1214, 13)	0.66	0.60	0.74	0.75	0.72	0.70	0.64	0.59	0.61	0.60	0.57	0.61	0.54	0.29
Animal products	0.66	0.62	0.60	0.62	0.59	0.62	0.54	0.34	0.27	0.29	0.25	0.21	0.22	0.14
Live animals (01)	0.92	0.92	0.91	0.87	0.82	0.84	0.78	0.71	0.76	0.82	0.73	0.47	0.20	0.50
Pig (0103)	1.00	1.00	1.00	0.99	0.99	0.99	0.98	0.97	0.97	0.98	0.98	0.98	0.98	1.00
Poultry (0105)	0.70	0.78	0.82	0.79	0.78	0.78	0.81	0.80	0.81	0.77	0.73	0.63	0.44	1.00
Meat and edible meat offal (02)	0.73	0.67	0.76	0.83	0.75	0.73	0.71	0.16	0.08	0.17	0.03	-0.08	0.20	0.12
Beef (0201, 0202)	0.81	0.68	0.71	0.79	0.85	0.89	0.87	0.63	0.53	0.69	0.20	0.11	0.50	1.00
Pork (0203)	1.00	1.00	1.00	0.99	0.99	0.98	0.93	0.47	0.07	0.53	0.44	0.49	0.79	1.00
Mutton (0204)	0.78	0.82	0.60	0.62	0.11	0.06	-0.11	-0.33	-0.40	-0.63	-0.55	-0.30	-0.01	na
Poultry (0207)	0.54	0.52	0.68	0.77	0.66	0.65	0.66	0.14	0.10	0.15	-0.03	-0.18	-0.03	1.00
Fish and aquatic products (03)	0.62	0.55	0.52	0.55	0.49	0.57	0.45	0.38	0.30	0.32	0.29	0.28	0.27	0.20
Dairy products, eggs and natural honey (04)	0.39	0.43	0.29	0.43	0.55	0.42	0.33	0.00	-0.07	-0.07	-0.17	-0.22	-0.31	-0.27
Dairy products (0401-0406)	-0.46	-0.30	-0.59	-0.37	-0.28	-0.21	-0.36	-0.57	-0.62	-0.69	-0.66	-0.76	-0.78	na
Product of animal origin not elsewhere specified (05)	0.82	0.80	0.84	0.82	0.75	0.72	0.73	0.68	0.65	0.58	0.55	0.54	0.59	0.64

	1992	1993	1994	1995	1996	1997	1998	1999	2000	2001	2002	2003	2004	2005
Animal fats (1501--1506, 1516--1518, 1520--1521)	-0.71	-0.71	-0.68	-0.81	-0.61	-0.22	-0.34	-0.71	-0.78	-0.79	-0.72	-0.65	-0.68	-0.82
Processed agricultural products	0.49	0.62	0.57	0.41	0.36	0.30	0.37	0.53	0.47	0.50	0.53	0.55	0.52	0.52
Products of milling industry (11)	0.11	0.39	0.36	0.17	0.51	0.46	0.32	0.01	0.18	0.14	0.11	0.02	-0.05	0.04
Preparations of meet, fish and aquatic products (16)	0.97	0.98	0.98	0.98	0.99	0.99	0.99	0.98	0.99	0.99	0.98	0.98	0.99	0.99
Sugars and sugar confectionery (17)	0.42	0.66	-0.10	-0.60	-0.17	-0.13	0.03	-0.13	-0.01	-0.41	-0.10	-0.05	-0.14	-0.04
Cocoa and cocoa preparations (18)	0.00	0.07	-0.17	-0.18	-0.09	-0.12	-0.19	-0.15	-0.42	-0.49	-0.38	-0.36	-0.32	-0.24
Preparations of cereals, flour (19)	0.78	0.67	0.68	0.80	0.87	0.88	0.89	0.72	0.67	0.63	0.51	0.56	0.54	0.52
Preparations of vegetables and fruits (20)	0.96	0.93	0.96	0.97	0.97	0.97	0.95	0.93	0.91	0.89	0.88	0.88	0.90	0.90
Miscellaneous edible preparations (21)	0.41	0.33	0.40	0.53	0.49	0.55	0.60	0.48	0.42	0.37	0.44	0.27	0.13	0.40
Beverages, spirits and vinegar (22)	0.81	0.73	0.82	0.83	0.81	0.74	0.71	0.58	0.51	0.59	0.60	0.54	0.48	0.27
Residues from food industry and animal feeds (23)	0.03	0.18	0.13	-0.11	-0.58	-0.73	-0.76	-0.48	-0.57	-0.37	-0.31	-0.26	-0.31	-0.46
Tobacco and manufactured tobacco substitutes (24)	0.37	0.52	0.91	0.47	0.36	0.44	0.69	0.58	0.19	0.18	0.28	0.23	0.28	0.17
Agricultural products as raw materials for textiles	-0.22	-0.10	-0.40	-0.65	-0.69	-0.69	-0.53	-0.26	-0.41	-0.65	-0.60	-0.77	-0.90	-0.92
Raw hides and skins, leather, fur skins														

and articles	-0.50	-0.66	-0.60	-0.77	-0.84	-0.81	-0.88	-0.94	-0.96	-0.97	-0.97	-0.98	-0.99	-0.99
Raw hides and skins (4101–4103)	0.41	-0.62	-0.56	-0.76	-0.84	-0.85	-0.91	-0.95	-0.98	-0.98	-0.98	-0.99	-1.00	-1.00
Raw fur skins (4301)	-0.72	-0.73	-0.84	-0.85	-0.77	-0.52	-0.68	-0.86	-0.85	-0.88	-0.85	-0.90	-0.90	-0.90
Silk	0.98	0.99	0.96	0.91	0.96	0.88	0.91	0.91	0.92	0.93	0.95	0.92	0.88	0.89
Cocoon (5001)	0.83	0.96	0.89	0.62	0.91	0.50	0.50	0.37	0.54	0.01	0.28	0.29	0.33	0.33
Raw silk (5002)	1.00	0.99	0.99	1.00	0.99	0.99	1.00	0.98	0.95	0.98	0.98	0.97	0.98	0.98
Waste silk (5003)	0.95	0.98	0.83	0.49	0.62	0.27	0.52	0.63	0.80	0.79	0.70	0.60	0.25	0.25
Wool, fine or coarse animal hair	-0.61	-0.71	-0.63	-0.80	-0.83	-0.78	-0.84	-0.94	-0.97	-0.96	-0.96	-0.93	-0.92	-0.92
Uncarded wool (5101)	-0.96	-0.96	-0.97	-0.96	-0.91	-0.84	-0.89	-0.96	-0.97	-0.96	-0.96	-0.93	-0.92	-0.92
Uncarded fine or coarse animal hair (5102)	0.79	0.73	0.64	0.13	0.22	-0.02	0.11	-0.59	-0.92	-0.80	-0.73	-0.96	-0.96	-0.96
Waste wool and fine or coarse animal hair (5103)	-0.60	-0.21	-0.50	-0.30	-0.65	-0.15	0.22	0.55	-0.10	-0.10	0.63	0.09	-0.02	-0.02
Cotton	-0.35	0.81	-0.70	-0.93	-0.98	-0.99	-0.72	0.55	0.56	-0.01	-0.05	-0.80	-0.99	-0.99
Uncarded cotton (5201)	-0.34	0.85	-0.71	-0.93	-0.98	-1.00	-0.71	0.62	0.61	0.06	-0.03	-0.79	-0.99	-1.00
Waste cotton (5202)	0.19	0.79	0.33	-0.40	-0.45	-0.33	0.20	-0.33	-0.54	-0.70	-0.36	-0.87	-0.88	-0.88
Carded cotton (5203)	-0.88	0.09	-0.82	-1.00	-1.00	-0.98	-1.00	-0.90	-0.84	-0.81	-0.84	-0.76	-0.94	-0.94
Other vegetable textile fibres (5301-5302)	0.14	-0.36	-0.63	-0.81	-0.77	-0.71	-0.79	-0.87	-0.88	-0.94	-0.95	-0.97	-0.97	-0.97

12 Agricultural trade between China and ASEAN

Dynamics and prospects

Jun Yang and Chunlai Chen

Bilateral trade between China and the Association of South-East Asian Nations (ASEAN) has expanded very quickly since 2001. China became ASEAN's third largest export market in 2005, after the United States and Japan. In particular, ASEAN's agricultural exports to China have increased rapidly, reaching US$5.9 billion in 2005. As a result, China became the third largest agricultural export market for ASEAN in 2005. With its rapid economic growth and structural change, slowing population growth, continuing income growth, rapid urbanisation and limited natural resources, China can be expected to import an increasing volume of agricultural products to meet its increasing food demand and the raw material demands of its high-growth industries (Chen 2006; Huang and Yang 2006; Chen 2004; Huang and Rozelle 2003). China's huge and fast growing purchasing power will provide great opportunities for agricultural exporting countries (Huang and Yang 2006).

The trade relationship between China and ASEAN has been strengthened by the ASEAN–China Free Trade Agreement (ACFTA), signed in 2002. In 2005, the ASEAN–China free trade area was the world's largest free trade area, with a population of 1.86 billion, combined gross domestic product (GDP) of US$2.62 trillion (2000 constant US dollar) and total trade value of US$1.23 trillion (2000 constant US dollar). The free trade agreement will be implemented fully in 10 years (by 2010). As a first step, the so-called early harvest program (EHP) was launched successfully in 2004. According to the EHP, from 2004 to 2006 the import tariffs on about 500 agricultural commodities traded between China and the original ASEAN members[1] were to be reduced to zero.

Many studies have been undertaken to explore the possible impacts of the agreement. Some indicate that China's export structure is quite similar to ASEAN in many aspects; therefore, integration will increase the competitive pressure on ASEAN economies (Tongzon 2005; Holst and Weiss 2004; Wong and Chan 2002). Other studies find that the ACFTA could promote net trade gains, stimulate economic growth and greatly improve social welfare in the partner economies (Suthiphand 2002; Chia 2004). The majority of studies, however, focus on impacts in the industrial sectors.

There are questions worthy of study with respect to agricultural trade between China and ASEAN. What has happened to agricultural trade in recent years? Have China and ASEAN become more competitive or more complementary in agriculture? Has trade integration helped the two economies make adjustments towards their respective comparative advantages? What are possible challenges in agriculture as a result of the free trade agreement?

This chapter is structured as follows: in the next section, we describe the classification of agricultural commodities adopted for the exercise. Section three highlights the characteristics and changing trends in ASEAN-China agricultural trade. Sections four and five calculate and analyse revealed comparative advantage and trade complementarity between the two economies. Section six summarises the main findings.

Classification of agricultural commodities

We use the Standard International Trade Classification (SITC Revision 3) to classify agricultural products. Agricultural products are defined to include: SITC0 (food and live animals), SITC1 (beverages and tobacco), SITC4 (animals and vegetable oils, fats and waxes) and some sub-groups[2] of SITC2 (crude materials, inedibles, except fuels). The agricultural trade data are from United Nations Statistics Division, Commodity Trade Statistics Database (COMTRADE). All values of agricultural trade are in 2000 constant US dollar prices.

In order to demonstrate more clearly changes in agricultural trade between China and ASEAN, the agricultural products are classified further into nine commodity groups: animal products, fish, cereals, vegetables and fruits, sugar, oil seeds, raw materials, rubber, processed products and vegetable oils. As we also analyse changes in terms of the factor intensities of the commodities, agricultural products are classified into labour-intensive and land-intensive commodities. The two sets of classifications at the two-digit level are shown in Table A12.1.

Agricultural trade between China and ASEAN

Changes in China's imports, exports and net exports with ASEAN

Bilateral trade in agricultural products has expanded in recent years. As Figure 12.1 shows, China's imports from ASEAN increased in 1999 after a short period of decline after 1995, and accelerated after 2002. The annual import growth rate jumped from 17.3 per cent during 1999-2001 to 27.3 per cent during 2001-05. The import value of agricultural products from ASEAN to China in 2005 reached US$5 billion—more than 2.6 times the level in 2001.

Exports of agricultural products from China to ASEAN fluctuated slightly during 1992-2001 and increased continuously after 2001. The average annual export growth rate reached 17 per cent in the period 2001-05—a rapid rate but much lower than that of imports. Therefore, China's trade deficit with ASEAN in agricultural products has been increasing, reaching US$2.8 billion in 2005.

Shares of bilateral agricultural trade between China and ASEAN

The Chinese market is becoming more and more important for ASEAN's agricultural exports. As Figure 12.2 shows, there was a short-term decline in 1998 and 1999 in the share of ASEAN's total agricultural exports to China. This could have been a temporary effect of the East Asian financial crisis; however, the share resumed its strong growth after 1999. The share of exports to China in ASEAN's total agricultural exports increased quickly from 4.8 per cent in 1999 to 10.2 per cent in 2005. In contrast, the share of agricultural exports to ASEAN in China's total agricultural exports fluctuated during 1992-2005 around a declining trend, falling from 10.1 per cent in 1999 to 8.8 per cent in 2005.

ASEAN is an important source of China's agricultural imports, as Figure 12.2 shows. On average, imports from ASEAN accounted for 15.6 per cent of China's total agricultural imports during 1992-2005 and there has been a rising trend in recent years, increasing from 14.7 per cent in 2000 to 16.6 per cent in 2005. China's share in ASEAN's agricultural imports is not as important—in 2005, it was only 8.8 per cent.

China–ASEAN import and export structure

We examine the structure of the bilateral agricultural trade in two ways. First, we use the data from 2005 to analyse the relative importance of agricultural commodities in bilateral trade. Second, we use data for the period 1992-2005 to investigate the trends in bilateral agricultural trade between the two partners.

Figure 12.1   China's imports, exports and net exports with ASEAN (US$ million)

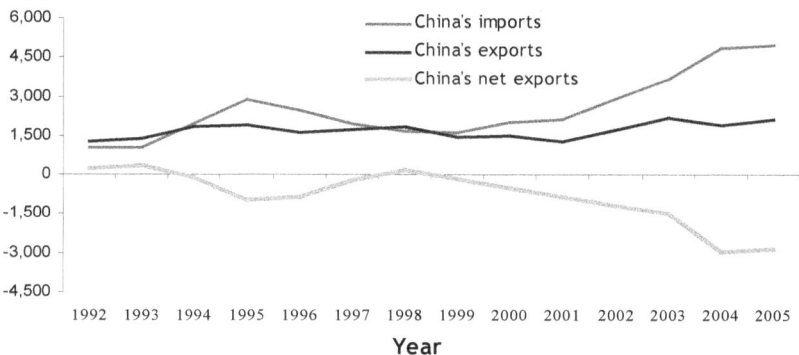

Source: COMTRADE.

Figure 12.2   Shares of bilateral agricultural trade in total agricultural trade, 1992–2005 (per cent)

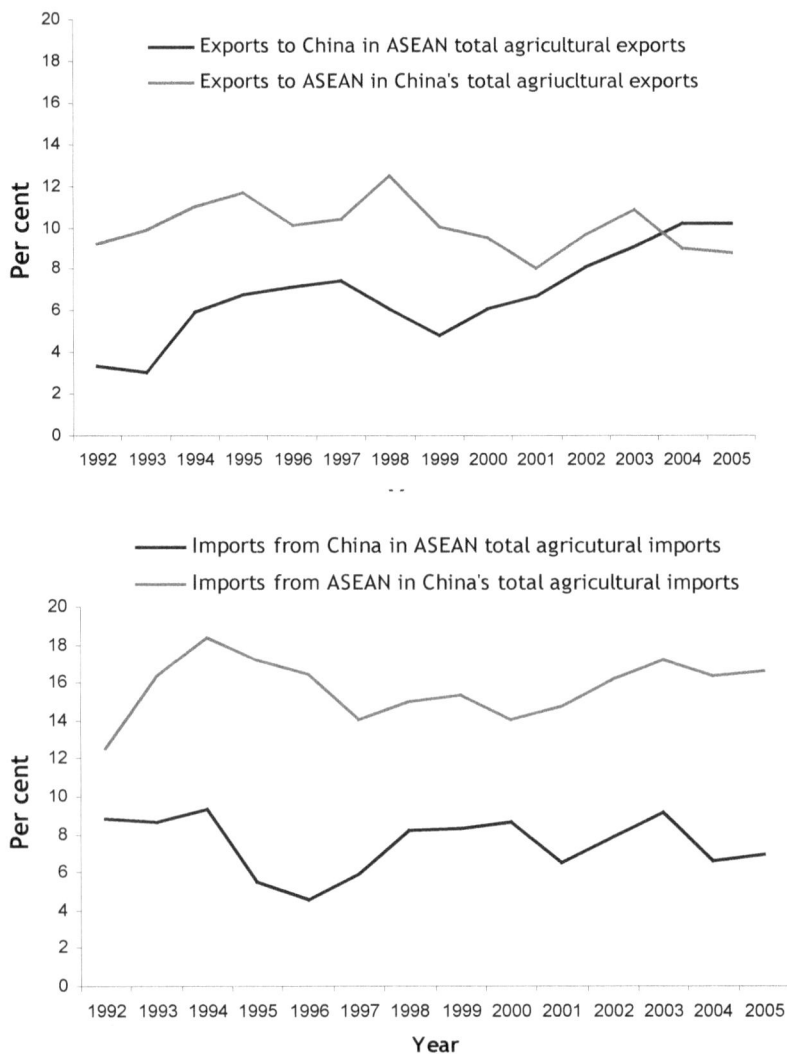

350

China-ASEAN agricultural export and import structure, 2005. Figure 12.3 shows China's export and import shares of agricultural products with ASEAN in 2005. China's agricultural exports to ASEAN are concentrated mainly in three groups of commodities: vegetables and fruits, processed food, and fish. The combined share of the three commodity groups accounted for 77 per cent of total agricultural exports to ASEAN. Vegetables and fruits are the largest export commodity group, accounting for 40 per cent.

China's agricultural imports from ASEAN are also concentrated in three different commodity groups. The most important imported agricultural commodity is vegetable oils, taking up 36 per cent. Decomposing this category, we find that palm oil accounted for 98.8 per cent of total vegetable oil imports from ASEAN in 2005. The second most important commodity is rubber—accounting for 33 per cent of total agricultural imports from ASEAN—and the third is vegetables and fruits, accounting for 14 per cent. The combined share of the three commodity groups accounted for 83 per cent of total agricultural imports from ASEAN.

Changes in China's agricultural trade structure with ASEAN, 1992-2005. The export share of vegetables and fruits in China's total agricultural exports to ASEAN increased steadily from 12.2 per cent in 1992 to 40.9 per cent in 2005 (Table 12.1). Vegetables and fruits became the largest group of agricultural exports from China to ASEAN in 2002 and its status has been strengthened by strong export growth since then. The remarkable improvement might have resulted from the EHP tariff-reduction program launched between China and ASEAN in 2004.

The export share of fish was very small (no more than 3 per cent) before 2002; however, it increased quickly from 4.7 per cent in 2002 to 13.8 per cent in 2005. In 2004, the export share of fish from China to ASEAN jumped to 16 per cent from 7.5 per cent in 2003. The large change was due mainly to two factors: first, the United States placed anti-dumping duties on China's prawn exports in 2004 and some products were diverted to the ASEAN market. Second, the EHP launched in 2004 provided good opportunities for China's fish products to access ASEAN's market (Agricultural Information Centre of Guangxi Autonomous Region 2004).

As cereals belong to 'sensitive products' related directly to food security, exports of cereals have always been affected significantly by China's trade policy. For example, exports of the largest export component in cereals (that is, maize) have sometimes been heavily subsidised by the Chinese government (Tian et al. 2005). Therefore, the export share of cereals has fluctuated with the changes in China's domestic production and trade policies.

Processed agricultural products were the largest export commodity group until cereals took over in 1998. This share has, however, been on a declining trend, although it is still the second largest commodity group in agricultural exports from China to ASEAN. The export share of animal products varied slightly around 5 per cent during the period. For the rest of the commodities, their combined share fell from 24 per cent in 1992 to 10.7 per cent in 2005. In general, China's agricultural exports to ASEAN have become concentrated more and more in vegetables and fruits and processed agricultural products.

Vegetable oils and rubber dominated China's agricultural imports from ASEAN during 1992-2005 (Table 12.2). The combined import share of the two commodities accounted for 52-73 per cent of China's total agricultural imports from ASEAN. Moreover, their combined share has risen in recent years, increasing from 60 per cent during 1992-2001 to 70 per cent during 2002-05.

Figure 12.3 China-ASEAN agricultural trade structure in 2005

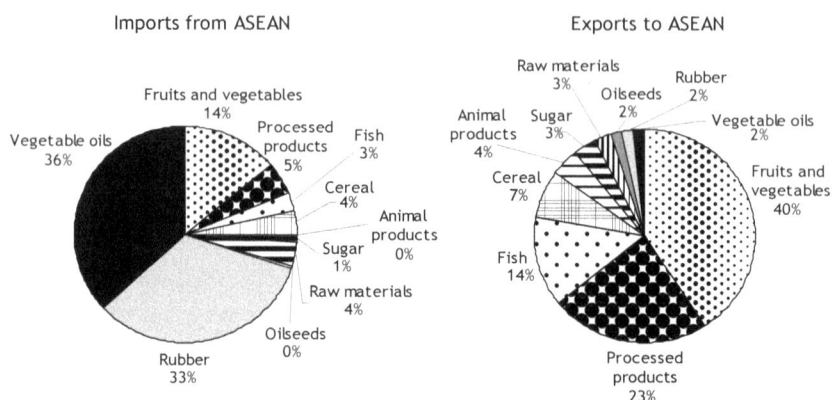

Imports from ASEAN

Exports to ASEAN

Source: COMTRADE.

352

Table 12.1   Shares in total agricultural exports from China to ASEAN, by commodity (per cent)

	1992	1993	1994	1995	1996	1997	1998	1999	2000	2001	2002	2003	2004	2005
Animal products	5.6	4.0	3.2	4.3	4.6	5.6	3.5	6.1	5.7	8.5	5.6	4.0	4.7	4.2
Fish	0.8	1.7	2.0	1.7	2.0	2.6	1.7	2.4	3.0	4.2	6.3	7.5	16.0	13.8
Cereals	23.3	16.2	17.1	3.0	2.7	22.7	46.3	34.1	34.5	19.4	25.9	29.6	7.5	7.2
Vegetables and fruits	12.2	13.2	15.9	17.7	18.7	18.6	12.3	19.0	19.5	26.9	27.4	27.6	38.1	40.9
Sugar	2.3	3.1	1.6	6.1	4.4	1.7	4.7	3.0	2.4	4.6	3.1	2.3	2.9	3.2
Oil seeds	11.1	5.8	12.4	6.8	3.9	1.4	1.1	3.1	2.7	3.2	2.9	2.8	1.7	1.8
Raw materials	10.0	11.9	9.2	4.4	3.6	4.5	2.9	8.1	11.1	2.7	4.0	4.8	1.5	2.5
Rubber	0.1	0.2	0.3	0.7	0.8	1.1	0.4	0.6	0.8	1.4	1.1	0.9	1.4	1.7
Processed products	34.1	43.1	37.9	54.8	56.6	38.4	24.7	22.8	19.9	28.8	23.2	20.1	25.3	23.2
Vegetable oils	0.6	0.8	0.3	0.6	2.7	3.5	2.2	0.8	0.4	0.4	0.5	0.5	0.8	1.5

Table 12.2   Shares in total agricultural imports from ASEAN to China, by commodity (per cent)

	1992	1993	1994	1995	1996	1997	1998	1999	2000	2001	2002	2003	2004	2005
Animal products	0.7	1.2	0.7	0.5	0.6	0.9	1.3	3.3	1.7	0.9	0.6	0.4	0.1	0.1
Fish	6.4	5.5	4.2	2.1	1.9	1.5	1.6	1.6	4.7	3.8	2.9	2.4	2.5	2.8
Cereals	4.8	4.5	8.7	17.0	12.7	8.4	8.2	5.8	6.0	5.2	3.5	2.6	4.8	3.7
Vegetables and fruits	6.4	6.9	3.5	4.9	5.8	8.5	10.8	10.1	9.9	16.9	13.4	11.5	12.4	13.8
Sugar	2.6	1.1	7.3	14.3	6.4	3.5	1.3	1.6	1.3	5.4	2.5	1.0	1.3	0.9
Oil seeds	0.2	0.2	0.5	0.1	0.1	0.2	0.2	0.0	0.0	0.1	0.2	0.2	0.2	0.1
Raw materials	7.5	5.1	5.6	4.2	6.5	7.0	6.3	4.0	4.6	4.2	3.9	3.7	3.7	3.8
Rubber	25.7	24.0	18.8	16.4	31.8	25.8	20.7	20.0	30.7	28.5	29.9	30.6	29.0	33.1
Processed products	14.3	13.3	5.9	4.6	6.0	6.3	7.0	9.5	11.9	8.4	6.2	4.9	5.8	4.9
Vegetable oils	31.4	38.2	44.8	36.0	28.3	38.0	42.6	44.1	29.4	26.6	37.0	42.7	40.2	36.9

Source: COMTRADE.

353

The third largest group of agricultural commodities imported by China from ASEAN is vegetables and fruits. The share of this group increased during 1992–2005, rising from an average of 8.8 per cent during 1992–2001 to 12.8 per cent during 2002–05. China also imports certain raw materials and processed products from ASEAN; however, on average, the share of these commodities has been declining. In general, China's agricultural imports from ASEAN have become concentrated more and more on palm oil, rubber and vegetables and fruits.

China–ASEAN agricultural trade patterns

Aggregation of the commodities into labour-intensive and land-intensive groups presents a different view of the trends in trade between China and ASEAN. As Figure 12.4 shows, imports of labour-intensive commodities from ASEAN to China were stable from 1992 to 1999 and began to increase rapidly after 2000. The average annual growth rate of these kinds of commodities during 2001–05 was 14.6 per cent.

China's exports of labour-intensive agricultural commodities to ASEAN increased between 1992 and 1995 and then declined to the 1992 level between 1996 and 2000. Exports of these kinds of commodities began to increase strongly after 2000, achieving an annual growth rate of 21.2 per cent between 2001 and 2005.

China has been enjoying a trade surplus with ASEAN in labour-intensive agricultural commodities (Figure 12.4). The changing trend in net exports is quite similar to that of China's exports. As the growth rate of exports was higher than that of imports after 2000, the net export value of labour-intensive agricultural commodities increased and the trade surplus reached US$0.67 billion in 2005.

The picture of China's exports and imports of land-intensive agricultural products with ASEAN is different from that of labour-intensive agricultural products. As Figure 12.5 shows, exports of land-intensive agricultural products from China to ASEAN were relatively stable between 1992 and 2003; however, they declined in 2004 and 2005. In contrast, China's imports of land-intensive agricultural products from ASEAN have increased strongly since 2001. The annual import growth rate of land-intensive agricultural products was 27.5 per cent between 2001 and 2005.

China has been in a trade deficit position with ASEAN with respect to land-intensive agricultural products (Figure 12.5). This situation strengthened rapidly after 2001. The trade deficit in land-intensive products rose from US$1.5 billion in 2001 to US$3.95 billion in 2005.

On the whole, China's agricultural trade with ASEAN has been in deficit since 1998 and the deficit has been increasing (Figure 12.1); however, China maintains a trade surplus with ASEAN in labour-intensive agricultural products. Moreover, the trade surplus in labour-intensive agricultural products has increased quickly in recent years. Therefore, China's rising trade deficit in agriculture is the result of the rapid increase in imports of land-intensive agricultural products, mainly palm oil and rubber. These imported commodities are used to meet the changing consumption preference for high-quality food with income growth and the demand for raw materials by China's high-growth industry.

Revealed comparative advantage

In this section, we examine comparative advantage in China and ASEAN as revealed through their agricultural trade and see how it is reflected in trade between the two partners. First, we present the definition of the revealed comparative advantage (RCA) index adopted in this study. Second, we compare the RCA by commodity in the two economies in 2005. Finally, we examine changes in the RCA to assess the export potential of the two economies.

Definition of revealed comparative advantage (RCA)

It is difficult if not impossible to measure comparative advantage directly. The most common approach to indirect estimation draws on the principle of RCA proposed by Balassa (1965). The logic of this principle is that the trade of a country is generated by its underlying comparative advantage, therefore, its real exports and imports can be used to infer the underlying pattern of comparative advantage. Following this principle, several indicators of RCA have been developed. In this exercise, we adopt the widely used net export ratio (NER), which is defined as

Figure 12.4 China's exports, imports and net exports with ASEAN in labour-intensive commodities, 1992-2005 (US$ million)

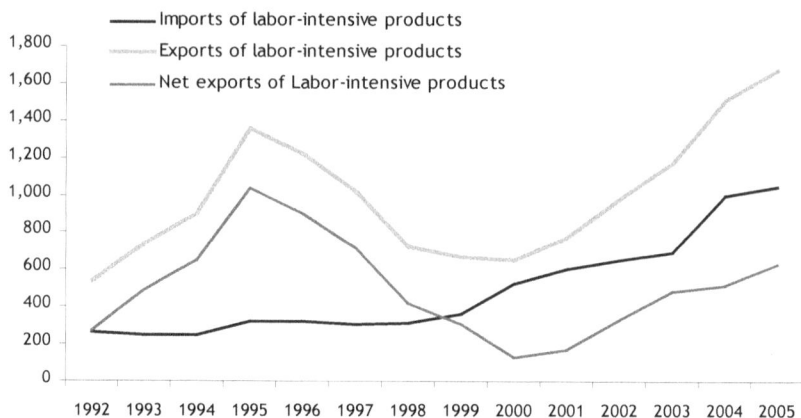

Source: COMTRADE.

Figure 12.5 China's exports, imports and net exports with ASEAN in land-intensive commodities, 1992-2005 (US$ million)

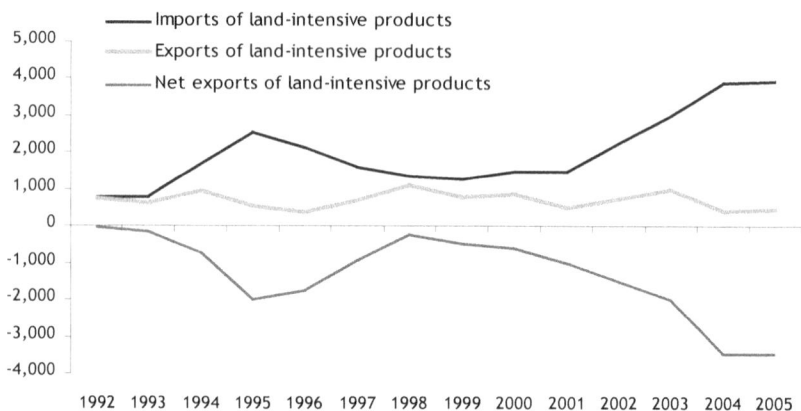

Source: COMTRADE.

$$RCA_{ic} = \frac{X_{ic} - M_{ic}}{X_{ic} + M_{ic}}$$ (1)

where RCA_{ic} = revealed comparative advantage index of commodity C for country i; X_{ic} = export value of commodity C by country i; M_{ic} = import value of commodity C by country i.

The rationale behind the index is that country i has RCA in good C if it exports more of it than it imports. Country i has a comparative advantage in exporting commodity C when RCA_{ic} is positive. If RCA_{ic} is negative, country i has a comparative disadvantage in producing the commodity. The larger RCA_{ic} is, the stronger is the comparative advantage. The index ranges between -1 and +1.

A comparison of RCA in China and ASEAN in 2005

In 2005 China had strong RCA in vegetables and fruits, fish, animal products, and processed products, relatively weak RCA in sugar and cereals, and clear revealed comparative disadvantage in raw materials, oil seeds, vegetable oils and rubber (Figure 12.6).

ASEAN has RCA in vegetables and fruits, fish, vegetable oils and rubber, lower RCA in processed products and sugar, and revealed comparative disadvantage in raw materials, oil seeds and cereals.

There are some overlaps between the two economies. China and ASEAN show significant comparative advantage in vegetables and fruits, and fish. Usually, we think that there will be competition if both economies have comparative advantage in the same commodity. Therefore, as these two economies become more integrated, some adjustments are inevitable.

With further decomposition of the vegetables and fruits group, we find that competition is not as likely as the aggregate RCA would suggest. Taking the fresh and dry fruits category (SITC057) as an example,[3] China has RCA in temperate fruits (for example, apples, citrus and pears) and revealed comparative disadvantage in tropical fruits (for example, bananas and mangoes). The combined share of apples, citrus and pears accounted for 90.5 per cent of China's total exports of fruits to ASEAN in 2005. In contrast, ASEAN has RCA in tropical fruits and revealed comparative disadvantage in temperate fruits. The share of bananas, mangoes and other fresh fruits (SITC05798) accounted for 84.1 per cent of China's total imports of fruits from ASEAN in 2005. Therefore, the two economies have good potential for complementarity in fruit trade. This is apparently the reason for the rapid increase in bilateral trade in fruits in recent years.

With regard to processed products, China has stronger RCA than ASEAN. China's RCA value for processed products was 0.25 in 2005—much higher than ASEAN's RCA value of 0.03. Comparing the RCA in past years, we find that China has been maintaining strong RCA in processed products (Table 12.3). In contrast, ASEAN's RCA in processed products has been declining and has approached zero in recent years (Table 12.4). It is expected, therefore, that China will export more processed products to ASEAN if tariffs on processed products are removed through the ACFTA.

Many commodities—for example, animal products, vegetable oils and rubber—enjoy complementarity between China and ASEAN, and it will be

Figure 12.6   Comparisons of revealed comparative advantage in agricultural products in China and ASEAN, 2005

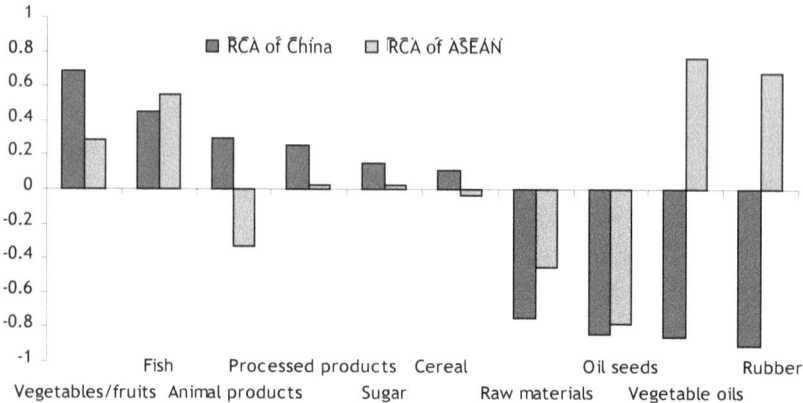

Source: COMTRADE.

easier for these sectors to be integrated, as both sides will benefit. Because China and ASEAN do not possess comparative advantage in raw materials and oil seeds, these are areas for export opportunity for other countries.

Changes in comparative advantage in China and ASEAN, by commodity

It is useful to examine the historical changes in comparative advantage in China and ASEAN. The RCA indices in Table 12.3 show that China had comparative advantage in the production of the following commodities during 1992-2005: animal products, fish, vegetables and fruits, and processed products.

During this period, the values of China's RCA indices for these products declined; however, the relatively large positive values of the RCA indices imply that China's comparative advantage in these commodities will continue for some time.

The values of China's RCA indices for cereals and sugar have fluctuated sharply and were negative in some years. As cereals are related to food security and sugar is very important for farmers' income, especially poor farmers, the production and trade of these commodities are among the most highly distorted (Tian et al. 2005; Huang et al. 2005). With imbalances between domestic supply and demand as a result of policy changes, large fluctuations in exports are expected. Based on the country's endowment of natural resources, however, it is impossible for China to export large quantities of these commodities to ASEAN in the future. With regard to raw materials, rubber and vegetable oils, the values of China's RCA indices have been negative since 1992, implying that China is and will remain heavily dependent on the world market to meet its growing domestic demand for these commodities.

As shown in Table 12.4, during 1992-2005 ASEAN had strong comparative advantage in the production of vegetable oils, rubber, vegetables and fruits, and fish. The high positive RCA indices for these agricultural products suggest that ASEAN will be able to sustain its exports. The values of ASEAN's RCA indices in processed products, though slightly positive, are close to zero, which suggests that imports will soon exceed exports.

ASEAN had comparative disadvantage in animal products, raw materials, cereals and oil seeds during 1992-2005. The high negative values of the RCA indices for animal products, raw materials and oil seeds indicate that ASEAN has been heavily dependent on the world market to meet its demand for these commodities and is likely to remain so.

Changes in comparative advantage in China and ASEAN, by commodity group

To further investigate the changing pattern in comparative advantage in agricultural trade in China and ASEAN, we calculated the RCA indices for labour-intensive and land-intensive commodity groups. As Figure 12.7 shows, China does not have comparative advantage in land-intensive commodities.

China has comparative advantage in labour-intensive agricultural commodities, which was declining in the 1990s but has been stable since 2000. Chen et al. (2006) suggest, however, that the sanitary and phytosanitary (SPS) regulations in industrialised countries could have significant negative impacts on China's exploitation of its comparative advantage in labour-intensive agricultural commodities (see also Sun et al. 2005).

ASEAN's RCA indices reveal that it has comparative advantage in land-intensive and labour-intensive agricultural commodities. As shown in Figure 12.7, the RCA for ASEAN's labour-intensive agricultural commodities declined during 1992-97, but jumped sharply in 1998. This sudden change was caused by the East Asian financial crisis, which began in 1997. With ASEAN's currencies depreciating rapidly, imports became much more expensive than domestic goods (Philippe 1998). As a result, agricultural imports fell by 24 per cent in 1998. As the level of exports changed much less, the RCA value rose abruptly. Since then, the RCA value has declined continuously, and since 2003 it has been below the 1997 level.

The change in the RCA for ASEAN's land-intensive agricultural commodities is very impressive. During 1992-2001, the RCA value fluctuated around a declining trend; however, beginning in 2001, the RCA index increased sharply. The fastest growth happened during 2002 and 2003, when exports of land-intensive agricultural commodities to China increased by 130 per cent (Figure 12.7). Meanwhile, exports of land-intensive agricultural

Table 12.3 China's RCA, by commodity, 1992-2005

	1992	1993	1994	1995	1996	1997	1998	1999	2000	2001	2002	2003	2004	2005
Animal products	0.74	0.74	0.72	0.70	0.69	0.69	0.64	0.28	0.25	0.33	0.23	0.14	0.22	0.30
Fish	0.65	0.62	0.60	0.65	0.65	0.69	0.60	0.54	0.50	0.50	0.49	0.48	0.48	0.45
Cereal	-0.03	0.22	0.12	-0.85	-0.66	0.23	0.39	0.40	0.50	0.31	0.56	0.70	-0.35	0.11
Vegetables and fruits	0.91	0.91	0.93	0.90	0.83	0.80	0.79	0.78	0.73	0.69	0.70	0.71	0.68	0.69
Sugar	0.49	0.70	0.04	-0.42	0.05	0.12	0.29	0.17	0.28	-0.08	0.09	0.28	0.13	0.16
Oil seeds	0.88	0.88	0.83	0.65	0.19	-0.52	-0.62	-0.61	-0.75	-0.75	-0.79	-0.81	-0.85	-0.84
Raw materials	-0.42	-0.34	-0.48	-0.70	-0.70	-0.67	-0.63	-0.41	-0.52	-0.65	-0.67	-0.68	-0.79	-0.75
Rubber	-0.93	-0.90	-0.90	-0.89	-0.90	-0.84	-0.90	-0.92	-0.92	-0.91	-0.92	-0.92	-0.92	-0.90
Processed products	0.46	0.53	0.64	0.54	0.27	0.18	0.26	0.40	0.25	0.32	0.32	0.32	0.28	0.25
Vegetable oils	-0.57	-0.39	-0.56	-0.69	-0.62	-0.42	-0.64	-0.81	-0.77	-0.73	-0.89	-0.93	-0.94	-0.85

Table 12.4 ASEAN RCA, by commodity, 1992-2005

	1992	1993	1994	1995	1996	1997	1998	1999	2000	2001	2002	2003	2004	2005
Animal products	-0.31	-0.35	-0.35	-0.42	-0.43	-0.44	-0.30	-0.31	-0.32	-0.24	-0.21	-0.18	-0.36	-0.33
Fish	0.51	0.58	0.62	0.62	0.60	0.61	0.66	0.65	0.67	0.64	0.61	0.59	0.56	0.55
Cereal	-0.05	-0.07	-0.11	-0.19	-0.29	-0.03	-0.06	-0.08	-0.06	-0.01	-0.02	-0.02	0.06	-0.04
Vegetables and fruits	0.41	0.33	0.25	0.22	0.25	0.22	0.27	0.33	0.29	0.29	0.29	0.29	0.28	0.29
Sugar	0.39	0.28	0.33	0.24	0.15	0.14	0.00	-0.11	0.10	0.10	0.15	0.20	0.20	0.03
Oil seeds	-0.60	-0.77	-0.75	-0.76	-0.87	-0.78	-0.69	-0.79	-0.80	-0.82	-0.82	-0.82	-0.77	-0.78
Raw materials	-0.74	-0.70	-0.66	-0.64	-0.65	-0.62	-0.61	-0.61	-0.60	-0.68	-0.62	-0.56	-0.51	-0.45
Rubber	0.69	0.66	0.69	0.70	0.70	0.63	0.61	0.57	0.56	0.56	0.63	0.67	0.68	0.68
Processed products	0.10	0.06	0.08	0.08	0.06	0.05	0.18	0.15	0.07	0.00	0.00	0.01	0.02	0.03
Vegetable oils	0.64	0.68	0.77	0.79	0.83	0.82	0.83	0.79	0.81	0.80	0.82	0.83	0.78	0.76

Source: Authors' calculations from COMTRADE data.

Figure 12.7 RCA of labour-intensive and land-intensive agricultural commodities in China and ASEAN, 1992-2005

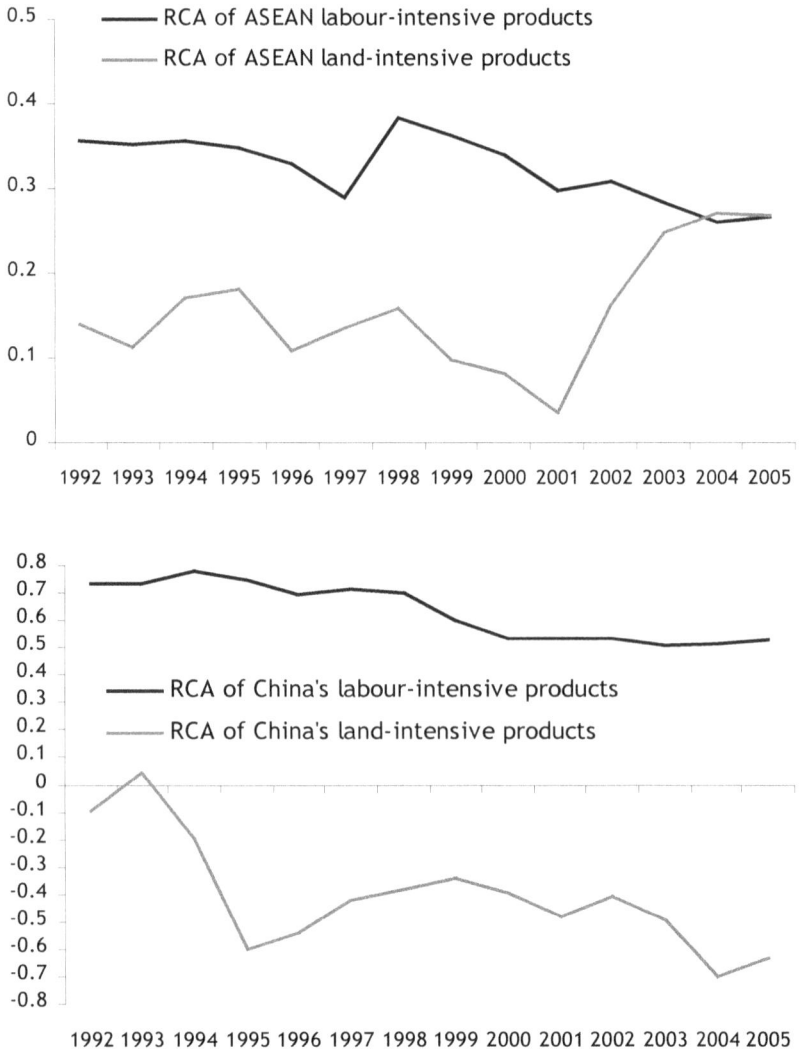

Source: Authors' calculations from COMTRADE data.

commodities to other regions rose by 56 per cent. Consequently, the share to China in ASEAN's total exports of land-intensive agricultural commodities increased from 8.9 per cent in 2001 to 13.2 per cent in 2003, and increased further to 16.4 per cent in 2005. Therefore, we could argue that China's huge growth in demand contributed significantly to the change. It remains to be seen whether the change will be maintained.

Complementarity in agricultural trade between China and ASEAN

We use the trade complementarity index (TCI) to measure how well the structure of ASEAN–China exports matches the structure of China–ASEAN imports. We calculate the TCI for food and live animals (SITC0) and for all agricultural commodities. By analysing the changes in the TCI, we can predict potential adjustments in agricultural structures in the process of economic integration between China and ASEAN.

Definition of the trade complementarity index (TCI)

The TCI measures the degree to which one country's relative export share structure corresponds with another's across certain commodities (Vollrath and Johnston 2001). The TCI assesses the market match between two economies—that is, is one country selling what the other country wants to buy? The formula for calculating TCIs is as follows

$$TCI_{ij}^{s} \equiv \sum_{k \in s} \left[\theta^{k} * RXS_{i}^{k} * RMS_{i}^{k} \right] \tag{2}$$

where

$$RXS_{i}^{k} \equiv \frac{X_{iw}^{k} / X_{iw}^{s}}{X_{ww}^{k} / X_{ww}^{s}} \equiv \frac{\text{share of } k \text{ in country } i\text{'s exports of } s \text{ goods}}{\text{share of } k \text{ in the world's exports of } s \text{ goods}}$$

$$RMS_{j}^{k} \equiv \frac{M_{jw}^{k} / M_{jw}^{s}}{M_{ww}^{k} / M_{ww}^{s}} \equiv \frac{\text{share of } k \text{ in country } j\text{'s imports of } s \text{ goods}}{\text{share of } k \text{ in the world's imports of } s \text{ goods}}$$

363

$$\theta^{k} \equiv \frac{X_{ww}^{k}}{X_{ww}^{s}} \equiv \text{share of } k \text{ in global exports of } s \text{ goods}$$

RXS_i^K is Balassa's revealed comparative advantage. RMS_j^k has the same structure, except that import rather than export data are used. In other words, the index can be interpreted as being a trade-weighted measure for sector s of the degree to which exporter i's profile of comparative advantages corresponds with the profile of comparative disadvantages for importer j. That is, this index depicts how specialisation in the commodity composition of country i's exports matches the specialisation in the commodity composition of country j's imports.

A TCI equal to one represents a threshold, with a value greater (less) than one showing a greater (lesser) level of complementarity in the composition of what exporter i exports and what importer j imports than what occurs between the average pair of countries. Further, an upward-sloping TCI suggests that the structural change is consistent with more efficient use of partner and global resources. Such a change is likely to be welfare enhancing.

Changing trends in trade complementarity between China and ASEAN

Figure 12.8 presents TCIs between ASEAN and China in food and live animals (SITC0) for the period 1992-2005. As the figure shows, the value of the TCI of ASEAN's exports and China's imports is always greater than that of China's exports and ASEAN's imports. After a short period of decline during 1994-97, the TCI value for ASEAN's exports and China's imports increased steadily from 1 in 1998 to 1.23 in 2005, indicating that there not only exists complementarity in ASEAN's exports and China's imports, but that the complementarity has been increasing since 1998.

In contrast, the TCI value for China's exports and ASEAN's imports became less than 1 in 1994; since then, it has been fluctuating about 0.8 and 0.9. This implies that there is less complementarity in China's exports and ASEAN's imports in food and live animals.

Figure 12.9 shows the TCI between China and ASEAN in all agricultural commodities for the period 1992-2005. The TCI value for all agricultural products in ASEAN's exports and China's imports is larger than that of food and live animals (SITC0). This is because the TCI of all agricultural commodities adds commodities (for example, rubber and vegetable oils) in which ASEAN has comparative advantage and China has comparative disadvantage. Therefore, agricultural products as a whole show more complementarity between ASEAN's exports and China's imports.

The TCI value of all agricultural products in ASEAN's exports and China's imports declined continuously during 1994-2001, but the decline has been reversed since 2001 (Figure 12.9). The TCI value increased quickly from 1.07 in 2001 to 1.48 in 2005.

The TCI value of China's exports and ASEAN's imports for all agricultural commodities has, however, been less than 1 since 1995. As Figure 12.9 shows, the TCI value fluctuated within the range 0.85-0.93 between 1995 and 2005. This indicates that China's exports and ASEAN's imports have less complementarity in this area. Moreover, this situation has not been changed by the implementation of the free trade agreement.

Figure 12.8   TCIs for China and ASEAN in food and live animals (SITC0), 1992-2005

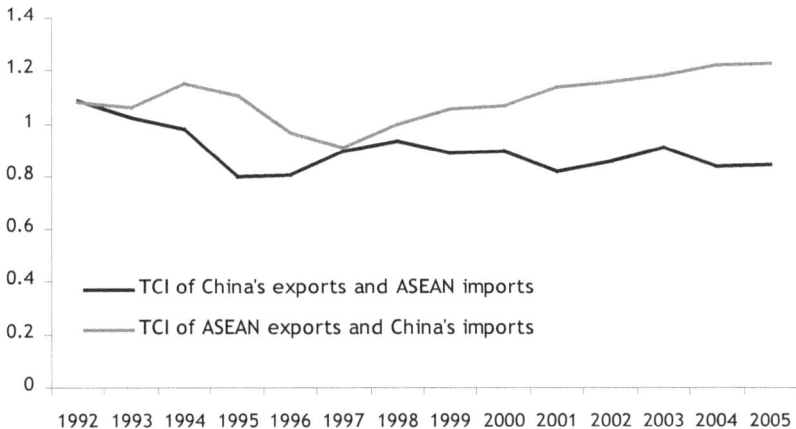

Source: Authors' calculations from COMTRADE data.

From the increasing TCIs in recent years for ASEAN's exports and China's imports in food and live animals (SITC0) and in all agricultural commodities, we could argue that ASEAN's exports to China have been experiencing some structural adjustments based on China's market demand. Moreover, such adjustments are more apparent and started earlier in the food and live animals sectors (SITC0). Such adjustments seemed, however, not to be happening in China. The increasing TCI of ASEAN's exports and China's imports corresponds with the rising share of exports to China in ASEAN's total agricultural exports. With China's market becoming more and more important for ASEAN's agricultural exports, ASEAN's agricultural production structure has started to adjust to match Chinese demand.

The changes in the TCI have another implication. The larger and increasing value of the TCI for ASEAN's exports and China's imports reveals that ASEAN is selling what China wants to buy and this match is becoming stronger. Therefore, exports from ASEAN will meet less resistance in entering China's domestic market. Consequently, ASEAN farmers will not only enjoy more

Figure 12.9 TCIs for China and ASEAN in all agricultural products, 1992–2005

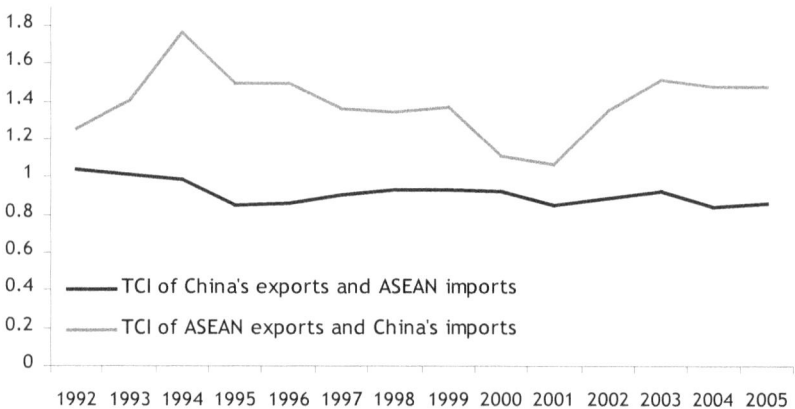

Source: Authors' calculations from COMTRADE data.

opportunities provided by the ACFTA, they will have easier access to the Chinese market. In comparison, the lower value of the TCI for China's exports and ASEAN's imports implies that a lot of effort will be needed for Chinese exporters to grasp the opportunities in the ASEAN markets.

Conclusions

Since China's WTO accession and the implementation of the ACFTA, bilateral agricultural trade between China and ASEAN has increased very rapidly and new trends have emerged during the process of economic integration. The following are the main findings of the preceding analysis.

Bilateral agricultural trade between China and ASEAN has increased rapidly in recent years, especially since the negotiation and implementation of the ACFTA and the launch of the EHP. ASEAN's agricultural exports to China have increased rapidly, reaching US$5 billion in 2005. China's agricultural exports to ASEAN have also increased but at a slower pace, reaching US$2.2 billion in 2005. ASEAN has been enjoying a surplus with China in agricultural trade and this surplus has grown.

China's domestic market is becoming more and more important for ASEAN's agricultural exports. The share of exports to China in ASEAN's total agricultural exports increased rapidly from 4.8 per cent in 1999 to 10.2 per cent in 2005. China became the third largest export market for ASEAN's agricultural products in 2005. With the full implementation of the ACFTA, the share can be expected to rise further. China's status as an export destination for ASEAN's agricultural products will be further enhanced; however, such a trend has not been witnessed in China.

China can be expected to export more labour-intensive agricultural products to ASEAN and import more land-intensive agricultural products from ASEAN. As China's RCA in labour-intensive agricultural products is higher than that of ASEAN, it should be possible for China to increase its exports to ASEAN in labour-intensive agricultural products—that is, fruits and vegetables, processed products, animal products, and fish. Compared with China, ASEAN has an overwhelming comparative advantage in certain land-intensive agricultural products (such as rubber and palm oil). Therefore, it will be better for both sides to exploit their comparative advantage in agricultural sectors by deeper integration of their economies.

The agricultural production structure in ASEAN has experienced some adjustments to match Chinese market demand. The TCI for ASEAN's exports and China's imports in food and live animals (SITC0) rose from 1 in 1998 to 1.23 in 2005. The TCI for ASEAN's exports and China's imports of all agricultural products also increased quickly—from 1.07 in 2001 to 1.48 in 2005. These trends demonstrate that the complementarity of ASEAN's exports and China's imports has been increasing in recent years. This implies that ASEAN could have undergone a structural adjustment in its agricultural sectors in response to China's rising status as an important export destination for ASEAN's agricultural products. Such an adjustment has not, however, been witnessed in China—at least, not one as significant as in ASEAN.

It should be relatively easy for ASEAN to gain access to the Chinese market during the integration of the two economies. The high and increasing value of the TCI for ASEAN's exports and China's imports reveals the strong market match between ASEAN and China: ASEAN is selling what China wants to buy. Therefore, the structural adjustment in agricultural production in ASEAN (shown by the rising TCI) should improve ASEAN's capacities to grasp the opportunities provided by China's huge market.

The integration of the two economies also provides opportunities for other agricultural exporting countries to increase their exports to China and ASEAN. The ACFTA will increase its member countries' competitiveness in many commodities. There are, however, some agricultural commodities in which China and ASEAN have no comparative advantage: for example, cereals, milk, beef and raw materials. With income growth induced by the ACFTA, demand for these commodities will rise, providing opportunities for other countries. As for countries with the same comparative advantage as China or ASEAN, they will confront certain challenges with possibly shrinking export shares in the Chinese and ASEAN markets.

There also are some challenges for the trading partners in the short term. Although the ACFTA will assist both sides to exploit their comparative advantage, it puts great pressure on those sectors with comparative disadvantage. Some adjustments will be inevitable. For example, as China's imports from ASEAN in tropical fruits increased quickly in recent years, many Chinese farmers producing tropical fruits in coastal areas (that is,

Guangdong, Guangxi, Fujian and Yunnan) found that they were losing profits and domestic market shares (Newspaper of Southern Agriculture 2006). As a result, many fruit trees have been destroyed. Therefore, certain policies should be taken to assist the transition to different farming activities or to help farmers move to non-agricultural sectors.

Notes

1 The original ASEAN members include Brunei, Singapore, Thailand, Malaysia, Indonesia and the Philippines.
2 Some sub-groups include the following commodities: SITC21 (hides, skins and fur, raw), SITC22 (oil seeds/oil fruits), SITC23 (crude rubber), SITC26 (textile fibres) and SITC29 (crude animal and vegetable materials).
3 SITC057 is one of the important components of fruit exports in China and ASEAN, accounting for 35 per cent and 54.6 per cent of total fruit exports, respectively, in 2005. Moreover, this category dominated fruit exports from ASEAN to China in 2005, making up 94.8 per cent of the total.
4 Other fresh fruits (SITC05798) include tropical fruits such as durian, longan and mangosteen, which China imports in large quantities from ASEAN.

References

Agricultural Information Centre of Guangxi Autonomous Region, 2004. Available from http://www.gxny.gov.cn/2005/1207/154414-1.html.

Balassa, B., 1965. 'Trade liberalization and "revealed" comparative advantage', *The Manchester School of Economic and Societal Studies*, 33:99–123.

Chen, C., 2006. 'Changing patterns in China's agricultural trade after WTO accession', in R. Garnaut and L. Song (eds), *The Turning Point in China's Economic Development*, Asia Pacific Press, The Australian National University, Canberra:227-55.

Chen, C.L., Yang, J. and Christopher, F., 2006. Measuring the effect of food safety standards on China's agricultural exports, Paper presented at IAAE conference, Gold Coast, Australia.

Chen, Y.F., 2004. *China's Food: demand, supply and projections*, China Agricultural Press, Beijing.

Chia, S.Y., 2004. ASEAN-China free trade area, Paper presented at the AEP conference, Hong Kong.

Holst, D. and Weiss, J., 2004. 'ASEAN and China: export rivals or partners in regional growth?', *World Economy*, 27(8):1,255–74.

Huang, J.K., Rozelle, S., Xu, Z.G. and Li, N.H., 2005. *Impacts of trade liberalization on agriculture and poverty in China*, Agricultural Policy Paper, No.18, Centre for Applied Economics and Policy Studies, Department of Applied and International Economics, Massey University, Palmerston North, New Zealand.

Huang, J.K. and Rozelle, S., 2003. 'Trade reform, the WTO and China's food economy in the twenty-first century', *Pacific Economic Review*, 8(2):143–56.

Huang, J.K., and Yang, J., 2006. 'China's rapid economic growth and its implications for agriculture and food security in China and the rest of world', *Management World*, 1:67–76.

Newspaper of Southern Agriculture, 2006. 'Pressures by ASEAN fruit import', 18 April.

Philippe, D., 1998. *Asian Crisis: the implosion of the banking and finance systems*, Kyodo Printing Company, Singapore.

Sun, D., Zhou, J. and Yang, X., 2005. 'Study on the effect of Japan's technical barrier on China's agricultural exports', *Journal of Agricultural Technical Economics*, 5:6–12.

Suthiphand, C., 2002. 'ASEAN-China free trade area: background, implications and future development', *Journal of Asian Economics*, 13:671–86.

Tian, W.M., Yang, Z.H., Xin, X. and Zhou, Z.Y., 2005. 'China's grain trade: recent developments', in Z.Y. Zhou and W.M. Tian (eds), *Grains in China: food grain, feed grain and world trade*, Aldershot, Ashgate:131–47.

Tongzon, J., 2005. 'ASEAN-China free trade area: a bane or boon for ASEAN countries?', *World Economy*, 28(2):191–210.

United Nations Statistics Division, Commodity Trade Statistics Database, COMTRADE, United Nations, New York. Available online at http://unstats. un.org/unsd/comtrade/default.aspx.

Vollrath, T.L. and Johnston, P.V., 2001. *The Changing Structure of Agricultural Trade in North America Pre- and Post-CUSTA/NAFTA: what does it mean?*, Economic Research Service, USDA, Washington, DC. Available from http:// www.ers.usda.gov/Briefing/NAFTA/PDFFiles/Vollrath2001AAEAPoster. pdf

Wong, J. and Chan, S., 2002. 'China's emergence as a global manufacturing centre: implications for ASEAN', *Asia Pacific Business Review*, 9(1):79–94.

Appendix

Table A12.1 Classification of agricultural commodities, SITC Revision 3

Commodities	Two digital classifications in SITC Revision 3
Animal products	00, 01, 02, 41
Fish	03
Cereals	04
Vegetables and fruits	05
Sugar	06
Oil seeds	22
Raw materials	21,26
Rubber	23
Processed products	11, 12, 29, 08, 07, 09
Vegetable oil	42, 43
Labour intensive	00, 01, 03, 05, 07, 09, 11, 12, 29
Land intensive	02, 04, 06, 08, 21, 22, 23, 26, 41, 42, 43

Source: COMTRADE.

13 The economic impact of the ASEAN-China Free Trade Area
A computational analysis with special emphasis on agriculture

Jun Yang and Chunlai Chen

Trade between China and the Association of South-East Asian Nations (ASEAN) has increased very rapidly in the past decade. Total trade (imports plus exports) between China and ASEAN expanded 6.7 times, from US$19.3 billion in 1996 to US$105.1 billion in 2005—an annual growth rate of 21 per cent. Currently, China is ASEAN's third largest trading partner, and ASEAN is China's fourth largest trading partner. As the growth rate of China's imports from ASEAN is higher than that of its exports to the regional group, China's trade status with ASEAN has changed from a trade surplus to a trade deficit, and the deficit has been rising in recent years. In 2005, China's trade deficit with ASEAN reached US$19.8 billion.

The trading relationship between China and ASEAN is expected to become closer because of the establishment of the ASEAN-China Free Trade Agreement (ACFTA). The Framework Agreement on ASEAN-China Comprehensive Economic Cooperation was signed in 2002, and represented a milestone in cooperation between the two economies. The ACFTA is scheduled to enter into force in 2010.[1] With a view to accelerating the implementation of the agreement, the parties agreed to implement an Early Harvest Program (EHP) with a package of agricultural and industrial products. The EHP committed the participating countries to the elimination of tariffs on these products between 2004 and 2006.[2]

Recent studies have shown that economic integration between China and ASEAN will bring numerous opportunities as well as challenges for the participants. Some studies show that China and ASEAN will experience net trade gains from the ACFTA and that it will promote economic growth in both economies (Chirathivat 2002; ASEAN Joint Experts Group 2001). In contrast, some studies find that China and ASEAN will be more likely to compete with, rather than complement, each other (Tongzon 2005; Holst and Weiss 2004; Voon and Yue 2003; Wong and Chan 2002). As the ACFTA will likely increase the competitive pressures on ASEAN producers in third-country markets and in ASEAN domestic markets, some special and differential treatment has been seen as necessary for the poorer ASEAN economies (Wattanapruttipaisan 2003). Studies have, however, pointed out that China's market liberalisation under the ACFTA would provide ASEAN countries with promising economic opportunities (Tongzon 2005; Holst and Weiss 2004). Moreover, both economies would gain large benefits from becoming more competitive and attracting foreign investment into their integrated market in the long run (Wong and Chan 2003).

These different views certainly raise issues worthy of further study. Moreover, many empirical studies have shown that regional free trade agreements contribute to member countries' growth through the accumulation of physical and human capital, productivity growth and accelerated domestic reforms (for example, Ethier 2000; Fukase and Winters 2003), but that it cannot be assumed that all participants and sectors will benefit equally. Some could, in fact, be hurt by the liberalisation. Moreover, the realisation of the ACFTA is a complex procedure involving several steps. It is useful, therefore, to study the effects under different policy arrangements during the different stages. There has been no study focusing on the dynamics of the agreement. Therefore, one of the main aims of this chapter is to explore the different economic effects of the EHP during 2004–06 and the full implementation of the ACFTA in 2010.

Trade liberalisation in China and ASEAN

Because the major issue in a free trade agreement is tariff reduction, accurate estimation of the tariffs used by the participants is very important. In this section, we analyse the trade liberalisation schedules in China and ASEAN, examining China's World Trade Organization (WTO) commitments, ASEAN Free Trade Area (AFTA) commitments and the ACFTA commitments.

China's WTO commitments

After 15 years of arduous negotiations, China acceded to the WTO at the end of 2001. As a result of the negotiations, China has agreed to undertake a series of commitments to liberalise its trade policy in order to better integrate into the world economy and offer a predictable environment for trade and foreign investment in accordance with WTO rules.[3] Under its WTO commitments, China will further reduce its import tariffs on goods and reduce or eliminate trade barriers on services. Other prohibitions, quantitative restrictions or protective measures used against imports that are inconsistent with WTO agreements will be phased out or otherwise dealt with in accordance with mutually agreed terms.

Import tariffs will be reduced gradually between 2001 and 2010. Based on the Harmonised Commodity Description and Coding System (HS) tariff schedules of the protocol of China's WTO accession and weighted by 2001 import data from the United Nations Statistics Division, Commodity Trade Statistics Database (COMTRADE), China's average import tariff will be reduced gradually from 8.79 per cent in 2001 to 5.43 per cent in 2010.[4] The scheduled import tariff reductions—by commodity and year—are provided in Table A13.1. As there is no tariff line in the services sector, it is difficult to estimate the liberalisation directly. In the empirical analysis, we adopt the estimate of Tongeren and Huang (2004) and Francois and Spinager (2004) that the import tariff equivalent of China's services sector will be reduced from 19 per cent to 9 per cent.

China will confront considerable challenges in its liberalisation of the agricultural sector. In addition to the agreed tariff reductions, China committed to removing quantitative restrictions, phasing out all export subsidies and reducing product-specific support to 8.5 per cent.[5] Although China's WTO agreement allows the government to manage the trade of 'national strategic products'[6] through a tariff rate quota (TRQ), the quotas under low tariffs will be expanded while the shares of state-owned enterprises (SOEs) will be reduced gradually.

374

The details of China's TRQs in agricultural products are shown in Table 13.1. The within-quota tariffs are quite low, while the out-of-quota tariffs are almost prohibitive. For example, the in-quota tariff for sugar is 20 per cent, while it is 9 per cent for edible oils and only 1 per cent for rice, wheat, maize and wool. The quantities imported at these low tariff levels are limited; however, the in-quota volumes were to grow over a four-year period (2002–05) at annual rates ranging from 4 per cent to 19 per cent. At the same time, tariffs on out-of-quota imports and import shares for SOEs would be reduced substantially between 2002 and 2005.

AFTA's tariff reduction schedule

The ASEAN Free Trade Area (AFTA) was established in January 1992 with the objective of eliminating tariff barriers among ASEAN member countries. The Agreement on the Common Effective Preferential Tariff (CEPT) Scheme for the AFTA requires that tariff rates levied on a wide range of products traded within the region will be reduced to no more than 5 per cent. Quantitative restrictions and other non-tariff barriers are to be eliminated. The free trade agreement covers all manufactured and agricultural products; however, 734 tariff lines on the General Exception List—representing 1.09 per cent of all tariff lines in ASEAN—are permanently excluded from the agreement for reasons of national security (ASEAN Secretariat 2002).

ASEAN member countries have made good progress in lowering intra-regional tariffs. More than 99 per cent of the products in the CEPT Inclusion List of ASEAN+6 have been brought down to the 0-5 per cent tariff range (Figure 13.1). ASEAN's new members have also reduced their import tariffs, with almost 80 per cent of their products having been moved into their respective CEPT inclusion lists. Of these items, about 66 per cent already have tariffs within the 0-5 per cent band. Vietnam had until 2006 to bring down tariffs of products on its inclusion list to no more than 5 per cent; Laos and Myanmar have until 2008, and Cambodia until 2010.

According to the amending protocol[7] signed by ASEAN member states in 2003, import duties on products on the inclusion lists of Brunei, Indonesia, Malaysia, the Philippines, Singapore and Thailand will be eliminated no later than 1 January 2010. Import duties on products on the inclusion lists for Cambodia, Laos, Myanmar and Vietnam will be eliminated no later than 1 January 2015, with flexibility allowing for import duties on some sensitive products to be eliminated no later than 1 January 2018. The tariff reduction schedule for sensitive products is governed by the Protocol on the Special Arrangement for Sensitive and Highly Sensitive Products; however, all sensitive products will have final tariff rates of 0-5 per cent by the deadlines agreed for each country.[8]

Table 13.1   China's market access commitments on farm products subject to TRQs

	Share of SOE (%)		Quotas by year			
	2002	Terminating year	2002	2003	2004	2005
Wheat	90	90	8.468	9.052	9.636	--
Corn	71	60	5.850	6.525	7.200	--
Rice	50	50	1.995	2.328	2.660	--
Soybean oil	42	10	2.518	2.818	3.118	3.587
Palm oil	42	10	2.400	2.600	2.700	3.168
Rape-seed oil	42	10	0.879	1.019	1.127	1.243
Sugar	70	70	1.764	1.852	1.945	--
Wool	n.a.	n.a.	0.265	0.276	0.287	--
Cotton	33	33	0.819	0.856	0.894	--

	In-quota tariff	Out-of-quota tariff			
		2002	2003	2004	2005
Wheat	1	71	68	65	--
Corn	1	71	68	65	--
Rice	1	74	71	65	--
Soybean oil	9	75	71.7	68.3	65
Palm oil	9	75	71.7	68.3	65
Rape-seed oil	9	75	71.7	68.3	65
Sugar	20	90	72	50	--
Wool	1	38	38	38	--
Cotton	1	54.4	47.2	40	--

n.a. there is no information in the proposal
-- the TRQ regime was phased out in 2004
Source: World Trade Organization, 2001. *Accession of the People's Republic of China, decision of 10 November 2001*, World Trade Organization, Geneva.

The free trade agreement between ASEAN and China (ACFTA)

The ACFTA was proposed initially by then Chinese Premier, Zhu Rongji, at the ASEAN-China summit in November 2000. The framework agreement on comprehensive economic cooperation between China and ASEAN nations was signed on 4 November 2002, and represented a milestone in cooperation between the two parties. According to the time frame provided by the agreement, the ACFTA covering trade in goods will be established by 2010 for Brunei, China, Indonesia, Malaysia, the Philippines, Singapore and Thailand, and by 2015 for the new ASEAN member states, Vietnam, Laos, Cambodia and Myanmar.

With a view to accelerating the implementation of the framework agreement, the parties agreed to implement the EHP for a package of agricultural and industrial products. Starting in 2004, the EHP committed the countries to the elimination of tariffs on these products by 2006. As shown in Table 13.2, the EHP comprises agricultural products under HS Chapters 1-8. The original ASEAN members and China reduced the import tariffs on these commodities to zero before 1 January 2006. The newer ASEAN members enjoy a longer period before they are to eliminate tariffs on these commodities.[9]

According to the trade-in-goods agreement, participating countries will eliminate tariffs and non-tariff barriers substantially for all products traded over time. For ASEAN+6 and China, the schedule for the bulk of the goods subject to tariff elimination ranges from 2005-10. The newer ASEAN members have until 2015 to remove all import tariffs. Countries have the flexibility to protect a limited number of products that are considered sensitive for their economies; however, the tariffs on most of these products will be reduced to 0-5 per cent by 2018.

Methodology and policy scenarios

The main analytical tool used in this study is a global trade model based on the Global Trade Analysis Project (GTAP). Following a brief introduction of the model, efforts to enhance GTAP's database and parameters for China are discussed. Finally, the baseline and two policy scenarios are defined for the purposes of evaluating the effects of the EHP during 2004-06 and the full implementation of the ACFTA during 2006-10.

Figure 13.1 Percentage of tariff lines at 0-5 per cent in the tentative 2004 CEPT package

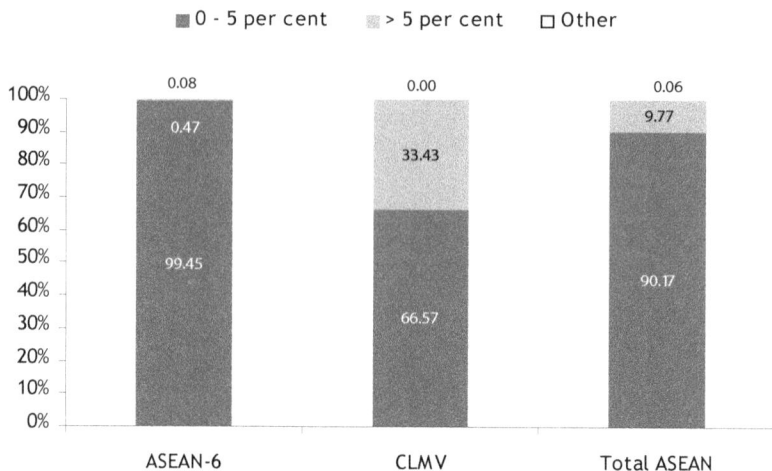

■ 0 - 5 per cent   ▨ > 5 per cent   ☐ Other

	ASEAN-6	CLMV	Total ASEAN
Other	0.08	0.00	0.06
> 5 per cent	0.47	33.43	9.77
0 - 5 per cent	99.45	66.57	90.17

Source: ASEAN statistics, 2005. Available online at http://www.aseansec.org

Table 13.2 Product coverage in the Early Harvest Program (EHP)

Chapter	Description
01	Live animals
02	Meat and edible meat offal
03	Fish
04	Dairy products
05	Other animal products
06	Live trees
07	Edible vegetables
08	Edible fruits and nuts

Source: ASEAN-China FTA Framework Agreement (2002). Available online at http://www.bilaterals.org

Brief introduction to the GTAP model

GTAP is a multi-region, multi-sector, computable general equilibrium model, with perfect competition and constant returns to scale. The model is described fully in Hertel (1997). It has been used widely to analyse the impacts of changes in trade policy.

In the GTAP model, each country or region is depicted within the same economic structure. The consumer side is represented by the country or regional household to which are assigned the income of factors, tariff revenues and taxes. The country or regional household allocates its income to three expenditure categories: private household expenditure, government expenditure and savings. For the consumption of the private household, the non-homothetic Constant Difference of Elasticities (CDE) function is applied. Firms combine intermediate inputs and primary factors, land, labour (skilled and unskilled) and capital. Intermediate inputs are composites of domestic and foreign components, and the foreign component is differentiated by region of origin (the so-called Armington assumption). With respect to factor markets, the model assumes full employment, with labour and capital being fully mobile within regions but immobile internationally. Labour and capital remuneration rates are determined endogenously at equilibrium. In the case of crop production, farmers make decisions on land allocation. Land is assumed to be imperfectly mobile between crops, and hence the model allows for endogenous land rent differentials. Each country or region is equipped with one country or regional household that distributes income across savings and consumption expenditure to maximise its utility.

The GTAP model includes two global institutions. All transport between regions is carried out by the international transport sector. The trading costs reflect the transaction costs involved in international trade as well as the physical activity of transportation itself. Using transport inputs from all regions, the international transport sector minimises its costs under Cobb-Douglas technology. The second global institution is the global bank, which takes the savings from all regions and purchases investment goods in all regions depending on the expected rates of return. The global bank guarantees that global savings are equal to global investments.

The model does not have an exchange rate variable; however, by choosing as a *numeraire* the index of global factor prices, each region's change of factor prices relative to the *numeraire* directly reflects a change in the purchasing power of the region's factor incomes on the world market. This change can be interpreted directly as a change in the real exchange rate. Welfare changes are measured by the equivalent variation, which can be computed from each region's household expenditure function.

Taxes and other policy measures are represented as *ad valorem* tax equivalents. These create wedges between the undistorted prices and the policy-inclusive prices. Production taxes are placed on intermediate or primary inputs, or on output. Trade policy instruments include applied most-favoured nation tariffs, anti-dumping duties, countervailing duties, export quotas and other trade restrictions. Additional internal taxes can be placed on domestic or imported intermediate inputs, and can be applied at differential rates that discriminate against imports. Taxes could also be placed on exports and on primary factor income. Finally, where relevant, taxes are placed on final consumption, and these can be applied differentially to the consumption of domestic and imported goods.

Data improvement

The GTAP database contains detailed bilateral trade, transport and protection data characterising economic linkages among regions. Regions are linked with individual country input–output databases, which account for inter-sectoral linkages among the 57 sectors in each of the 87 regions. The database provides detailed sectoral classifications for agriculture, with 14 primary agricultural sectors and seven agricultural-processing sectors. The base year for the version used in this study (Version 6) is 2001. For the purpose of the study, the database has been aggregated into 13 regions and 22 sectors. The regional and sectoral aggregations are summarised in Tables A13.2 and A13.3.

Before applying the GTAP Version 6, we carefully examined the database and parameters for China and made substantial improvements in several aspects related to agricultural input and output ratios, demand parameters, trade policies and production values. The main improvements to GTAP Version 6 include the following.

Input-output tables in the agricultural sector. We overcame some of the shortcomings in the database by taking advantage of data that were collected by the National Development and Reform Commission (NDRC) and government organisations. The NDRC collects data on the costs of production of all of China's major crops and livestock industries. The data set contains information on quantities and total expenditure of labour and material inputs as well as expenditure on a large number of miscellaneous costs such as tax, transportation and marketing. Each farmer reports output and the total revenue earned from crops or livestock. The data were used previously in analyses of China's agricultural supply and input demand (Huang and Rozelle 1996; World Bank 1997). Similar methods have been used for other studies (Huang and Yang 2006; Tongeren and Huang 2004). In this way, we ensure the balance and consistency of input-output relationships among sectors.

Improving own-price and income elasticities for China. We incorporated the most up-to-date estimates for price and income elasticities of demand for various foods in China (Fan et al. 1995; Huang and Bouis 1996; Huang and Rozelle 1998). Table A13.4 summarises the major adjustments that have been made. In addition, we assume that income elasticities of demand for various commodities will change as incomes increase. This is an essential assumption for long-term simulations. Based on other empirical studies (Huang and Bouis 1996; Huang and Rozelle 1998), we assume that food income elasticities decline with income growth (Table A13.4). Using information on uncompensated income elasticities and own-price elasticities, we recalibrate the expansion and substitution parameters for the CDE by the method introduced by Liu et al. (1998) and Yu et al. (2003).

Trade distortions. Various studies have estimated the magnitude of agricultural price distortions using available series on domestic and international prices. Unfortunately, the results obtained have varied widely. Huang et al. (2004) adopted a new approach, which estimated the policy impacts from detailed interviews with participants in China's agricultural marketing and trading activities. This approach provides a much clearer indication of the implications of agricultural trade policies than would otherwise be possible. Their results have been used in several recent studies on the impacts of WTO accession on China's economy (Bhattasali et al. 2004; Anderson et al. 2004; Ianchovichina and Martin 2004). We adjust import and export tariff equivalents of agricultural commodities in the base year (2001) using results from the study by Huang et al. (2004). Details of the adjustments are provided in Table A13.5.

Assumptions for the different scenarios

The central goal of this study is to assess the economic impacts of the ACFTA during its various stages of implementation. Towards this end, three scenarios have been developed: I) the baseline scenario; II) the EHP policy scenario in 2006; and III) the full implementation of the ACFTA by 2010. **The baseline scenarios.** In this study, we compare two trade liberalisations over two different periods: the EHP during 2004-06 and the full implementation of the ACFTA during 2006-10. We construct two baselines (I and II) to evaluate the effects of policy changes in the two periods. Baseline (I) is constructed for the period 2001-06 to capture the effects of the EHP; Baseline (II) incorporates the effects of the EHP during 2001-06 and projects to 2010 in order to isolate the effects of the liberalisation in the second stage.

Both baselines are constructed using a recursive dynamic approach to reflect the changes over time in the endowments of the countries. This procedure has been used in several other studies (for example, Hertel and Martin 1999; Tongeren and Huang 2004). The growth in endowments (GDP, population, skilled and unskilled labour, capital and natural resources) is taken mainly from other similar studies (Huang and Yang 2006; Tongeren and Huang 2004; Walmsley et al. 2000).

The baseline projection also includes a continuation of existing policies and the effectuation of important policy events related to international trade, as they are known to date. The important policy changes are: implementation of the remaining commitments from the General Agreement on Tariffs and Trade (GATT) Uruguay Round agreements; China's WTO accession commitments between 2001 and 2010; the global phase-out of the Multifibre Agreement under the WTO Agreement on Textiles and

Clothing (ATC) by January 2005; European Union enlargement to include Central and Eastern European countries (CEECs); and the implementation of the AFTA among ASEAN member countries.

The economic effects of the EHP during 2004-06. Under the EHP scenario, all assumptions under the Baseline (I) scenario are held except for the import tariffs on commodities listed in the EHP between China and ASEAN member countries. According to the protocol of the EHP, the import tariffs were to be eliminated in China and the original ASEAN members before 1 January 2006. Therefore, in this simulation, tariffs between China and the old ASEAN members are reduced to zero on the commodities listed in the EHP.

As for the new ASEAN members, there is a longer period for them to reduce their tariffs on commodities listed in the EHP. As the tariffs on many commodities were to be reduced to about 5 per cent by 2006 (Shang 2005), we adopt the simple and reasonable assumption that the import tariffs of commodities listed in the EHP for China's exports to the newer ASEAN members were reduced to 5 per cent in 2006. The import tariffs for the newer ASEAN members' exports to China will be eliminated.

The full implementation of the ACFTA by 2010. Under the scenario of the full implementation of the ACFTA, all assumptions for Baseline (II) are maintained except for the import tariffs between China and ASEAN. According to the trade-in-goods agreement, participating countries will eliminate substantially tariffs and non-tariff barriers for all products traded. For ASEAN+6 and China, the schedule for the bulk of the goods subject to tariff elimination is before 2010. Therefore, tariffs between China and old ASEAN members will be reduced to zero on all commodities that are not included in the EHP.

As for the newer ASEAN members, they have five additional years (until 2015) to remove all import tariff lines, including tariffs on commodities listed in the EHP. Therefore, in this simulation, we assume newer ASEAN members will make no liberalisation except for implementing their commitments in the EHP, but their exports to China will face zero import tariffs.

Comparison of the simulation results with those for the first stage of the EHP will help us understand the impacts of the second stage of liberalisation.

Simulation results and explanations

In this section, we present the economic impacts of the EHP during 2004-06 and the full implementation of the ACFTA for other commodities during 2006-10 separately. Some of the economic forces underlying the impacts are analysed.

The economic impacts of the EHP

The welfare effects of the EHP are presented in Table 13.3, and the participant countries capture the benefits. The ASEAN+5 countries are the biggest winners in terms of the absolute increase in social welfare (US$56.55 million). Newer ASEAN members are next, with welfare increasing by US$25.8 million. The newer ASEAN members are, however, the biggest winners in relative terms. China's welfare increase is modest at US$13.65 million. As for other regions,[10] their total welfare declines by US$77.03 million because of trade diversion effects. As a whole, global welfare rises by US$18.97 million.

The returns to primary inputs—that is, land, capital and labour—increase in all the EHP countries. As trade liberalisation through the EHP promotes production in all countries, the demand for primary factors increases. Therefore, the price of primary factors rises in the new equilibrium under the assumption of the fixed supply of primary factors. The price of land increases by 0.029 per cent in China, by 0.71 per cent in ASEAN+5 and by 1.57 per cent in the newer ASEAN members. As relatively more unskilled labour is employed in the agricultural sector, the wage increases of unskilled labour are greater than for skilled labour in EHP countries. The returns to primary factors fall in other regions.

Trade increases in all of the ACFTA signatory countries. As shown in Table 13.4, total exports and imports increase in China by US$276 million and US$297 million respectively, by US$121 million and US$83 million in ASEAN+5 and by US$59 million and US$49 million in new ASEAN member countries. In aggregate, the EHP increases global exports (or imports) by US$286 million. Moreover, the exports and imports of the commodities listed in the EHP increase in all ACFTA signatory countries and their growth rates are larger than those of other commodities.

There are trade diversion effects for non-member regions. This can be seen more clearly in Table 13.5, which shows the changes in bilateral trade flows associated with the EHP. China's exports to ASEAN increase by US$317 million but its exports to other regions fall by US$41 million; as a result, China's total exports increase by US$276 million. China's imports increase by US$297 million, of which those from ASEAN increase by US$592 million; but China's imports from other regions decline by US$296 million. The changes in exports and imports with ASEAN also reflect the importance of China-ASEAN trade. The exports of other regions to China and ASEAN+5 decline by US$350 million. Although exports from other regions to new ASEAN member countries increase only marginally (by US$4 million), the total exports of other regions fall by US$169 million.

In general, the output prices of commodities in ACFTA signatory countries rise because of the increasing cost of production. As the standard GTAP model assumes perfect competition in markets, firms have zero profits and the output price is equal to the production cost. Therefore, the rising prices of primary input factors pull up the output prices. As shown in Table 13.6, prices increase in all the participant countries; however, in China, output prices for processed food, fish, textiles and apparel decline marginally. This is mainly because the tariff reductions lower the import prices of fruits and vegetables, pork and poultry products, which are the

Table 13.3   The macro impacts of the EHP

	China	Old ASEAN	New ASEAN	Other regions
EV (US$ million)	13.65	56.55	25.80	-77.03
EV/GNP (%)	0.001	0.012	0.021	0.000
GDP (%)	0.000	0.031	0.079	-0.002
GDP price (%)	-0.001	0.030	0.078	-0.002
Price of land (%)	0.029	0.706	1.570	-0.030
Wages of unskilled labour (%)	0.012	0.040	0.063	-0.002
Wages of skilled labour (%)	0.011	0.007	0.021	-0.001
Price of capital (%)	0.010	0.009	0.019	-0.001

Note: EV - Equivalent variation
Source: Results of author's simulation.

Table 13.4 The impact of the EHP on imports and exports

	Percentage change in exports				Percentage change in imports			
	China	Old ASEAN	New ASEAN	Other regions	China	Old ASEAN	New ASEAN	Other regions
Rice	0.35	-0.27	-0.73	0.12	-0.58	0.12	2.50	-0.03
Wheat	0.00	0.00	0.00	-0.01	0.00	0.00	0.00	-0.01
Coarse grain	0.00	0.00	-4.76	0.01	0.14	0.34	0.00	-0.01
Vegetables and fruits	4.13	3.26	11.04	-0.14	10.01	2.09	16.81	-0.02
Oil seeds	0.00	0.00	-1.56	0.02	0.03	0.12	0.00	0.00
Sugar	0.00	0.00	0.00	0.03	0.00	0.15	0.00	0.00
Cotton	0.00	0.00	0.00	0.02	0.11	0.00	0.00	-0.01
Vegetable oil	0.72	-0.08	0.00	0.02	-0.16	0.00	0.00	-0.01
Other crops	6.65	0.92	2.03	-0.23	4.38	1.49	11.36	-0.04
Cattle and mutton	1.11	3.23	0.00	-0.01	0.18	0.17	0.00	-0.01
Pork and poultry	3.01	9.50	6.45	-0.42	3.76	5.38	1.11	-0.02
Milk	0.00	0.97	0.00	-0.01	0.41	0.06	0.30	-0.01
Fish	0.16	0.31	1.33	-0.03	1.14	0.00	0.00	-0.01
Processed food	0.10	-0.05	-0.29	0.00	-0.05	0.10	0.14	0.00
Textiles and apparel	0.03	-0.02	-0.16	-0.01	-0.01	-0.01	0.00	0.00
Other industries	-0.03	-0.05	-0.10	0.00	0.01	-0.02	0.05	0.00

	Change in exports (US$ million)				Change in imports (US$ million)			
	China	Old ASEAN	New ASEAN	Other regions	China	Old ASEAN	New ASEAN	Other regions
Rice	2	-5	-5	6	-1	1	1	-2
Wheat	0	0	0	-1	0	0	0	-1
Coarse grain	0	0	-1	1	1	2	0	-2
Vegetables and fruits	78	51	72	-62	99	27	20	-7
Oil seeds	0	0	-1	3	2	1	0	0
Sugar	0	0	0	2	0	1	0	0
Cotton	0	0	0	2	2	0	0	-1
Vegetable oil	1	-4	0	2	-1	0	0	-1
Other crops	89	56	16	-75	58	32	10	-16
Cattle and mutton	1	2	0	-2	1	1	0	-2
Pork and poultry	141	206	6	-175	118	65	1	-8
Milk	0	4	0	-2	2	1	1	-2
Fish	1	2	1	-2	1	0	0	-1
Processed food	7	-7	-5	6	-2	7	2	-6
Textiles and apparel	47	-5	-12	-40	-4	-1	0	-6
Other industries	-88	-175	-13	168	21	-56	15	-88
Total	276	121	59	-170	297	83	49	-143

Source: Results of authors' simulation.

important intermediate inputs of processed food. Therefore, because the cost reduction on intermediate inputs is dominant, the combined effects of the EHP reduce the production costs of these commodities in China.

The changes in production reflect the combined changes in sectoral exports and imports and domestic consumption resulting from the removal of trade barriers. The driving force underlying such change is the comparative advantage in each region. It is clear that the EHP will shift the primary input factors into the agricultural sectors experiencing tariff reductions. As shown in Table 13.6, all the industrial and other agricultural sectors without tariff reductions will shrink in ASEAN. It is, however, a little different for China. The production of processed food, textiles and apparel expands marginally due to enhanced competitiveness arising from their falling output price.

Increases in the output of the commodities listed in the EHP are not, however, assured. Two factors will determine changes in production: the first is competition from China's trading partners. Taking vegetables and fruits in China, for example, although production will increase by 0.3 per cent due to ASEAN's import tariff reduction, the increasing competition induced by China's tariff reduction will reduce production by 0.4 per cent. Therefore, the total effect on China's production of vegetables and fruits is a decline of 1 per cent. The second factor is competition among sectors for limited resources. If more production factors are drawn into sectors experiencing strong expansion, the production of other sectors could be undermined. Taking the milk sector in China, for example, as ASEAN

Table 13.5   Changes in bilateral trade flows (US$ million)

	China	Old ASEAN	New ASEAN	Other regions	Exports
China		266	51	-41	276
Old ASEAN	465	-118	-6	-220	121
New ASEAN	127	-12	0	-58	59
Other regions	-296	-54	4	176	-169
Imports	297	83	49	-143	

Source: Results of authors' simulation.

Table 13.6　Changes in supply price and output by the EHP (per cent)

	Supply price	Output	Contribution to output by China Tariff reduction	Old ASEAN Tariff reduction	New ASEAN Tariff reduction
China					
Vegetables and fruits	0.01	-0.01	-0.04	0.02	0.01
Other crops	0.28	0.74	-0.82	1.29	0.27
Cattle and mutton	0.01	0.02	0.03	-0.02	0.00
Pork and poultry	0.01	0.01	-0.07	0.08	0.00
Milk	0.00	-0.03	0.00	-0.02	-0.01
Fish	-0.01	0.01	0.00	0.01	0.00
Processed food	-0.04	0.03	0.03	0.00	0.00
Textiles and apparel	-0.01	0.03	0.06	-0.03	-0.01
Other agricultural products	0.01	0.02	0.01	0.01	0.00
Other industries	0.00	-0.01	0.00	-0.01	0.00
Old ASEAN					
Vegetables and fruits	0.31	0.11	0.25	-0.11	-0.03
Other crops	0.30	0.09	0.18	-0.09	0.00
Cattle and mutton	0.04	0.02	-0.01	0.03	0.00
Pork and poultry	0.26	0.82	1.16	-0.34	0.01
Milk	0.05	0.23	0.16	0.07	0.00
Fish	0.01	0.02	0.02	0.00	0.00
Processed food	0.04	-0.01	-0.05	0.04	0.00
Textiles and apparel	0.00	-0.01	-0.11	0.10	0.00
Other agricultural products	0.11	-0.04	-0.07	0.03	0.00
Other industries	0.01	-0.03	-0.03	0.01	0.00
New ASEAN					
Vegetables and fruits	1.37	1.04	1.32	-0.01	-0.27
Other crops	0.52	0.04	0.09	-0.01	-0.05
Cattle and mutton	0.17	-0.07	-0.09	0.00	0.02
Pork and poultry	0.27	0.11	0.11	-0.04	0.04
Milk	0.10	-0.08	-0.11	0.00	0.04
Fish	0.01	0.01	0.00	0.00	0.01
Processed food	0.12	-0.10	-0.13	-0.01	0.04
Textiles and apparel	0.02	-0.10	-0.15	0.01	0.03
Other agricultural products	0.29	-0.15	-0.18	-0.01	0.04
Other industries	0.03	-0.02	-0.03	0.00	0.00

Source: Results of authors' simulation.

tariff reductions will promote production in some sectors quite strongly, production factors will be drawn away from the milk sector; as a result, the production of milk will fall by 0.03 per cent due to the relocation of production factors.

The economic impacts of the full implementation of the ACFTA

The macro effects of the full implementation of the ACFTA are much larger than those of the EHP. The increases in real GDP and social welfare in the ACFTA signatories are much larger than those in the EHP (Table 13.7). Welfare in all ACFTA signatories increases by US$1.8 billion, with US$451 million in China, US$1.25 billion in ASEAN+5 and US$92 million in the new ASEAN members. ASEAN+5 is the largest beneficiary in absolute and relative terms. The negative impacts on the rest of the world are also more significant than those of the EHP. The social welfare of other regions declines by US$1.9 billion. The global welfare loss due to the creation of the ASEAN-China Free Trade Area is US$115 million.

The impacts on the returns to primary factors in ACFTA signatories are substantial. As shown in Table 13.7, all the returns to primary factors increase in China and in the new ASEAN members. As for ASEAN+5, returns to labour and capital increase while returns to land decline. Land is the sluggish factor in the GTAP model, so its price can vary across sectors (Hertel 1997). The land prices reported in Table 13.7 reflect the aggregate effects of policy changes on land use. Because the non-agricultural sectors in ASEAN+5 grow so strongly and draw labour and capital away from agricultural sectors, the demand for land declines. As a result, land prices drop after the full implementation of the ACFTA. The returns to primary factors in the other regions decline marginally.

The results indicate that there will be trade gains for all ACFTA signatories. Trade creation will easily offset trade diversion. As shown in Tables 13.8 and 13.9, total exports and imports increase by US$6.5 billion and US$6.8 billion, respectively, for China, by US$4.7 billion and US$4.9 billion for ASEAN+5 and by US$153 million and US$203 million for the new ASEAN members. As a whole, the implementation of the ACFTA by 2010 will promote global exports of US$8.2 billion.

The effects on exports and imports vary remarkably across sectors. While there are declines in cattle and mutton, pork and poultry, manufacturing and services, exports of other commodities from China increase. As shown in Table 13.8, the most significant growth in absolute terms is in electronic products in the industrial sector and in processed food in the agricultural sector. Although the growth of sugar in China is very impressive, its increase in absolute terms is limited as its initial export value is very small. As for ASEAN+5 and the new ASEAN members, exports of natural resource-related industrial products in the industrial sector and vegetable oils in the agricultural sector increase most significantly in absolute terms. The growth in the export of sugar is also quite remarkable—in ASEAN+5 and in the new ASEAN members.

Imports also rise among the ACFTA signatories. China's imports of sugar and vegetable oil rise significantly, increasing by 10.5 per cent (US$40 million) and 28.5 per cent (US$174 million) respectively. As China does not have comparative advantage in these two agricultural commodities, its imports of vegetable oils (mainly palm oil) from ASEAN have risen dramatically in recent years. The elimination of import tariffs will further stimulate the importation of vegetable oil from ASEAN. China's imports of natural resource-related industrial products, electronics and metal and machinery also increase significantly. The increase in these three commodities accounts for 85.5 per

Table 13.7   The macro effects of the implementation of the ACFTA, up to 2010

	China	Old ASEAN	New ASEAN	Other regions
EV (US$ million)	451	1,254	92	-1,912
EV/GNP (%)	0.025	0.244	0.062	-0.006
GDP (%)	0.141	0.610	0.351	0.034
GDP price (%)	0.136	0.576	0.338	-0.033
Price of land (%)	0.200	-0.071	0.156	0.001
Wages of unskilled labour (%)	0.288	0.877	0.310	-0.002
Wages of skilled labour (%)	0.311	0.848	0.317	-0.001
Price of capital (%)	0.306	0.889	0.333	-0.001

Note: EV - Equivalent variation
Source: Results of authors' simulation.

cent of the total increase in imports of industrial products.

For ASEAN+5, among agricultural products, imports of processed food increase most significantly (by 2.83 per cent or US$212 million). Imports of all industrial products increase, with the highest growth rate of 6.75 per cent (US$683 million) in textiles and apparel, and the largest increase in value of US$1.05 billion in metal and machinery (a 1.98 per cent increase). Because the new ASEAN members will continue to implement the tariff reduction

Table 13.8   Impacts on exports after implementation of the ACFTA, up to 2010

	Change in exports (US$ million)				Percentage change in exports			
	China	Old ASEAN	New ASEAN	Other regions	China	Old ASEAN	New ASEAN	Other regions
Rice	6	-14	3	21	1.15	-0.77	0.40	0.42
Wheat	2	0	0	-9	6.61	-0.70	1.24	-0.06
Coarse grain	2	0	0	3	0.49	-0.33	-0.23	0.02
Vegetables and fruits	6	-13	0	19	0.34	-0.73	-0.04	0.04
Oil seeds	8	-1	0	-18	2.68	-0.90	-0.57	-0.09
Sugar	12	81	3	-43	64.51	12.33	11.55	-0.55
Cotton	2	2	2	-4	0.84	2.92	14.34	-0.03
Vegetable oil	4	102	18	51	2.43	2.29	83.90	0.45
Other crops	3	-34	1	34	0.25	-0.55	0.07	0.10
Cattle and mutton	-1	-2	0	3	-0.75	-2.75	0.04	0.01
Pork and poultry	-37	-43	-1	85	-0.70	-2.03	-0.81	0.20
Milk	0	-8	0	9	0.77	-1.57	-1.02	0.03
Fish	0	0	0	4	0.06	0.01	-0.06	0.06
Processed food	474	-22	8	-193	6.21	-0.16	0.46	-0.11
Natural resources	48	-273	-11	696	2.63	-1.45	-0.19	0.20
Textiles and apparel	642	478	-59	-84	0.38	2.14	-0.71	-0.03
Natural industry	708	2,736	200	-1,297	1.18	4.59	9.50	-0.12
Metal and machinery	1,489	1,522	-16	-1,198	1.27	2.79	-0.90	-0.09
Transportation	550	9	-3	-362	3.91	0.11	-1.38	-0.05
Electronics	2,723	1,309	33	-2,429	2.06	0.75	5.93	-0.42
Manufactures	-48	28	4	135	-0.08	0.35	0.64	0.10
Services	-132	-1142	-28	1,462	-0.52	-1.74	-0.80	0.11
Total	6,463	4,717	153	-3,115	1.08	1.06	0.56	-0.05

Source: Results of authors' simulation.

schedule for the commodities listed in the EHP during 2006-10, imports of these agricultural commodities increase more than imports of other commodities. As there is no liberalisation in industrial sectors, the import growth of industrial products in the new ASEAN members is very small.

There is significant trade diversion from the non-member regions during the full implementation of the ACFTA. China's exports to ASEAN+5 and the new ASEAN members increase by 35.6 per cent (US$8.8 billion) and 0.9 per

Table 13.9   Impacts on imports of the implementation of the ACFTA, up to 2010

	Change in imports (US$ million)				Percentage change in imports			
	China	Old ASEAN	New ASEAN	Other regions	China	Old ASEAN	New ASEAN	Other regions
Rice	7	16	0	-7	3.26	1.54	0.44	-0.09
Wheat	2	-8	0	-2	0.45	-0.61	0.02	-0.01
Coarse grain	2	2	0	0	0.28	0.36	0.27	0.00
Vegetables and fruits	4	8	7	-9	0.31	0.60	4.45	-0.02
Oil seeds	-21	8	0	3	-0.28	0.89	1.07	0.02
Sugar	40	15	0	-2	10.52	2.17	0.33	-0.02
Cotton	6	-4	0	-1	0.32	-0.34	0.41	-0.01
Vegetable oil	174	10	1	-10	28.45	1.93	0.53	-0.06
Other crops	3	5	8	-12	0.18	0.21	6.75	-0.03
Cattle and mutton	2	7	0	-9	0.42	1.17	1.04	-0.03
Pork and poultry	16	13	3	-26	0.41	0.93	2.55	-0.06
Milk	2	6	1	-8	0.35	0.32	0.40	-0.03
Fish	0	1	0	3	0.10	0.30	1.54	0.04
Processed food	116	212	6	-67	2.25	2.83	0.28	-0.04
Natural resources	40	356	1	63	0.15	2.27	1.01	0.02
Textiles and apparel	601	683	8	-314	1.81	6.75	0.27	-0.07
Natural industry	1,978	785	36	-452	3.08	1.76	0.53	-0.04
Metal and machinery	1,641	1,048	26	-919	1.90	1.35	0.38	-0.07
Transportation	179	264	11	-260	0.98	1.39	0.45	-0.04
Electronics	1,691	699	13	-766	2.20	0.68	0.78	-0.11
Manufactures	123	155	3	-161	3.53	3.00	0.65	-0.08
Services	238	662	79	-819	0.36	1.15	0.56	-0.06
Total	6,845	4,944	203	-3,773	1.71	1.40	0.52	-0.06

Source: Results of authors' simulation.

cent (US$44 million), respectively (Table 13.10). In contrast, China's exports to other regions decline by 0.4 per cent (US$2.4 billion). The same pattern occurs in ASEAN+5 and the new ASEAN countries, even with the reduction in trade among ASEAN members. As for the other regions, exports to China and ASEAN+5 fall by 0.9 per cent (US$5.2 billion) and by 0.5 per cent (US$1.6 billion), respectively. Exports to the new ASEAN members, however, and to the other regions (individually) rise by 1 per cent (US$201 million) and 0.1 per cent (US$3.5 billion) respectively. Overall, exports of the rest of the world decline by 0.1 per cent (US$3.1 billion).

Table 13.11 shows the changes in supply prices and output in China. Except for sugar, vegetable oil and electronics, supply prices rise due to the increasing costs of the primary factors. Because land is the sluggish factor, land rents in the sugar and vegetable oil sectors in China fall because of the dramatic decline in the production and prices of sugar and vegetable oil caused by the large imports of these commodities from ASEAN. The price decline in electronics is caused by the cost reduction of the intermediate inputs more than offsetting the increases in prices of the primary factors. The elimination of import tariffs will reduce the

Table 13.10   Changes in bilateral trade after the implementation of the ACFTA, up to 2010

	China	ASEAN+5	New ASEAN	Other regions	Exports
Value (US$ million, in world prices)					
China	0	8,800	44	-2,380	6,463
Old ASEAN	11,539	-2,147	-42	-4,632	4,717
New ASEAN	465	-97	0	-215	153
Other regions	-5,159	-1,611	201	3,454	-3,115
Imports	6,845	4,944	203	-3,773	
Percentage					
China		35.6	0.9	-0.4	1.1
Old ASEAN	41.4	-2.7	-1.0	-1.7	1.4
New ASEAN	9.6	-1.0	0	0.8	0.4
Other regions	-0.9	-0.5	1	0.1	-0.1
Imports	1.7	1.4	0.5	-0.1	

Source: Results of authors' simulation.

import prices of finished and semi-finished products of electronics from ASEAN. Imported semi-finished electronic products account for a significant share of the total production cost of electronics. As a result, the prices of electronics in China fall.

In China, the ACFTA will promote the production and increase the output of rice, vegetables and fruits, processed foods and fish in the agricultural sector and metal and machinery, transportation and electronics in the industrial sector (Table 13.11). In order to distinguish between the effects of trade liberalisation on output in agriculture and industry, we further decompose the total impacts on output into three sources: the contribution of tariff reductions on agricultural commodities listed in the EHP for the new ASEAN members; the contribution of tariff reductions on other agricultural commodities; and the contribution of tariff reductions on industrial products.

The decomposition of the total impacts on output reveals the direct effect of trade liberalisation and the effect of resource relocation. Taking the processed-food sector in China, for example, the tariff reduction in the agricultural sector increases its production by 0.26 per cent, but the liberalisation of industry draws resources out of the agricultural sector and reduces processed-food production by 0.03 per cent. The combined impact is to increase processed-food production by 0.24 per cent. As vegetables and fruits are the most important intermediate inputs of the processed-food industry, the increase in output of the processed-food sector promotes the production of vegetables and fruits.[11]

China's huge domestic market will provide great opportunities for ASEAN countries, but the effects on output will be determined by their comparative advantages and by competition among sectors. As Table 13.12 shows, the supply prices of all commodities in ASEAN+5 rise; however, this is not the case for output. In agricultural sectors, the output of sugar and vegetable oil increases but other agricultural sectors shrink. In the industrial sector, the output of natural resource-related industry, electronics and metal and machinery increases but the others decline. The full implementation of the ACFTA should, therefore, help ASEAN+5 members to exploit their comparative advantages.

According to the ACFTA, the new ASEAN members will have a transitional period beyond 2010 to eliminate most import tariffs, but they will continue to reduce the tariff lines on commodities listed in the EHP until 2010. For commodities not listed in the EHP, the new ASEAN members will enjoy the opportunities of reduced tariffs in China's market without any liberalisation on their part before 2010. As shown in Table 13.13, the tariff reductions on the commodities in the EHP reduce the production of those agricultural commodities, except for fish, in the new ASEAN members.

Table 13.11   Changes in China in supply prices and output from the implementation of the ACFTA, up to 2010

	Supply price	Output	Contribution to output by		
			Tariff reductions from EHP	Tariff reductions in agriculture	Tariff reductions in industry
Rice	0.216	0.002	0.000	0.023	-0.021
Wheat	0.160	-0.026	0.000	0.042	-0.068
Coarse grain	0.166	-0.058	0.000	0.009	-0.068
Vegetables and fruits	0.240	0.032	0.003	0.058	-0.029
Oil seeds	0.109	-0.332	-0.002	-0.203	-0.127
Sugar	-0.139	-1.778	-0.002	-1.710	-0.067
Cotton	0.105	-0.173	-0.004	-0.030	-0.139
Vegetable oil	-0.039	-2.968	-0.001	-3.107	0.140
Other crops	0.205	-0.099	0.198	-0.051	-0.246
Cattle and mutton	0.190	-0.126	-0.001	-0.010	-0.115
Pork and poultry	0.200	-0.074	0.001	-0.013	-0.062
Milk	0.179	-0.133	0.011	-0.003	-0.141
Fish	0.080	0.009	0.001	-0.013	0.021
Processed food	0.171	0.236	-0.001	0.262	-0.025
Natural resources	0.086	-0.121	-0.001	-0.005	-0.115
Textiles and apparel	0.152	-0.113	-0.003	-0.021	-0.090
Natural industry	0.117	-0.262	-0.001	-0.007	-0.255
Metal and machinery	0.129	0.017	-0.001	-0.015	0.033
Transportation	0.138	0.473	-0.001	-0.009	0.482
Electronics	-0.097	0.912	-0.001	-0.022	0.935
Manufactures	0.162	-0.219	-0.001	-0.019	-0.198
Services	0.193	0.022	0.000	-0.001	0.023

Source: Results of authors' simulation.

The output of vegetable oil, sugar, cotton, processed food, oil seeds and rice is, however, expected to increase. In the industrial sector, the output of natural resource-related industry, electronics, and manufactures will increase, but others will shrink.

Table 13.12   Changes in ASEAN+5 in supply prices and output after the implementation of the ACFTA, up to 2010

	Supply price	Output	Contribution to output by		
			Tariff reductions from EHP	Tariff reductions in agriculture	Tariff reductions in industry
Rice	0.514	-0.250	-0.002	-0.013	-0.235
Wheat	0.054	-0.689	0.009	-0.086	-0.612
Coarse grain	0.222	-0.473	0.002	-0.126	-0.348
Vegetables and fruits	0.464	-0.027	-0.011	-0.028	0.011
Oil seeds	0.384	-0.133	0.004	0.222	-0.358
Sugar	0.784	1.037	0.001	1.470	-0.434
Cotton	0.232	-0.405	0.007	0.209	-0.621
Vegetable oil	0.674	0.875	0.000	2.177	-1.303
Other crops	0.198	-0.428	-0.001	-0.069	-0.358
Cattle and mutton	0.536	-0.199	0.001	0.016	-0.216
Pork and poultry	0.468	-0.237	-0.004	0.055	-0.288
Milk	0.340	-1.134	-0.009	0.212	-1.337
Fish	0.191	-0.094	0.000	-0.047	-0.047
Processed food	0.387	-0.647	0.001	-0.207	-0.440
Natural resources	0.296	-0.396	0.000	-0.014	-0.382
Textiles and apparel	0.112	-0.666	0.002	-0.064	-0.603
Natural industry	0.361	1.670	0.001	-0.038	1.707
Metal and machinery	0.213	1.287	0.001	-0.080	1.366
Transportation	0.268	-0.678	0.000	-0.031	-0.647
Electronics	0.158	0.488	0.001	-0.090	0.577
Manufactures	0.254	-0.728	0.000	-0.060	-0.669
Services	0.638	-0.225	0.000	-0.010	-0.215

Source: Results of authors' simulation.

Conclusions

This chapter assesses the economic effects of the ACFTA in its two stages up to 2010. The analysis is based on an improved recursive GTAP model. The data are based on Version 6 of the GTAP database for 2001, together with data derived from other sources. There are two distinguishing characteristics of this study. The first is that, in addition to the commitments in the ACFTA, the study incorporates trade liberalisation in China (China's

Table 13.13   Changes in the new ASEAN members in supply prices and output after the implementation of the ACFTA, up to 2010

	Supply price	Output	Contribution to output by		
			Tariff reductions from EHP	Tariff reductions in agriculture	Tariff reductions in industry
Rice	0.274	0.054	0.018	-0.010	0.046
Wheat	0.221	-0.030	0.386	-0.074	-0.342
Coarse grain	0.225	-0.023	0.024	0.040	-0.087
Vegetables and fruits	0.168	-0.072	-0.066	-0.003	-0.003
Oil seeds	0.300	0.091	0.052	0.107	-0.068
Sugar	0.365	0.394	0.007	0.364	0.023
Cotton	0.693	0.676	0.046	0.989	-0.359
Vegetable oil	0.307	3.560	0.005	3.264	0.291
Other crops	0.036	-0.275	-0.083	-0.039	-0.153
Cattle and mutton	0.271	-0.030	0.003	-0.008	-0.024
Pork and poultry	0.242	-0.022	-0.030	-0.002	0.010
Milk	0.233	-0.319	-0.012	-0.069	-0.238
Fish	0.279	0.048	-0.001	0.035	0.014
Processed food	0.262	0.073	0.016	0.139	-0.082
Natural resources	0.222	-0.063	0.000	-0.035	-0.029
Textiles and apparel	0.234	-0.595	0.021	-0.094	-0.521
Natural industry	0.250	1.019	0.012	-0.031	1.038
Metal and machinery	0.255	-0.537	0.001	-0.067	-0.472
Transportation	0.281	-0.130	0.000	-0.014	-0.116
Electronics	0.249	0.541	0.001	-0.043	0.584
Manufactures	0.245	0.002	0.030	-0.032	0.004
Services	0.294	-0.015	0.000	-0.002	-0.013

Source: Results of authors' simulation.

WTO commitments) and trade liberalisation in ASEAN (ASEAN free trade commitments). The second is that we have separated and explored the different effects of the two-stage implementation of the ACFTA. The following are the key findings of the study.

All member countries will gain from the ACFTA: it will increase social welfare and promote real GDP in the EHP phase from 2004-06 and in the fuller implementation during 2006-10. As the EHP includes only a small package of agricultural commodities, the gains during the fuller implementation of the ACFTA will be much larger in all member countries.

There is a large trade creation effect among the ACFTA signatories; their total exports will increase. A trade diversion effect is, however, also apparent. Trade between ACFTA signatories and other regions can be expected to decline due to the creation of the agreement. Because the trade creation effect is much larger than the trade diversion effect, global trade will be increased by the ACFTA, especially in the fuller implementation stage of the agreement.

The ACFTA will bring about substantial structural changes in China and in ASEAN countries. Trade liberalisation will improve the exploitation of comparative advantages in ACFTA signatories. The structural changes will take place in the agricultural and industrial sectors. Our results also show that the different policy arrangements stemming from the two-stage trade liberalisation will have different impacts on the shifts in economic structure during the process of implementation.

The rest of the world will have to face the challenges brought about by the ACFTA. Because the agreement will enhance the competitiveness of China and ASEAN in each other's markets, exports from non-member countries will be substituted. Social welfare and real GDP will decline in the non-member countries as a result of the creation of the ACFTA.

The results provide some useful insights into the impacts of the ACFTA on trade and economic relations between China and ASEAN; however, some limitations of the exercise should be mentioned. First, as many studies have observed, there are serious disguised unemployment problems in agricultural sectors. This reality is not modelled. Therefore, instead of the increases in wages the ACFTA gives rise to in the simulations, it is more likely that the ACFTA will create job opportunities. If we take this factor into account, China's gain from the agreement could be much larger and the changes in sectoral output could be different to the results of this study. Second, because of the lack of information about the barriers to trade in services, this study does not capture the impact of liberalisation in the services sectors. Finally, no allowance has been made for possible increases in capital formation and improvements in productivity that the ACFTA could engender. It is possible that the dynamic growth and productivity gains of the ACFTA could turn out to be very significant.

Notes

1   The ACFTA will be established in 2010 for ASEAN+6 and China; it will include the newer ASEAN member states of Cambodia, Laos, Myanmar and Vietnam after 2015.
2   This schedule holds for China, Brunei, Singapore, Thailand, Malaysia and Indonesia. Cambodia, Vietnam, Myanmar and Laos will complete their EHP in 2010. The Philippines has not concluded its negotiations.
3   A more complete description of the terms of China's WTO accession is available from http://www.wto.org/english/news_e/pres01_e/pr243_e.htm
4   In this study, all import tariffs—by commodity and country—are calculated by this method.
5   Despite its status as a developing country, China's de minimis exemption for product-specific support is only 8.5 per cent of the value of production of each agricultural product. In comparison, a 10 per cent rate has been agreed with other developing countries.
6   These products include rice, wheat, maize, edible oils, sugar, cotton and wool.
7   The full name for the amending protocol is the Protocol to Amend the Agreement on the Common Effective Preferential Tariff (CEPT) Scheme for the ASEAN Free Trade Area (AFTA) for the Elimination of Import Duties. The agreement can be downloaded from http://www.aseansec.org/14183.htm
8   Member states—except Cambodia, Laos, Myanmar and Vietnam—will eliminate all other non-tariff barriers on sensitive and highly sensitive products by 1 January 2010. Vietnam will eliminate all other non-tariff barriers on sensitive products by 1 January 2013, Laos and Myanmar by 1 January 2015 and Cambodia by 1 January 2017.
9   The deadlines for eliminating tariffs on commodities listed in the EHP vary among the newer ASEAN members: Vietnam before 1 January 2008; Laos and Myanmar before 1 January 2009; and Cambodia before 1 January 2010.
10 We focus mainly on the effects of the ACFTA on China and ASEAN countries. For other countries, we present only the total effects on the rest of the world (other regions).
11 As the GTAP model adopts the Leontief technology, an output increase in a sector promotes its demand for intermediate inputs by the same proportion.
12 As the GTAP Version 6 database has no detailed information on Brunei, only the other five members are included as representative of the original ASEAN members.

References

Anderson, K., Huang, J. and Ianchovichina, E., 2004. 'Will China's WTO accession worsen farm household income?', *China Economic Review*, 15:443-56.

ASEAN Joint Experts Group, 2001. *Forging closer ASEAN-China economic relations in the 21st century*, Report submitted by the ASEAN-China Expert Group on Economic Cooperation, October. Available from http://www.aseansec.org/newdata/asean_chi.pdf.

ASEAN-China FTA Framework Agreement, 2002. Available online at http://www.bilaterals.org/article.php3?id_article=2488.

ASEAN Secretariat, 2002. *Southeast Asia: a free trade area*. Available from http://www.aseansec.org/viewpdf.asp?file=/pdf/afta.pdf.

Bhattasali, D., Li., S. and Martin, W. (eds), 2004. *China and the WTO: accession, policy reform, and poverty reduction strategies*, The World Bank and Oxford University Press, Washington, DC.

Chirathivat, S., 2002. 'ASEAN-China Free Trade Area: background, implications and future development', *Journal of Asian Economics*, 13(5):671-86.

Ethier, W., 2000. 'The new regionalism', *Economic Journal*, 108:1,149-61.

Fan, S., Wales, E.J. and Crame, G.L., 1995. 'Household demand in rural China: a two-stage LES-AIDS model', *American Journal of Agricultural Economics*, 77:54-62.

Francois, J.F. and D. Spinanger, 2004. 'WTO accession and the structure of China's motor vehicle sector', in D. Bhattasali, S. Li and W. Martin (eds), *China and the WTO: accession, policy reform, and poverty reduction strategies*, The World Bank and Oxford University Press, Washington, DC.

Fukase, E. and Winters, L.A., 2003. 'Possible dynamic effects of AFTA for the new member countries', *World Economy*, 26:853-71.

Hertel, T.W. (ed.), 1997. *Global Trade Analysis: modelling and applications*, Cambridge University Press.

Hertel, T.W. and Martin, W., 1999. *Would developing countries gain from inclusion of manufactures in the WTO negotiations?*, GTAP Working Paper, Purdue University, West Lafayette.

Holst, D. and Weiss, J., 2004. 'ASEAN and China: export rivals or partners in regional growth?', *The World Economy*, 27(8):1,255-74.

Huang, J. and Bouis, H., 1996. *Structural changes in demand for food in Asia*, Food, Agriculture and the Environment Discussion Paper, International Food Policy Research Institute, Washington, DC.

Huang, J. and Rozelle, S., 1996. 'Technological change: rediscovery of the engine of productivity growth in China's rural economy', *Journal of Development Economics*, 49(2):337-69.

——, 1998. 'Market development and food consumption in rural China', China *Economic Review*, 9:25-45.

Huang, J. and Yang, J., 2006. 'China's rapid economic growth and its implications for agriculture and food security in China and the rest of world', *Management World*, 1:67-76.

Huang, J., Rozelle, S. and Chang, M., 2004. 'Tracking distortions in agriculture: China and its accession to the World Trade Organization', *World Bank Economic Review*, 18(1):59-84.

Ianchovichina, E. and Martin, W., 2004. 'Economic impacts of China's accession to WTO', in D. Bhattasali, S. Li. and W. Martin (eds), *China and the WTO: accession, policy reform, and poverty reduction strategies*, The World Bank and Oxford University Press, Washington, DC.

Liu, J., Surry, Y., Dimmaraman, B. and Hertal, T., 1998. 'CDE calibration', in R.A. McDougall, A. Elbehri and T.P. Truoung (eds), *Global Trade Assistance and Protection: the GTAP 4 Data Base*, Centre for Global Trade Analysis, Purdue University, West Lafayette.

Shang, G., 2005. *The Trial Field of China-ASEAN Free Trade Area: early harvest*. Available from http://big5.mofcom.gov.cn/gate/big5/www. mofcom.gov.cn/aarticle/Nocategory/200507/20050700180151.html

Tongzon, J., 2005. 'ASEAN-China Free Trade Area: a bane or boon for ASEAN countries?', *World Economy*, 28(2):191-210.

Tongeren, F. and Huang, J., 2004. *China's food economy in the early 21st century*, Report, No.6.04.04, Agricultural Economics Research Institute (LEI), The Hague.

United Nations Statistics Division, Commodity Trade Statistics Database, COMTRADE. Available online at http://unstats.un.org/unsd/comtrade/ default.aspx.

Voon, J. and Yue, R., 2003. 'China–ASEAN export rivalry in the US market: the importance of the HK-China production synergy and the Asian financial crisis', *Journal of Asia Pacific Economy*, 8(2):157-79.

Walmsley, T.L., Betina, V.D. and Robert, A.M., 2000. *A Base Case Scenario for the Dynamic GTAP Model*, Centre for Global Trade Analysis, Purdue University, West Lafayette.

Wattanapruttipaisan, T., 2003. 'ASEAN-China Free Trade Area: advantages, challenges, and implications for the newer ASEAN member countries', *ASEAN Economic Bulletin*, 20(1):31-48.

Wong, J. and Chan, S., 2002. 'China's emergence as a global manufacturing centre: implications for ASEAN', *Asia Pacific Business Review*, 9(1):79-94.

——, 2003. 'China-ASEAN Free Trade Agreement: shaping future economic relations', *Asian Survey*, 43(3):507-26.

World Bank, 1997. *China 2020: development challenges in the new century*, The World Bank, Washington, DC.

World Trade Organization, 2001. *Accession of the People's Republic of China, decision of 10 November 2001*, World Trade Organization, Geneva.

Yu, W., Hertel, T., Preckel, P. and Eales, J., 2003. 'Projecting world food demand using alternative demand systems', *Economic Modelling*, 21:99-100.

Appendix

Table A13.1   Tariff rates in China for its WTO accession, 2001-2010
(per cent)

	2001	2002	2003	2004	2005	2006	2007	2008	2009	2010
Rice	1.00	1.00	1.00	1.00	1.00	1.00	1.00	1.00	1.00	1.00
Wheat	1.00	1.00	1.00	1.00	1.00	1.00	1.00	1.00	1.00	1.00
Coarse grain	2.97	2.97	2.97	2.97	2.97	2.97	2.97	2.97	2.97	2.97
Vegetables and fruits	19.13	15.95	13.76	11.57	11.40	11.40	11.40	11.40	11.40	11.40
Oil seeds	3.20	3.20	3.20	3.20	3.20	3.20	3.20	3.20	3.20	3.20
Sugar	19.07	19.07	19.07	19.07	19.07	19.07	19.07	19.07	19.07	19.07
Cotton	1.24	1.24	1.24	1.24	1.24	1.24	1.24	1.24	1.24	1.24
Vegetable oil	11.79	11.44	11.08	10.75	10.70	10.61	10.61	10.61	10.61	10.61
Other crops	20.11	16.29	12.52	8.77	8.74	8.72	8.72	8.72	8.72	8.72
Cattle and mutton	13.09	12.05	11.02	9.99	9.99	9.99	9.99	9.99	9.99	9.99
Pork and poultry	10.96	10.13	9.33	8.53	8.53	8.53	8.53	8.53	8.53	8.53
Milk	14.46	12.92	11.37	9.81	8.70	8.70	8.70	8.70	8.70	8.70
Fish	15.93	14.21	12.10	11.15	10.44	10.44	10.44	10.44	10.44	10.44
Processed food	16.89	14.19	11.84	10.12	9.80	9.80	9.80	9.80	9.80	9.79
Natural resources	0.38	0.36	0.36	0.36	0.36	0.36	0.36	0.36	0.36	0.36
Textiles and apparel	17.42	15.00	12.74	10.61	9.24	9.24	9.24	9.24	9.24	9.23
Natural industry	10.85	9.44	8.67	8.00	7.66	7.38	7.14	6.87	6.87	6.87
Metal and machinery	8.76	7.14	6.24	5.75	5.69	5.69	5.69	5.69	5.69	5.69
Transportation	15.03	12.72	11.24	10.19	9.19	8.22	8.22	8.22	8.22	8.22
Electronics	5.10	3.89	3.24	3.12	3.11	3.11	3.11	3.11	3.11	3.11
Manufactures	18.93	17.63	16.56	15.67	14.88	14.88	14.88	14.88	14.88	14.88

Source: Based on the HS system tariff schedules of the protocol of China's WTO accession and weighted by 2001 import data from the COMTRADE database. The tariff rates for rice, wheat, other grains and plant-based fibres are in-quota rates.

Table A13.2 Regional aggregations

	Description	Original GTAP Version 6 regional aggregation
China	Mainland China	Mainland China
HK	Hong Kong, China	Hong Kong, China
TW	Taiwan, China	Taiwan, China
Japkor	Japan and South Korea	Japan, South Korea
ASEAN-old	ASEAN old members[12]	Indonesia, Malaysia, the Philippines, Thailand, Singapore
ASEAN-new	ASEAN new members	Vietnam, rest of Southeast Asia
OthAsia	Other Asia	India, Bangladesh, Sri Lanka, rest of East Asia, rest of South Asia
Australia	Australia and New Zealand	Australia
NAFTA	North American Free Trade Agreement area	Canada, United States, Mexico
SAM	South and Central America	Central America, Caribbean, Colombia, Peru, Venezuela, rest of Andean Pact, Argentina, Brazil, Chile, Uruguay, rest of South America, rest of Caribbean
EU15	European Union	Austria, Belgium, Denmark, Finland, France, Germany, United Kingdom, Greece, Ireland, Italy, Luxembourg, Netherlands, Portugal, Spain, Sweden
CEEC	Central European Associates	Hungary, Poland, Albania, Bulgaria, Croatia, Cyprus, Czech Republic, Malta, Romania, Slovakia, Slovenia, Estonia, rest of Europe
ROW	Rest of World	Switzerland, New Zealand, rest of EFTA, Turkey, rest of Middle East, Morocco, rest of North Africa, Malawi, Mozambique, Tanzania, Zambia, Zimbabwe, other southern Africa, Uganda, rest of Sub-Saharan Africa, former Soviet Union, Botswana, rest of SACU, Russia, rest of world

Table A13.3 Sector aggregation

	Original GTAP Version 6 sector aggregation
Rice	Paddy rice, processed rice
Wheat	Wheat
Coarse grain	Cereals, grains nec
Vegetables and fruits	Vegetables, fruit, nuts
Oil seeds	Oil seeds
Sugar	Sugar cane, sugar beet, sugar
Cotton	Plant-based fibres
Other crops	Crops-nec
Vegetable oil	
Cattle and mutton	Cattle, sheep, goats, horses and their meat
Pork and poultry	Animal products nec, wool, silk-worm cocoons, meat products
Milk	Raw milk, dairy products
Fish	Fish
Processed food	Food products nec, beverages, tobacco products
Natural resources	Forestry, coal, oil, gas, minerals nec
Textiles and apparel	Textiles, clothing apparel, leather products
Natural industry	Wood products, paper products and publishing, petroleum, coal products; chemical, rubber and plastic products; mineral products
Metal and machinery	Ferrous metals, metals nec, metal products, machinery and equipment nec
Transportation	Motor vehicles and parts, transport equipment nec
Electronics	Electronic equipment
Manufactures	Manufactures nec
Services	Electricity, gas manufacture, distribution, water, construction, trade, transport nec, sea transport, air transport, communication, financial services nec, insurance, business services nec, recreation and other services, public administration/defence/health/education, dwellings

Note: nec - not elsewhere classified

Table A13.4 Adjusted own-price and income elasticities for China

	Own-price elasticity		Income elasticity		
	GTAP 2001	Adjusted 2001	GTAP 2001	Adjusted 2001	2006
Rice	-0.08	-0.27	0.4	0.04	0.03
Wheat	-0.07	-0.29	0.4	0.06	0.05
Coarse grain	-0.06	-0.26	0.4	-0.35	-0.35
Vegetables and fruits	-0.12	-0.65	0.4	0.53	0.53
Oil seeds	-0.06	-0.57	0.4	0.42	0.41
Sugar	-0.07	-0.60	0.42	0.55	0.5
Cotton	-0.22	-0.50	1.06	1.06	1.06
Other crops	-0.18	-0.57	0.87	0.42	0.41
Vegetable oil	-0.06	-0.65	0.4	0.53	0.53
Cattle and mutton	-0.25	-0.78	1.23	0.66	0.65
Pork and poultry	-0.34	-0.65	1.23	0.56	0.55
Milk	-0.25	-0.89	1.23	1.05	1.04
Fish	-0.28	-0.67	1.23	0.8	0.79
Processed food	-0.28	-0.55	0.87	1.12	1.04
Natural resources	-0.26	-0.26	1.26	1.26	1.26
Textiles and apparel	-0.29	-0.29	1.06	1.06	1.06
Natural industry	-0.33	-0.32	1.25	1.25	1.25
Metal and machinery	-0.29	-0.29	1.25	1.25	1.25
Transportation	-0.28	-0.28	1.26	1.26	1.26
Electronics	-0.28	-0.28	1.25	1.25	1.25
Manufactures	-0.28	-0.28	1.25	1.25	1.25
Services	-0.49	-0.48	1.15	1.15	1.15

Source: Estimated by the Centre for Chinese Agricultural Policy (CCAP).

Table A13.5 Summary of tariff equivalents for China, 2001-20

	Import tariff equivalents (%)				Export tariff equivalents (%)			
	GTAP	2001	2006	2010	GTAP	2001	2006	2010
Rice	1.0	1.0	1.0	1.0	0	-9	-5	-3
Wheat	1.0	1.0	1.0	1.0	0	0	0	0
Coarse grain	87.8	3.0	3.0	3.0	0	31	0	0
Vegetables and fruits	24.8	19.1	11.4	11.4	0	-11	-6	-4
Oil seeds	101.0	3.2	3.2	3.2	0	0	0	0
Sugar	18.6	19.1	19.1	19.1	0	0	0	0
Cotton	1.6	1.2	1.2	1.2	0	14	0	0
Other crops	12.8	11.8	10.6	10.6	0	0	0	0
Vegetable oil	17.0	20.1	8.7	8.7	0	0	0	0
Cattle and mutton	15.3	13.1	10.0	10.0	0	-8	-5	-3
Pork and poultry	10.6	11.0	8.5	8.5	0	-21	-11	-6
Milk	19.9	14.5	8.7	8.7	0	0	0	0
Fish	11.5	15.9	10.4	10.4	0	-20	-10	-7
Processed food	21.6	16.9	9.8	9.8	0	-10	-6	0
Natural resources	0.3	0.4	0.4	0.4	0	0	0	0
Textiles and apparel	19.4	17.4	9.2	9.2	-5	0	0	0
Natural industry	12.3	10.9	7.4	6.9	0	0	0	0
Metal and machinery	11.4	8.8	5.7	5.7	0	0	0	0
Transportation	20.5	15.0	8.2	8.2	0	0	0	0
Electronics	10.1	5.1	3.1	3.1	0	0	0	0
Manufactures	17.4	18.9	14.9	14.9	0	0	0	0
Services	0	19.0	9.0	9.0	0	0	0	0

Source: The import tariff equivalents, excluding services, were calculated by the authors; the estimates for services are from Tongeren, F. and Huang, J., 2004. *China's food economy in the early 21st century*, Report, No.6.04.04, Agricultural Economics Research Institute (LEI), The Hague; Francois, J.F. and D. Spinanger, 2004. 'WTO accession and the structure of China's motor vehicle sector', in D. Bhattasali, S. Li and W. Martin (eds), *China and the WTO: accession, policy reform, and poverty reduction strategies*, The World Bank and Oxford University Press, Washington, DC; the export tariff equivalents are based on the estimates by the Centre for Chinese Agricultural Policy (CCAP).

Index